XVII International Workshop on Magneto-Plasma Aerodynamics 2018

Journal of Physics: Conference Series Volume 1112

Moscow, Russia
17 - 19 April 2018

ISBN: 978-1-5108-7716-0
ISSN: 1742-6588

Printed from e-media with permission by:

Curran Associates, Inc.
57 Morehouse Lane
Red Hook, NY 12571

Some format issues inherent in the e-media version may also appear in this print version.

Institute of Physics
Dirac House, Temple Back
Bristol BS1 6BE UK

Phone: 44 1 17 929 7481
Fax: 44 1 17 920 0979

techtracking@iop.org

Additional copies of this publication are available from:

Curran Associates, Inc.
57 Morehouse Lane
Red Hook, NY 12571 USA
Phone: 845-758-0400
Fax: 845-758-2633
Email: curran@proceedings.com
Web: www.proceedings.com

TABLE OF CONTENTS

IOP Conf. Series: Journal of Physics: Conf. Series **1112** (2018) 011001 doi:10.1088/1742-6596/1112/1/011001

We are proud to present the Proceedings of 17th International Workshop for Magneto-Plasma Aerodynamics (The Workshop) which was held in Moscow on April 17 – 19, 2018 at Joint Institute for High Temperatures of Russian Academy of Sciences (JIHT). The Workshop was organized by JIHT under Scientific Council for Direct Energy Conversion, Scientific Council for Fluid Mechanics and Scientific Council for Low Temperature Plasma Physics of Russian Academy of Sciences. The Workshop was sponsored by Russian Foundation for Basic Research (RFBR).

As in previous years, the objective of the 17th Workshop was to provide forum for presentation and discussion of recent progress in science and technology in magneto-plasma aerodynamics. The results of analytical and experimental studies of MHD energy conversion, physics of gas discharge, physics of shock waves, plasma assisted ignition and combustion, physical gas dynamics of high-speed airflows and their applications were discussed at the meeting.

The Workshop was attended by more than 70 registered participants from Russia and United States, 53 oral presentation and 15 posters were presented. There was plenary section and several profiled sections: shock waves, electrical discharges, plasma assisted combustion, plasma aerodynamics, MHD flow control and related problems.

This year is special for us because this is the first time when we publish our Proceedings in Journal of Physics: Conference Series and we are immensely grateful to IOP Publishing for cooperation. Any registered participant who presented an oral or poster presentation in our Workshop could submit a paper for the proceedings. A total of 32 original papers were submitted, of which 27 were eventually accepted for publication after reviewing process.

Valentin Bityurin, PhD, DS, chairman and program committee head
Anastasia Dobrovolskaya, organizing committee

IOP Conf. Series: Journal of Physics: Conf. Series **1112** (2018) 011002 doi:10.1088/1742-6596/1112/1/011002

Peer review statement

All papers published in this volume of *Journal of Physics: Conference Series* have been peer reviewed through processes administered by the proceedings Editors. Reviews were conducted by expert referees to the professional and scientific standards expected of a proceedings journal published by IOP Publishing.

IOP Conf. Series: Journal of Physics: Conf. Series **1112** (2018) 012001 doi:10.1088/1742-6596/1112/1/012001

Measurement of heat flux to model surface in the wind tunnel SMGDU

V A Bityurin[1], A N Bocharov[1], S S Bychkov[1*], T N Kuznetsova[1], N V Tretyakova[1], N I Batura[2] and N G Zhurkin[2]

[1]JIHT RAS, Izhorskaya 13 Bd 2, Moscow 127412, Russia
[2]TsAGI, Zhukovsky, Russia

*serg-bychkov@yandex.ru

Abstract. The results of studies of heat fluxes that can occur in the gas supersonic flow created in the wind tunnel with a magnetogasdynamic accelerator are presented. For two operation modes of the installation, measurements of heat fluxes to the surface of a calorimetric model were performed. The dependences of these parameters on induction of magnetic field on axis of the MHD accelerator were obtained for the first time. It was found that their values reached a level of 6 MW/m^2. The experimental data were compared with the results of a numerical simulation of a longitudinal flow past cylinder with a flat face in the real conditions of the wind tunnel, that allowed to calculate other characteristics of the gas flow. As a result, it is shown that this setup is enabled to create high-speed gas flows with high enthalpy that can be used in the development of prospective hypersonic aircrafts.

The problem of the creation of gas flow corresponding to high-altitude flight conditions in the atmosphere is of particular importance to date. It is equally important to know the characteristics of the flow formed in the wind tunnel. A standard set of diagnostic tools used at the experimental setup with a magnetogasdynamic accelerator (SMGDU) [1-5] usually included the measurement of the total pressure behind the shock wave and the video shooting of the flow. Spectra and shadow pattern were recorded in certain runs, also. However, to solve a number of actual problems, in particular, for testing of thermal protection materials, this information is not enough. It was necessary to carry out calculations simulating the flow under real experimental conditions to obtain a more complete data set in details. The purpose of the work is to expand the list of parameters registered in the air stream generated in this experimental setup and their comparison with calculations in order to determine the total enthalpy of hypersonic flow. The results of a series of SMGDU runes to determine the heat flux to the model surface are presented below.

The energy supply to the gas was carried out by two devices in the wind tunnel SMGDU. First, the air (pressure was 0.2 MPa) was warmed up by the electric-arc heater in the plenum. After the addition of the K-Na seed (weight fraction was about 1%), the gas entered to the Faraday segmented MHD accelerator through the primary supersonic nozzle (the Mach number was 2). In the accelerator channel, amount of energy supplied outside was spent for additional acceleration of the gas. Finally, the hypersonic gas flow in the working part of the installation was formed by a secondary nozzle.

A special calorimetric model was used to measure heat fluxes. The model was a copper cylinder with a flat end facing the oncoming gas flow. The model diameter was 20 mm. At its end there were symmetrically arranged two heat flux sensors at 6 mm distance from the axis. Sensitive elements of the sensors were copper calorimeters of cylindrical shape with 1.5 mm diameter and 5 mm length. The sensors were well insulated from the body of the model, and their inner ends were connected to the chromel-copel thermocouples. The hole in the center was intended for gas sampling in order to measure the pressure behind the direct shock wave. Its diameter was 1 mm.

The model was placed on the axis of the supersonic flow at a distance of 10 mm from the outlet of the secondary nozzle. The nozzle had a rectangular cross-section of 100 mm high and 70 mm wide. The signals recorded by the sensors of the calorimetric model during the run are shown in figure 1. The fragment of the oscillograms recording can be divided into three stages in time. The initial stage I was characterized by a steady-state total pressure $p0'$ about 6 kPa, and a slow increase in temperature $T1$ and $T2$ of the two calorimetric sensors. At that time only the electric-arc heater worked. In the next stage II MHD accelerator was turned on. During its operation that continued 0.7 s there was a sharp rise in pressure to 12 kPa and a much faster increase in temperature. Finally, the phase of gradual relaxation of signals came after the accelerated flow regime (stage III).

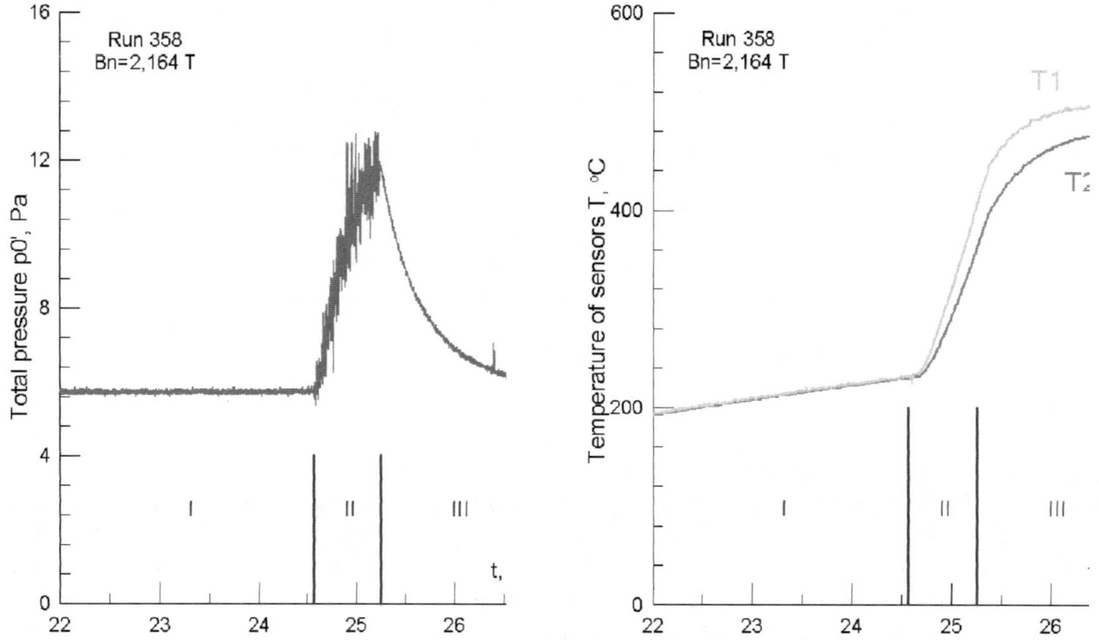

a) b)

Figure 1 Signals of the sensors of the calorimetric model:
a) total pressure behind the shock wave;
b) temperature of the copper calorimeters.

In the measurement method used, the average heat flux to the surface of the model q is proportional to change of the hot junction temperature dT_s during the time interval dt selected:

$$q = K_d \frac{dU}{dt} = K_d K_t \frac{dT_s}{dt},$$

where K_t is proportionality factor, U is the voltage measured between the cold ends of the thermocouple and $K_t = 0.067$ mV/°K is conversion factor for chromel-copel thermocouple. The proportionality factors for two calorimetric sensors were determined on a special pulse thermal installation that generated a stepped pulse of a calibrated convective heat flux. As a result of the calibration, the coefficients K_d measured was equal 35 W·s/cm^2·mV for the first sensor and 37 W·s/cm^{2}for the second one. The calibration accuracy was about 3%.

The results of this recalculation performed for the first and second stages of the run are shown in figure 2. The data are presented as a function of the external magnetic field induction on the axis of the

MHD accelerator. The blue color indicates the experimental points corresponding to the gas heating mode by the electro-arc heater only. Red points are obtained when the heater and accelerator are working together. All points were determined by averaging over the time intervals selected. For the stage II this interval was equal to the time of the MHD accelerator operation. The heat flux to the surface of the model did not depend on the magnetic field in the operation mode when only electro-arc heater operated, as expected. On the contrary, this value grew more than an order of magnitude with magnetic induction increasing and reached a value about 6 MW/m^2, if the MHD accelerator was additionally turned on. Such a growth in the heat flux associates with a rise in total enthalpy supplied to the MHD accelerator from the external electric circuit.

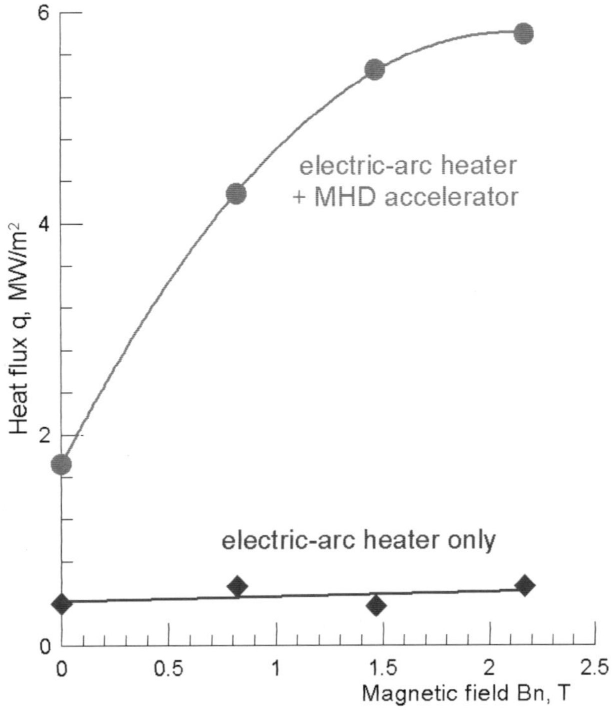

Figure 2 Heat fluxes measured at different stages of the run.

Other airflow characteristics can be estimated from the numerical simulation of ambient flow around the cylinder. For these purposes, the PlasmaAero software package [6,7] adapted to real conditions of the SMGDU wind tunnel was used. First, a two-dimensional axisymmetric flow in the facility duct was calculated. It was considered that the area of the circular cross-section was equal to the area of the real rectangular section. To calculate such an "effective" flow in the MHD channel, the bulk ponder motive force F and total power supplied W (source terms in the equations of motion and energy) were set like that the integral pulse and energy corresponded to the quantities defined earlier. These values were obtained as a result of calculation of the gas flow in the MHD channel consisting of 35 electrode pairs. A current of 35 A was set at each pair. The induction of the magnetic field was 2.4 T. In the example considered below the values were taken as $F = 55$ N and $W = 280$ kW, respectively. Then it was calculated the flow around the cylinder with a radius of 10 mm which end surface was 10 mm from the secondary nozzle outlet. Some results of the simulation are presented in figure 3. Also, there is an image of airflow around the calorimetric model in the Eiffel chamber of SMGDU in figure 3 on the left to compare calculated distributions with experimental data. The image was obtained during the video shooting with a high-speed camera. Exposure time of the frame was 500 μs.

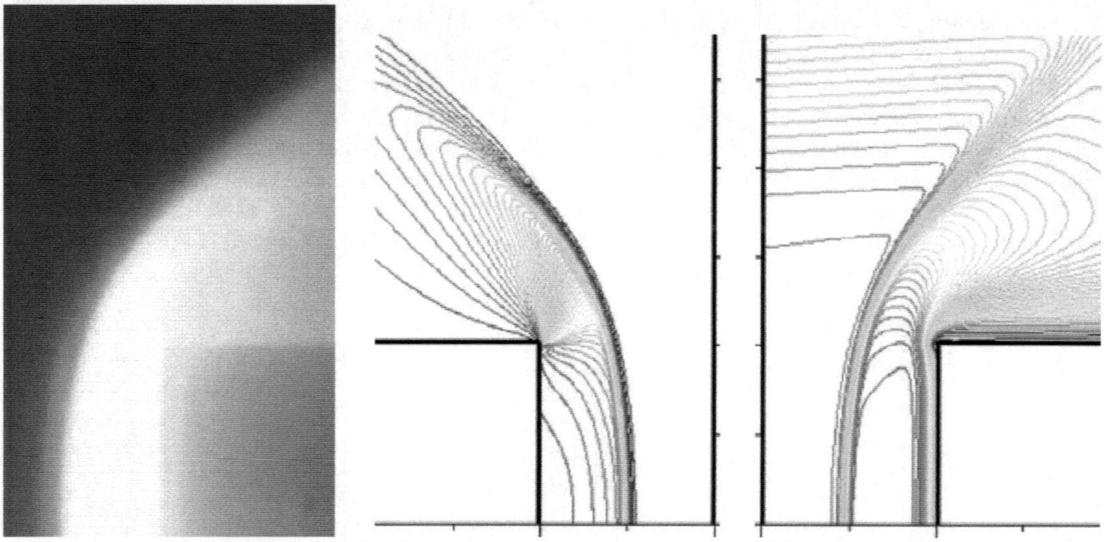

a) A frame of high-speed shooting of supersonic flow around the model.

b) Pressure distribution: $P_{min} = 147$ Pa, $P_{max} = 26530$ Pa

c) Temperature distribution: $T_{min} = 1550$°K, $T_{max} = 13100$°K

Figure 3 Longitudinal airflow around the copper cylinder with a flat end at SMGDU: a) an experimental pattern; b) and c) some results of simulating.

The next figure 4 demonstrates the radial distributions of the heat flux over the surface of the cylinder for two cases. The operating mode of the SMGDU when there was no additional pulse ($F = 0$) and energy ($W = 0$) input in the facility duct is shown in blue. This case corresponded to the turning on of the arc heater, only. For another mode, when also the MHD accelerator is turned on the distribution is shown in red.

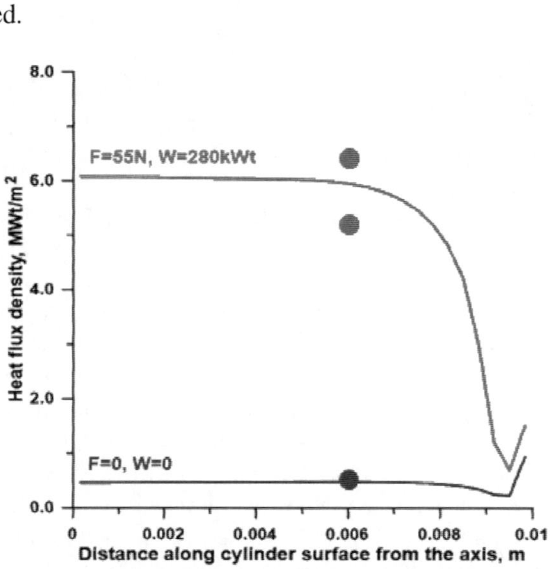

Figure 4 Heat flux distributions on the surface of the cylinder.

The same figure shows the points of experimental measurements of heat fluxes obtained under maximum value of magnetic field $B_n = 2.4$ T by means of two calorimetric sensors. The blue dot includes both measurements, since in this mode the readings of the sensors were practically the same.

When the MHD accelerator was turned on, the measured heat fluxes differed by approximately 20% (red dots). Apparently, this circumstance indicated the asymmetry of the gas flow since the calibration accuracy is much higher.

As the experimental data are in the good agreement with the results of numerical simulation, it is possible to determine other characteristics of the hypersonic flow in the test part of the SMGDU. Radial profiles of gas flow velocity (the Mach number on "effective" flow axis is 5.4) and specific total enthalpy in the outlet section of the secondary nozzle are shown in figure 5, as example.

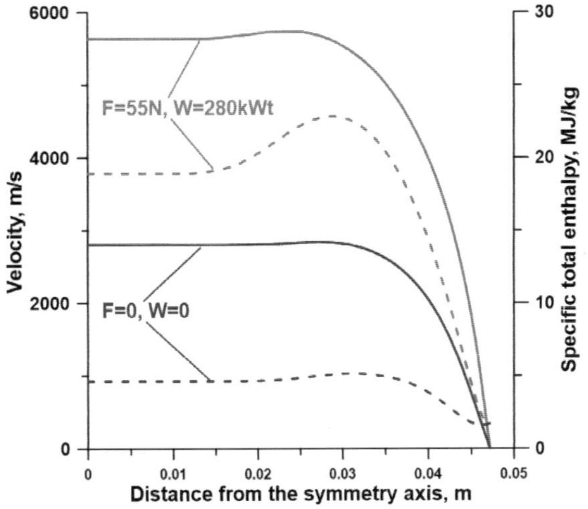

Figure 5 Radial distributions of flow characteristics at the nozzle exit cross-section: velocity (solid lines) and specific total enthalpy (dotted line).

Thus, it is possible to determine the characteristics of the supersonic gas flow formed by the wind tunnel SMGDU when measurements of heat fluxes are accompanied by mathematical modeling. Air flow with such parameters can be applied in the ground-based research needed to create new hypersonic devices.

Acknowledgments

The work was carried out with the partial support of the Russian Foundation for Basic Research (project 18-08-00890 A).

References

[1] Alferov V I 2000 Current status and potentialities of wind tunnels with MHD acceleration. *High Temp.* **38** 300

[2] Alferov V I, Bushmin A S and Egorov 2015 Experimental investigation of flow past simple model bodies in hypersonic wind tunnels at similar values of the Mach and Reynolds numbers but at different physical flow velocities *Fluid Dyn.* **50** 109

[3] Alfyorov V I, Dmitriev L M, Egorov B V, Markachev Yu E and Rudakova A P 2001 The possibility of simulating in wind tunnels the parameters of hypersonic flow and the conditions in the combustor of hypersonic scramjet engine *High Temp.* **5** 688

[4] Bityurin V, Bocharov A, Baranov D and Bychkov S 2007 Power Extraction Experiment with a Surface MHD Generator in Hypersonic Airflow *38th AIAA Plasmadynamics and Lasers Conf. In conjunction with the 16th Int. Conf. on MHD Energy Conversion, Miami, FL, June 25-28 2007* AIAA-2007-3882.

[5] Bityurin V A and Bocharov A N 2009 Hypersonic MHD: Features and Problems *Proceedings of the 17th Int. Conf. on MHD Energy Conversion, 14 – 17 September 2009, Shonan Village Center, Kanagawa, Japan* IK1-1.

IOP Conf. Series: Journal of Physics: Conf. Series **1112** (2018) 012001 doi:10.1088/1742-6596/1112/1/012001

[6] Bityurin V and Bocharov A 2006 Magnetohydrodynamic interaction in hypersonic air flow past a blunt body. *Fluid Dyn.* **41** 843
[7] Bityurin V A and Bocharov A N 2018 MHD heat flux mitigation in hypersonic flow around a blunt body with ablating surface *J. Phys*. D: *Appl. Phys.* **51** 264001

IOP Conf. Series: Journal of Physics: Conf. Series **1112** (2018) 012002 doi:10.1088/1742-6596/1112/1/012002

Influence of air flow speed on main characteristics of nonstationary pulsed discharge, created with help of stationary power source

V M Shibkov

Moscow State University, 119991 Moscow, Russia

shibkov@phys.msu ru

Abstract. The work is devoted to the investigation of a non-stationary pulsating transverse-longitudinal discharge created with the help of a stationary power source. The influence of the velocity of subsonic and supersonic airflows on main characteristics of the discharge is studied.

1. Introduction

Nowadays, low-temperature gas-discharge plasma in a flow is widely used in scientific research and various technological applications [1, 2]. To initiate and stabilize combustion of air–hydrocarbon fuels in a supersonic flow, we applied various types of discharge, namely, dc pulsed transverse discharges [3-6], freely localized microwave discharged produced by a focused electromagnetic radiation [7-9], and surface microwave discharges excited on a dielectric antenna [10, 11]. To optimize the burning of gaseous and liquid fuels under supersonic airflows it is necessary to know the parameters of the plasma used for the ignition and stabilization of the combustion of hydrocarbon fuels. The parameters of self-sustaining microwave discharges have been well studied. This work is devoted to research of the parameters of the discharge pulsating in a supersonic flow, which is essentially a nonstationary discharge generated using a DC voltage source.

2. Experimental setup

The experimental setup includes a vacuum chamber, a high-pressure air receiver, a system for generating a supersonic flow, a rectangular wind tunnel with an attached air duct, the high-voltage power supply to create the gas-discharge plasma, the synchronization system, and diagnostic equipment. The main part of the experimental setup is the 3-m-long evacuated metal cylindrical chamber with an inner diameter of 1 m. The wind tunnel was placed inside the pressure chamber. The longitudinal length of the channel was 50 cm. The supersonic flow in the wind tunnel was created by filling the pressure chamber with air through a specially shaped Laval nozzle. To prevent thermal choking a wind tunnel with a variable section was used; the ratio of the output section $S_2 = 38.1$ cm^2 to the input section $S_1 = 3$ cm^2 equals $S_2/S_1 = 12.7$.

A pulsed discharge was excited between two well streamlined electrodes, the cross section of which was an elongated rhomb with rounded vertices (the maximum thickness and width of the electrodes did not exceed 1 mm and 5 mm, respectively). The design of the electrode unit was described in [4, 5]. The minimum distance between the electrodes was varied from 0.1 to 3 mm. The electrodes were mounted inside the diverging wind tunnel. A special configuration of the electrodes made possible to easily excite a pulsed discharge without its additional initiation in a wide range of chamber pressures of $p = 10-760$ Torr. The main experiments were carried out in an open working chamber at atmospheric pressure. The discharge was produced using a power source with an output voltage of up to 5 kV and a discharge current of up to 20 A, the pulse duration being of up to 2 s. The air flow rate was varied from 50 to 140 g/s, flow speed varied from 150 to 550 m/s. The experiments were performed using a diagnostic complex consisting of digital spectrographs, a high-speed video camera, pressure gauges, digital photo cameras, oscilloscopes, and computers.

3. Experimental results

It was shown that non-stationary pulsating discharge produced by a stationary power source is a thin pulsating channel drawn in the form of a plasma loop down the direction of the high-speed airflow propagation. In the experiments, two discharge maintenance regimes were implemented. In the first case, the plasma loop reaches its maximum possible length without intermediate repeated breakdowns. In the second case, several additional repeated breakdowns are realized. In both modes, the process of drawing a discharge loop along the flow is repeated periodically.

Application of a dc voltage to the electrodes leads to air breakdown along the shortest path between them. The formed plasma channel begins to glide over the electrodes downstream the air flow. While moving in the interelectrode gap, the plasma channel undergoes deformation. The velocities at which the anode and cathode spots move along the electrodes are lower than the flow velocity in the central part of the channel. First, the anode spot reaches the tip of the first electrode and is fixed on it, and, then, the cathode spot is fixed at the tip of the second electrode. At the same time, the plasma channel continues to stretch out downstream the flow. Since the channel length grows, the voltage drop across the channel increases until it reaches the threshold for the second breakdown, which occurs along the shortest path between the electrodes and shunts the plasma channel. Then, the discharge glow begins to decrease and the discharge plasma gradually decays. By that time, a new plasma channel has already formed and the process repeats.

Figure 1 illustrated the dynamics of the development of a pulsating discharge created in a subsonic airflow $\upsilon = 275$ m/s. The video frame rate is 32 kHz, the exposure time of one frame is 2 μs. The air flow is directed from the top to down, the time increases from left to right. It is seen that under these conditions, the regime of discharge maintenance is implemented without intermediate breakdowns during the development of one loop.

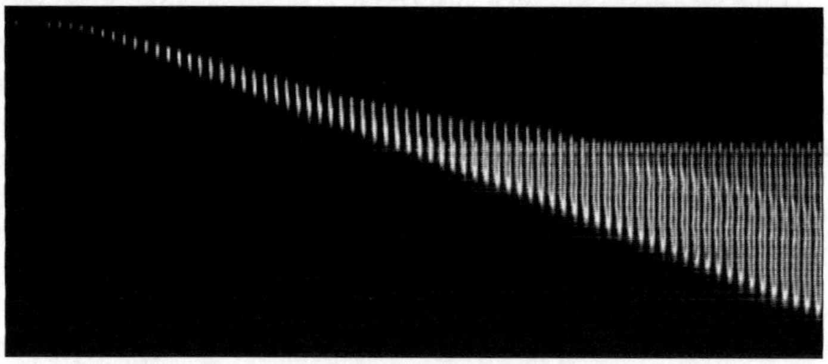

Figure 1 Dynamics of the development of a pulsating discharge created in a subsonic airflow

Discharge pulsations in a supersonic flow manifest themselves as oscillations of the gap voltage, discharge current, and plasma glow. Oscillations of the voltage and plasma glow are especially strong. The results of experimental studies carried out at various airflow velocities have shown that the magnitude of the current pulsation depends strongly on the flow velocity. Moreover, for fixed values of the output voltage of the power source, internal and ballast resistances, the maximum value of the discharge current i_{max} remains constant both for subsonic and supersonic air flow velocities. Then, the minimum value of ripple current i_{min} is highly dependent on the flow speed. The strongest pulsations, reaching 50% of the maximum value of the i_{max} current, appear at low subsonic airflow velocities in

IOP Conf. Series: Journal of Physics: Conf. Series **1112** (2018) 012002 doi:10.1088/1742-6596/1112/1/012002

the aerodynamic channel, whereas in supersonic airflows, the current ripples are smoothed, their magnitude decreases monotonically and $\Delta i = 5\%$ at $\upsilon = 550$ m/s. Figure 2 shows a fragment of the time dependences of the current and the voltage across the discharge gap. We see that the modulation of the voltage reaches 100%, while the modulation of the discharge current is about 45%, because, at any time, there is plasma bridging the electrodes. The length of the plasma bridge varies in time, but the bridge is destroyed only after the repeated breakdown and formation of a new plasma bridge. In the voltage waveform (figure 2, curve 2), the moments of repeated breakdowns are clearly seen.

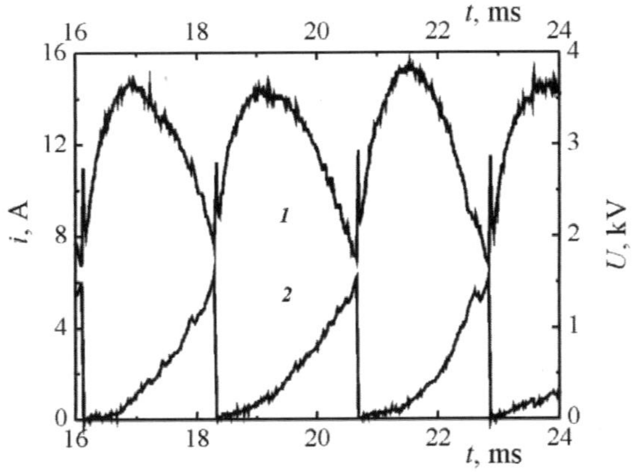

Figure 2 Fragment of the time dependences of the discharge current (*1*) and gap voltage (*2*) for a discharge with a duration of $\tau = 2$ s in a subsonic air stream. Flow velocity is 275 m/s

The repeated breakdown can also occur between the anode and cathode parts of the plasma channel or between the electrodes. In any case, the repeated breakdown leads to the shunting of the remaining part of the channel. Depending on the experimental conditions, one to five repeated breakdowns can occur during one cycle of the formation and development of the plasma loop. The process of stretching of the discharge loop downstream the air flow repeats periodically, i.e., the given form of a dc discharge produced in an air flow is, in fact, an unsteady pulsating discharge. In figure 3 shows the dynamics of the development of a pulsating discharge plasma loop created in a subsonic airflow by means of a stationary power source when the discharge is realized with two additional intermediate repeated breakdowns. Air speed is 200 m/s, maximum discharge current is 9.8 A.

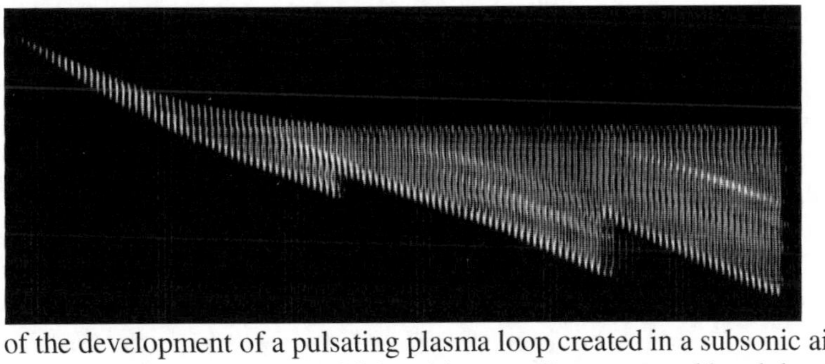

Figure 3 Dynamics of the development of a pulsating plasma loop created in a subsonic airflow when the discharge is realized with two additional intermedium repeated breakdowns

It was obtained that for any fixed values of the velocities of subsonic and supersonic air flows, the pulsation frequency is almost independent of the discharge current, varying within 100 Hz with a change in the discharge current from 2 to 16 A. In contrast, in the experiment observed a strong discharge pulsation frequency dependence on the air speed. The dependence of the pulsation frequency of the plasma loop on the airflow velocity at a discharge current of 15.5 A is shown in figure 4. It can be seen that an increase in the flow velocity from 100 m/s to 520 m/s results in a significant 40-time increase in the pulsation frequency from 50 Hz to 2 kHz.

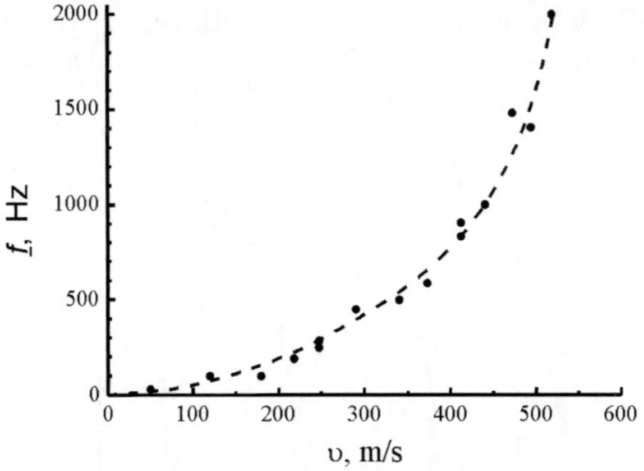

Figure 4 Pulsation frequency of the plasma loop as a function of the flow rate at $i_{max} = 15.5$ A

In figure 5 shows the dependence of the maximum achievable voltage U_{max} on the discharge gap on the airflow velocity. In the case of subsonic streams, as the velocity of the air flow increases, the discharge voltage remains approximately constant for any fixed value of the current, and the voltage in the discharge decreases with increasing discharge current. In going to a supersonic flow, U_{max} sharply decreases with increasing velocity, and the explicit dependence on $U(i)$ disappears.

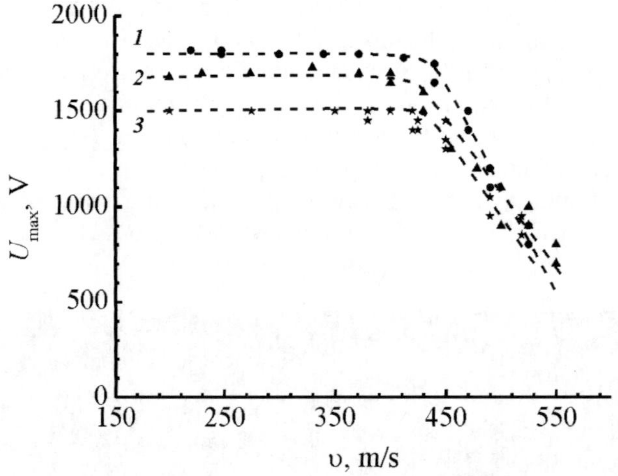

Figure 5 Dependence of the maximum attainable value of the voltage across the discharge gap on the air flow rate for different values of the pulsed discharge current i_{max}: 1 – 5.2, 2 – 9.8, 3 – 15.5 A

The dependence of the maximum distance on which the front boundary of the plasma loop extends down the air flow, on its velocity for different values of the discharge current, is shown in figure 6. It can be seen that as the flow rate increases, the maximum achievable length of the plasma loop decreases linearly, and at any speeds the distance to which the loop extends increases with increasing of discharge current. The total length of the plasma channel is equal to the sum of the lengths of the cathode and anode parts of the loop.

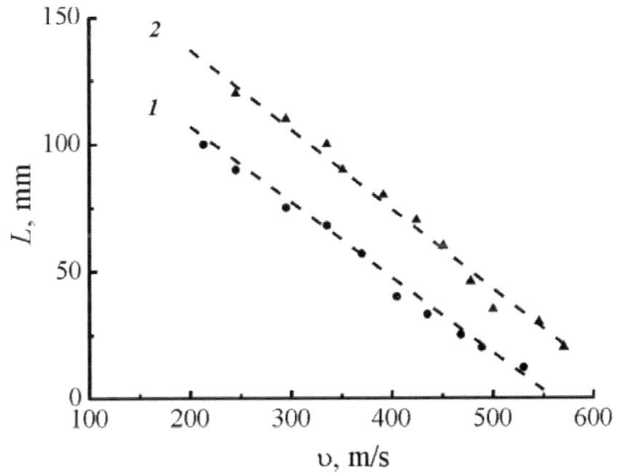

Figure 6 The distance covered by the front boundary of the plasma loop down the airflow, depending on its speed for different maximum values of the pulsed discharge current i_{max}: 1– 5.5, 2– 15.5 A.

The dependence of the electric field intensity averaged over time and the length of the channel in pulsating discharge plasma on the airflow velocity for different values of the discharge current is shown in figure 7.

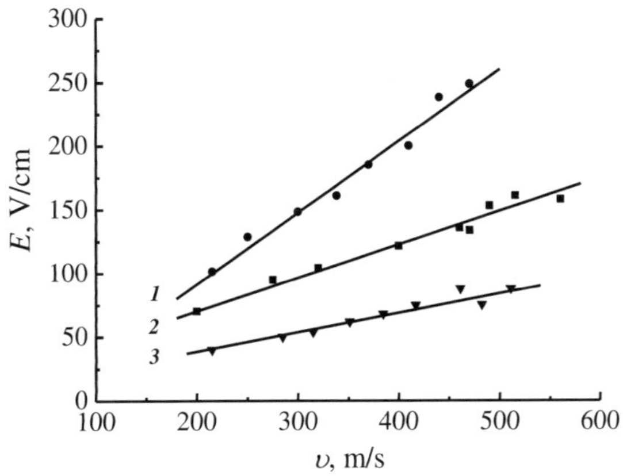

Figure 7 Electric field in pulsating discharge vs the airflow velocity. The maximum value of the discharge current i_{max}: 1– 5.5, 2– 9.8, 3– 15.5 A

It can be seen that the electric field increases with increasing flow velocity because the necessary condition for the plasma loop of a transverse-longitudinal discharge in a high-speed flow to remain closed onto the cathode is the increase in the drift velocity of positive ions, which is possible only if the electric field increases [12]. For small current the field grows faster than for large i. Thus, with an increase in the flow velocity from 200 to 500 m/s, the field doubles from 40 to 85 V/cm at i_{max} = 15.5 A, whereas for i_{max} = 5.5 A the electric field grows three times from 90 to 260 V/cm. For a fixed flow velocity, an increase in the discharge current leads to a decrease in the electric field in the discharge channel. The similar dependence $E(i)$ was obtained in [13].

4. Conclusions

It is shown that a pulsating discharge produced by a dc power source in a high-speed air flow operates in the form of a thin pulsating channel in the shape of a plasma loop stretched downstream the flow. Discharge exists in two regimes: without and with intermediate breakdowns during the development of one loop. The pulsation frequency of the plasma loop monotonically increases from 50 Hz to 2 kHz when flow velocity changes from 30 m/s to 520 m/s. The electric field strongly depends on the flow velocity and the magnitude of the discharge current, varying in the range of 30-260 V/cm.

Acknowledgments
This work was supported by the Russian Foundation for Basic Research, project # 18-02-00336-a.

References
[1] Starikovskaya S M 2006 *J. Phys.* D **39** R265.
[2] Shibkov M V, Aleksandrov A F, Chernikov V A, Ershov A P and Shibkova L V 2009 *J. Propul. Power* **25** 123 DOI: 10.2514/1.24803
[3] Ershov A P, Surkont O S, Timofeev I B, Shibkov V M and Chernikov V A 2004 *High Temp.* **42** 4 516 DOI: 10.1023/B:HITE.0000039979.89955.df
[4] Kopyl P V, Surkont O S, Shibkov V M and Shibkova L V 2012 *Plasma Phys. Rep.* **38** 6 503 DOI: 10.1134/S1063780X12050054
[5] Shibkov V M, Shibkova L V and Logunov A A 2017 *Plasma Phys. Rep.* **43** 3 373 DOI: 10.1134/S1063780X17030114
[6] Shibkov V M, Shibkova L V and Logunov A A 2017 *Moscow Univ. Phys. Bulletin* **72** 3 294 DOI: 10.3103/S0027134917030109
[7] Konstantinovskii R S, Shibkov V M and Shibkova L V 2005 *Kinet. Catal.* **46** 6 775 DOI: 10.1007/s10975-005-0136-2
[8] Shibkov V M, Aleksandrov A F, Ershov A P, Timofeev I B, Chernikov V A and Shibkova L V 2005 *Plasma Phys. Rep.* **31** 9 795 DOI: 10.1134/1.2048839
[9] Shibkov V M, Shibkova L V, Gromov V G, Karachev A A and Konstantinovskii R S 2011 *High Temp.* **49** 2 155 DOI: 10.1134/S0018151X11020143
[10] Shibkov V M, Ershov A P, Chernikov V A and Shibkova L V 2005 *Tech. Phys.* **50** 4 455 DOI: 10.1134/1.1901784
[11] Shibkov V M, Dvinin S A, Ershov A P and Shibkova L V 2005 *Tech. Phys.* **50** 4 462 DOI: 10.1134/1.1901785
[12] Pashchenko N T and Raizer Yu P 1982 *Sov. J. Plasma Phys* **8** 617
[13] Fridman A, Nester S, Kennedy L, Saveliev A and Mutaf-Yardimci O 1999 *Prog. Energy Combust. Sci.* **25** 211

Air ionization degree of the plasma in a nonstationary pulsed discharge in subsonic and supersonic flows

V M Shibkov

Moscow State University, 119991 Moscow, Russia

shibkov@phys.msu ru

Abstract. The degree of gas ionization in a nonstationary pulsating discharge created using a stationary power source in subsonic and supersonic air streams is determined. It was experimentally obtained that when the flow velocity is changed from 150 m/s to 520 m/s, the plasma electron density near the electrodes (z = 1 cm) varies from 10^{15} cm^{-3} to 3.7×10^{16} cm^{-3} at an unchanged discharge current of 15.5 A, and the gas temperature rises from 400 K to 1250 K. It is shown that in pulsating discharge plasma the degree of gas ionization is of the order of 10^{-4} at low subsonic airflow, and with an increase in the flow velocity it increases sharply and reaches a value of 10^{-2} at velocity of 500 m/s.

1. Introduction

The non-equilibrium low-temperature plasma of a pulsating discharge is optimal from the point of view of its application in different areas of supersonic plasma aerodynamics. Fundamental works in the field of plasma aerodynamics are studies of the properties of transverse electric dc discharges in supersonic air flows carried out by A.I. Alferov [1-2]. At present, intensive researches relating to the development of new ways to improve the efficiency of gaseous and liquid fuels combustion in a high-speed air flow are underway. In various laboratories it is proposed to use plasma technology for the purpose of reducing the fuel induction period, increasing the completeness and stabilizing supersonic combustion. That allows optimizing the combustion of air-hydrocarbon fuels in high-speed air flows without using mechanical stagnant zones. Different gas discharges are used to create the plasma in high-speed flow, among them: high-voltage nanosecond discharges [3-6], transverse and longitudinal electrode discharges [7-15], high-frequency discharges created either in the chamber volume or on the dielectric surface, spark, arc, glow, corona discharges, barrier, sliding arc [16] and laser discharges, as well as freely localized microwave discharges [17-22], microwave discharges created by a surface wave on a dielectric body [23-28].

The paper presents the results of studies of the main plasma parameters of a non-stationary pulsating transverse-longitudinal discharge created in subsonic and supersonic air flows using a stationary power source.

2. Experimental installation and diagnostic methods

The test setup includes a vacuum chamber, a high-pressure air receiver, a system for generating a supersonic flow, a rectangular wind tunnel with a variable section, high-voltage power supply, synchronization system, and diagnostic equipment. The wind tunnel was placed inside the 3-m-long evacuated metal cylindrical chamber with an inner diameter of 1 m. The ratio of the output section $S_2 = 38.1$ cm^2 to the input section $S_1 = 3$ cm^2 of wind tunnel equals $S_2/S_1 = 12.7$. The longitudinal length of the wind tunnel was 50 cm. The discharge was produced using a power source with an output voltage of up to 5 kV and a discharge current of up to 20 A, the pulse duration being of up to 2 s. The air flow rate was varied from 50 to 140 g/s, flow speed changed from 150 to 550 m/s. A pulsed discharge was excited between two well streamlined electrodes. The design of the electrode unit was described in [10, 11]. The minimum distance between the electrodes was varied from 0.1 to 3 mm.

The experiments were performed using the diagnostic complex consisting of digital spectrographs, high-speed video camera, pressure gauges, digital photo cameras, oscilloscopes, and computers. Parameters of high-speed flow (velocity, second mass flow consumption, temperature and density) were experimentally determined from the measured values of static p_1 and total p_0 pressures, respectively, in the wind tunnel and receiver containing compressed air. The main parameters of the pulsating discharge were determined using an automated system for collecting and processing information. This system allowed real-time recording of signals from electrical probes, pulse pressure sensors, thermocouple sensors; photoelectric multiplier; digital spectrograph; low resistance, the voltage drop in which is proportional to the magnitude of the discharge current; high-voltage divider, which makes it possible to measure the voltage across the discharge gap. The emission spectrum of the gas-discharge plasma was recorded with the Avantes double-channel spectrograph AvaSpec-2048-2-DT, with an inverse linear dispersion is equals of 0.05 nm/mm for the first spectral channel and 0.32 nm/mm for the second one. Averaging was performed for the exposure time $\tau = 20$ ms, the frame repetition rate was 20 Hz, i.e., for one start with duration of $t = 2$ s up to 40 spectra are consistently recorded. The gas temperature in the plasma channel of a pulsating discharge was determined by the spectral method from the bands of cyanogen and molecular nitrogen ion. At first, model distributions over the rotational levels of the CN and molecular nitrogen ion at different gas temperatures were calculated taking into account the instrumental function of the spectrograph. Various effects that lead to the broadening of spectral lines in the discharge plasma were also taken into account. Then, the experimentally obtained spectrum was compared with the calculated results. The gas temperature was considered to be equal to the temperature at which the best agreement between the calculated results and experimental data was achieved. The electron concentration was measured by the Stark broadening of the spectral lines H_α and H_β of the Balmer hydrogen series (for $n_e < 10^{16}$ cm^{-3} we used only H_β line, for $n_e > 10^{16}$ cm^{-3} we used H_β and H_α lines). The absolute sensitivity of the spectrographs was calibrated using special tungsten filament lamp, the radiation intensity which is determined by the temperature of the tungsten ribbon, depending on the magnitude of current flowing through the lamp.

3. Experimental results

The panoramic emission spectrum of a pulsating discharge in a supersonic air flow, recorded at the distance of $z = 1$ cm from the electrode tips is shown in figure 1 at the discharge current of $i = 15.5$ A. A distinctive feature of the spectrum is the presence of an intense continuum, which may electron bremsstrahlung from ions or atoms and/or recombination emission. Estimates show that, under these experimental conditions, the intensity of bremsstrahlung caused by electron–ion collisions exceed the intensity of bremsstrahlung caused by electron–neutral collisions. It is well known that the intensity of the recombination spectrum depends non-monotonically on the wavelength, whereas the measured intensity of the continuous spectrum depends monotonically on λ. The electron temperature in the plasma channel was determined by comparing the measured spectrum with the bremsstrahlung spectrum calculated for different electron temperatures. It has been shown that at the discharge current of $i = 10.8$ A the electron temperature decreases downstream the flow from 13000 K at distance from electrodes $z = 1$ cm to 7000 K at $z = 5$ cm. It has been also shown that at $z = 1$ cm the electron temperature increases from 10000 K to 15000 K as the discharge current rises from 2 to 16 A. It was shown experimentally that, upon adding a small amount of propane to the air flow, the electron temperature decreases abruptly to 3000 – 5000 K.

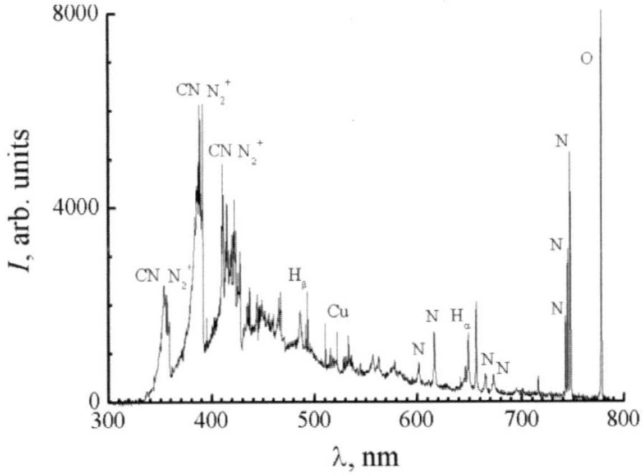

Figure 1 Panoramic emission spectrum of a dc pulsating discharge in a supersonic air flow, recorded at a distance of $z = 1$ cm from the electrodes

To measure the air ionization degree direct link electron concentration with half-width of a spectral line $\Delta\lambda$ is used. The line broadening occurs not only because of the Stark effect, but also due to the Doppler effect, the finite lifetime of the excited atom, the pressure effects, and the apparatus function of the spectral instrument. The apparatus function for the spectrograph was determined experimentally. From the experimentally obtained values $\Delta\lambda$, the Stark component was calculated in accordance with contour theory and empirical dependence [29].

It was experimentally obtained that the electron concentration at a distance of 1 cm from the electrodes increases almost forty times from 10^{15} cm^{-3} to 3.7×10^{16} cm^{-3} with an increase in the flow velocity from 150 m/s to 500 m/s (figure 2). As the distance from the electrodes is increased to 3 cm, n_e decreases by approximately 2-3 times at all flow velocities, but also increases linearly with increasing of air stream speed. A high level of electron concentration is also observed in the channel plasma of a microwave discharge at high air pressures [30].

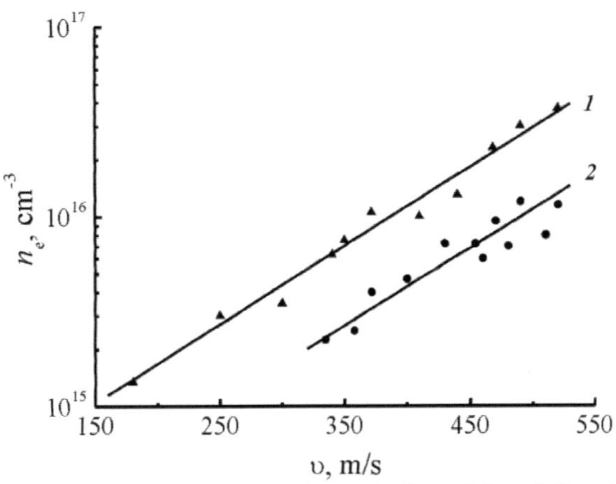

Figure 2 Dependence of the electron density on the velocity of the airflow in the pulsating channel discharge plasma at various distances from the electrodes $z = 1$ cm (*1*) and $z = 3$ cm (*2*). $i = 15.5$ A.

It has been experimentally shown that the electric field intensity linearly increases by 2-3 times with an increase in the flow velocity from 200 m/s to 500 m/s. The increase in the electric field strength with increasing flow velocity is explained by the fact that in order to close the plasma loop to the cathode under conditions of a transverse-longitudinal discharge in a high-speed flow, it is necessary to increase the drift velocities of positive ions, and this is possible only with an increase in the electric field. The growth in the electric field leads to an increase in the ionization frequency and accordingly increases the electron density.

In figure 3 shows the time evolution of the electron concentration measured at a distance of $z = 1$ cm downstream of the electrodes, with different values of the discharge current and a fixed value of the flow velocity $\upsilon = 500$ m/s. The exposure time of the spectrograph is 20 ms. Twenty spectra were recorded during the duration of the discharge current pulse $t = 2$ s. It can be seen that at a fixed flow rate the electron density increases with increasing discharge current and the instability of the concentration does not exceed 5-10% during the discharge current pulse.

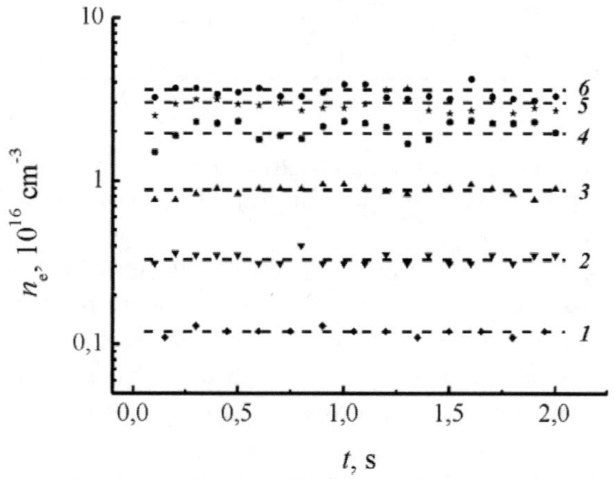

Figure 3 Electron concentration in a pulsating discharge plasma vs time for different values of the discharge current i, A: *1* - 2.3, *2* – 5.5, *3* – 9.8, *4* – 12.5, *5* – 14.5, *6* – 15.5 and fixed air velocity $\upsilon = 500$ m/s

To determine the degree of gas ionization, in addition to the electron density, it is also necessary to know the concentration of air molecules in the heated plasma channel. Initially, the dependence of the gas temperature on velocity of a high-speed flow was calculated without creating a discharge in it. The resulting relationship is shown in figure 4 (dotted curve 1).

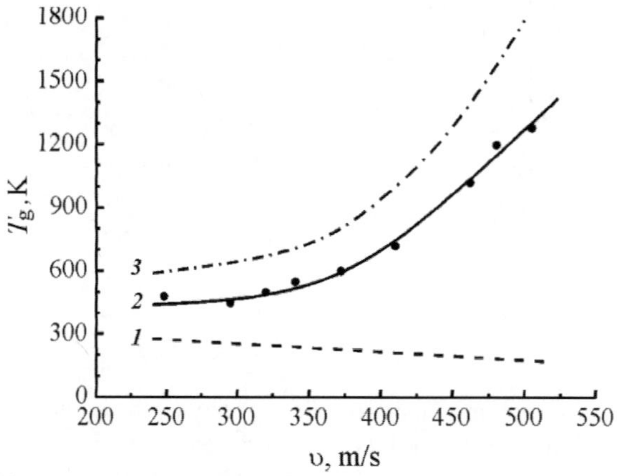

Figure 4 Dependence of the gas temperature on flow velocity without creating a discharge in it (dotted curve 1), in a pulsating discharge channel plasma (point 2) and calculated by the elastic heating mechanism (dot-dashed curve 3) with a discharge current $i = 15.5$ A. Distances from the electrodes $z = 1$ cm.

Then the gas temperature in the channel plasma of the pulsed discharge was measured at $z = 1$ cm and different flow rates. This dependence (points 2) is presented in figure 4. It can be seen that the gas temperature in the plasma channel near the electrodes increases from 450 K to 1250 K with an increase in the airflow velocity from 250 m/s to 500 m/s.

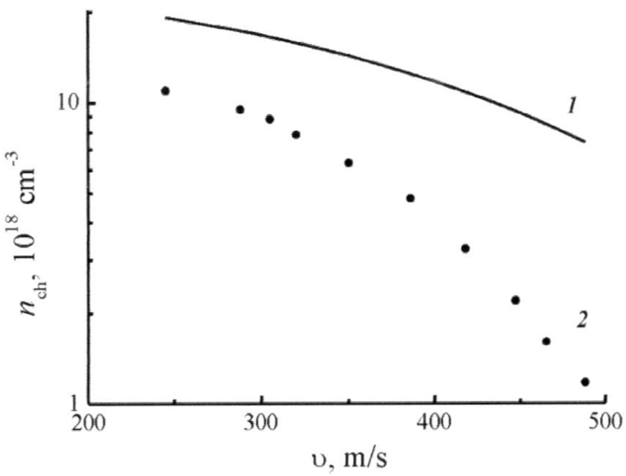

Figure 5 Dependence of the concentration of air molecules on flow velocity without creating a discharge in it (curve 1) and in a pulsating discharge channel plasma (point 2) at $i = 15.5$ A, $z = 1$ cm.

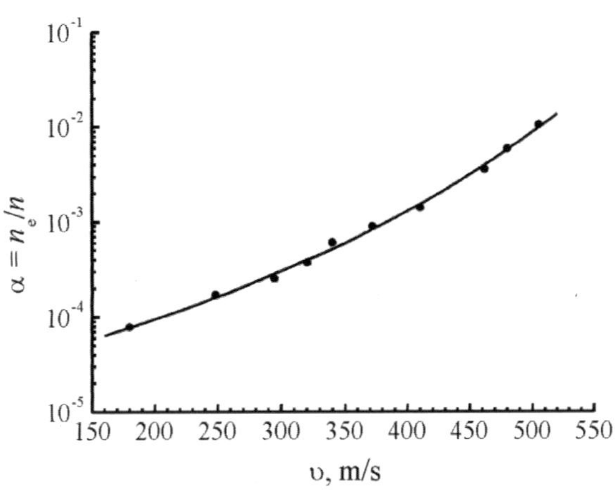

Figure 6 Dependence of the air ionization degree in a pulsating discharge plasma on the air flow velocity at $i = 15.5$ A, $z = 1$ cm.

The gas heating model took into account various processes leading to gas heating in weakly ionized plasma, such as energy transfer into translational degrees of freedom under elastic collisions of electrons with molecules, rotational-translational and vibrational-translational relaxation, vibrational-vibrational exchange, and quenching of electronically excited states of molecules. Estimates show that, in the highly ionized plasma of a pulsating discharge, the main mechanism for gas heating is heating due to elastic interactions [18]. Numerical simulations of gas heating via elastic collisions without allowance for the power loss due to the convective heat transfer in a high-speed flow shows that, at an electron density of 10^{15} cm^{-3}, the air heating rate is $dT_g/dt = 27$ K/μs. For an air flow velocity of 150 m/s, this corresponds to an increase in the gas temperature in the pulsating discharge by about 180 K. At velocity of 500 m/s and $n_e = 3 \times 10^{16}$ cm^{-3}, the air heating rate is $dT_g/dt = 810$ K/μs, which corresponds to an increase in the gas temperature in a pulsating discharge by about 1600 K. These results satisfactorily agree with the experimental data (figure 4).

With the help of these data, the concentration of air molecules in a heated plasma channel of a pulsating discharge was determined (figure 5).

Taking into account the obtained data on the concentrations of electrons and air molecules, the degree of ionization of the plasma in a nonstationary pulsating discharge created using a stationary power source in subsonic and supersonic air flows was determined. The dependence of the air ionization degree in the pulsating discharge plasma on the air flow velocity is shown in figure 6.

4. Conclusions

At low subsonic airflow rates, the degree of gas ionization in pulsating discharge plasma is of the order of 0.01 %, with the flow velocity increasing, the degree of ionization of the gas sharply increases and reaches a value of 1 % at flow velocity of 600 m/s. The obtained results show that near the electrodes the plasma channel of the nonstationary pulsating discharge is a strongly ionized medium.

Acknowledgments

The work was partially supported by the Russian Foundation for Basic Research (grant No. 18-02-00336-a).

References

[1] Alferov V I and Bushmin A S 1963 *Sov. Phys. JETP* **17** 1190
[2] Alferov V I, Bushmin A S and Kalachev B V 1967 *Sov. Phys. JETP* **24** 859
[3] Starikovskaya S M 2006 *J. Phys.* D **39** R265.
[4] Starikovskii A Y, Anikin N B, Kosarev I N, Mintoussov E I, Nudnova M M, Rakitin A E, Roupassov D V, Starikovskaia S M and Zhukov V P 2008 *J. Propul. Power* **24** 1182.
[5] Adamovich I V, Lempert W R Rich J W and Utkin Y G 2008 *J. Propul. Power* **24** 1198
[6] Jacobsen L, Carter C, Baurle R, Jackson T, Williams S, Barnett J, Tam C-J and Bivolaru D 2008 *J. Propul. Power* **24** 4 641
[7] Ershov A P, Surkont O S, Timofeev I B, Shibkov V M and Chernikov V A 2004 *High Temp.* **42** 4 516 DOI: 10.1023/B:HITE.0000039979.89955.df
[8] Ershov A P, Surkont O S, Timofeev I B, Shibkov V M and Chernikov V A 2004 *High Temp.* **42** 5 667 DOI: 10.1023/B:HITE.0000046519.53287.47
[9] Ershov A P, Kalinin A V, Surkont O S, Timofeev I B, Shibkov V M and Chernikov V A 2004 *High Temp.* **42** 6 865 DOI: 10.1007/s10740-005-0005-8
[10] Kopyl P V, Surkont O S, Shibkov V M and Shibkova L V 2012 *Plasma Phys. Rep.* **38** 6 503 DOI: 10.1134/S1063780X12050054
[11] Shibkov V M, Shibkova L V and Logunov A A 2017 *Plasma Phys. Rep.* **43** 3 373 DOI: 10.1134/S1063780X17030114
[12] Shibkov V M, Shibkova L V and Logunov A A 2017 *Moscow University Physics Bulletin* **72** 3 294 DOI: 10.3103/S0027134917030109
[13] Pashchenko N T and Raizer Yu P 1982 *Sov. J. Plasma Phys* **8** 617
[14] Dvinin S A, Ershov A P, Timofeev I B, Chernikov V A and Shibkov V M 2004 *High Temp.* **42** 2 171 DOI: 10.1023/B:HITE.0000026147.82949.36
[15] Akishev Yu S, Aponin G I, Grushin M E, Karal'nik V B, Pan'kin M V, Petryakov A V and Trushkin N. I. 2008 *Plasma Phys. Rep.* **34** 312
[16] Fridman A, Nester S, Kennedy L, Saveliev A and Mutaf-Yardimci O 1999 *Prog. Energy Combust. Sci.* **25** 211
[17] Aleksandrov A F, Kuzovnikov A A and Shibkov V M 1992 *J. Eng. Phys. Thermophys.* **62** 5 519 DOI: 10.1007/BF00862338
[18] Shibkov M V, Aleksandrov A F, Chernikov V A, Ershov A P and Shibkova L V 2009 *J. Propul. Power* **25** 123 DOI: 10.2514/1.24803
[19] Shibkov V M 1997 Heating of gas under conditions of free-localized microwave discharge in air:

Mathematical simulation *High Temp.* **35** 5 681

[20] Shibkov V M 1997 Heating of gas under the conditions of free-localized microwave discharge in air: Experiment *High Temp.* **35** 6 858

[21] Konstantinovskii R S, Shibkov V M and Shibkova L V 2005 *Kinet. Catal.* **46** 6 775 DOI: 10.1007/s10975-005-0136-2

[22] Shibkov V M, Aleksandrov A F, Ershov A P, Timofeev I B, Chernikov V A and Shibkova L V 2005 *Plasma Phys. Rep.* **31** 9 795 DOI: 10.1134/1.2048839

[23] Shibkov V M, Ershov A P, Chernikov V A and Shibkova L V 2005 *Tech. Phys.* **50** 4 455 DOI: 10.1134/1.1901784

[24] Shibkov V M, Dvinin S A, Ershov A P and Shibkova L V 2005 *Tech. Phys.* **50** 4 462 DOI: 10.1134/1.1901785

[25] Shibkov V M, Shibkova L V, Gromov V G, Karachev A A and Konstantinovskii R S 2011 *High Temp.* **49** 2 155 DOI: 10.1134/S0018151X11020143

[26] Shibkov V M and Shibkova L V 2009 *Tech. Phys.* **54** 10 1467 DOI: 10.1134/S1063784209100107

[27] Shibkov V M, Shibkova L V 2010 *Tech. Phys.* **55** 1 58 DOI: 10.1134/S106378421001010X

[28] Dvinin S A., Shibkov V M and Mikheev V V 2006 *Plasma Phys. Rep.* **32** 7 601 DOI: 10.1134/S1063780X06070099

[29] Olivero J J and Longbothum R L 1977 *J. Quant. Spectrosc. Radiat. Transfer* **17** 2 233

[30] Zlobin V V, Kuzovnikov A A and Shibkov V M 1988 *Moscow Univ. Phys. Bulletin* 42 1 98

Plasma Jet Interaction with Dielectric Surface

I Schweigert[1*], S Vagapov[1], L Lin[2] and M Keidar[2]

[1]Khristianovich Institute of Theoretical and Applied Mechanics, Novosibirsk 630090, Russia
[2]George Washington University, Washington D.C. 20052, USA

[*]ischweig@yahoo.com

Abstract. Propagation of cathode-directed streamer over the helium jet in ambient air is studied in the framework of fluid model and in experiments. The streamer is initiated by applying sinusoidal voltage to the powered electrode inside of dielectric tube and propagates to the direction of treated surface. The gas mixture is calculated with ANSYS Fluent software for helium supply of 4 sl/min. The experimental data on the speed of streamer are in good agreement with calculation results. The reflection of ionization wave from the surface was observed in the experiment and in simulations. It is shown that the reflected ionization wave follows a channel produced by the streamer.

1. Introduction

Cold atmospheric plasma jets are powerful instrument widely used for different applications in medicine [1], agriculture [2] and material surface manufacturing [3]. A streamer propagating from the powered electrode in a dielectric tube delivers the high density plasma to the treated surface, providing the local high electrical field, surface bombardment by ions and enhanced chemical reactions in the mixture of nitrogen, oxygen, water and etc. The large number of publications were devoted to the interaction of cold atmospheric plasma jet with different kinds of surfaces including the tissue. Nevertheless, some important mechanisms of influence of cold atmospheric plasma jet, for example, on human cells live cycle are still missing and are hot topics of recent experimental and theoretical studies. An interesting idea was proposed in Ref. [4]. Authors suggested that a change in permeability of the cell membrane is connected to cell compression due to an increase of osmosis pressure outside the cell as a result of presence ions created by plasma. Recently, the dynamics of different types of cold atmospheric plasma jets is intensively studied in the experiments and simulations (see for example, [5,6] for experiments and [7,8] for simulations).

In this work we study the dynamics of cold atmospheric plasma jet in helium in the experiment and in fluid model simulations and its interaction with a metal surface. The experimental setup is described in details in previous paper [6].

2. Model

The fluid model of discharge includes continuity equations for electrons, ions and mean electron energy U_e:

$$\frac{dn_e}{dt} = \mathrm{div}\big(\mu_e \mathbf{E} n_e + \nabla(D_e n_e)\big) + \alpha_i n_e n_g - \beta n_e n_i \,, \quad (1)$$

$$\frac{dn_i}{dt} = \mathrm{div}(\mu_i \mathbf{E} n_i + \nabla(D_i n_i) - V n_i) + \alpha_i n_e n_g - \beta n_e n_i \,, \quad (2)$$

IOP Conf. Series: Journal of Physics: Conf. Series **1112** (2018) 012004 doi:10.1088/1742-6596/1112/1/012004

$$\frac{\mathrm{d}(n_e U_e)}{dt} = \mathrm{div}\big(\mu_e \mathbf{E} n_e U_e + \nabla(D_u n_e U_e)\big) + e\mathbf{E}\big(\mu_e \mathbf{E} n_e + \nabla(D_e n_e)\big) - \nu_u n_e n_g \ , \quad (3)$$

and the Poisson equation for the electric potential φ and electric field $\mathbf{E}$:

$$\mathrm{div}\,\nabla\varphi = 4\pi e(n_e - n_i), \mathbf{E} = -\nabla\varphi \ , \quad (4)$$

where n_e, n_i are electron and ion densities, $\mathbf{V}$ is the gas flow velocity, n_g is the neutral gas density, e is the elementary charge, α_i, β are the coefficients of ionization and recombination.

The transport coefficients in the continuity equations for electrons and their mean energy are calculated from the electron energy distribution function (EEDF) $f_e(\varepsilon)$. The EEDF was calculated for a wide range of electric field strengths. The EEDF is found by solving of Boltzmann equation for a wide range of E/n_g. The transport coefficients are:

$$D_e = \frac{1}{n_e}\int \frac{v(\varepsilon)}{3n_g\sigma(\varepsilon)} f_e(\varepsilon)\mathrm{d}\varepsilon \ ,$$

$$\mu_e = \frac{-e}{n_e}\int \frac{v(\varepsilon)}{3n_g\sigma(\varepsilon)} \frac{df_e(\varepsilon)}{d\varepsilon}\mathrm{d}\varepsilon \ ,$$

$$D_u = \frac{1}{n_e}\int \frac{\varepsilon v(\varepsilon)}{3n_g\sigma(\varepsilon)} f_e(\varepsilon)\mathrm{d}\varepsilon \ ,$$

$$\mu_u = \frac{-e}{n_e}\int \frac{\varepsilon v(\varepsilon)}{3n_g\sigma(\varepsilon)} \frac{df_e(\varepsilon)}{d\varepsilon}\mathrm{d}\varepsilon \ ,$$

where $n_e = f_e(\varepsilon)\mathrm{d}\varepsilon$, $v(\varepsilon)$ is the electron velocity for the energy $\varepsilon, \sigma(\varepsilon)$ is the electron transport crossection.

For ions the transport coefficients are taken as:

$$\mu_i = \frac{3e}{16n_g\sigma_i}\left(\frac{\pi}{m_i kT}\right)^{1/2}\left(1 + \frac{9\pi^2 eE}{512 n_g kT\sigma_i}\right)^{-1/2} ,$$

$$D_i = \mu_i kT \ ,$$

where k is the Boltzmann constant, T is the neutral gas temperature, m_i is the ion mass and σ_i its transport crossection. The electron emission from the dielectric wall inside of the tube due to the electron and ion bombardments is taken into account. The surface charge on the walls of the dielectric tube inside and outside is calculated from the of calculation of electron and ion fluxes on the surface on every electron time step. This surface charge is included in the Poisson equation (4). The equations (1)-(4) we resolved self-consistently overcalculation domain shown in figure 1. The radius of calculation domain is $r_{max} = 4$ cm. The metal surface is at $z_{max} = 5.5$ cm.

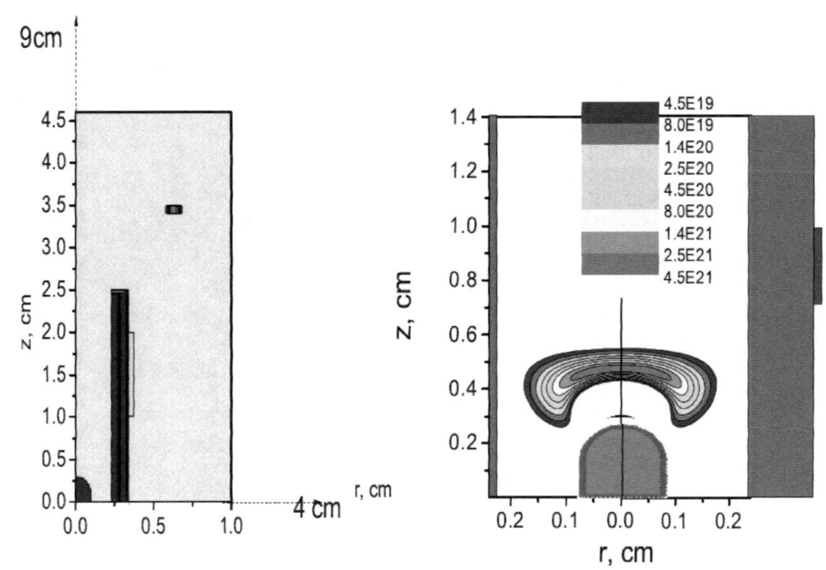

Figure 1 Calculation domain (left) and ionization rate distribution at $t = 3$ ns.

IOP Conf. Series: Journal of Physics: Conf. Series **1112** (2018) 012004 doi:10.1088/1742-6596/1112/1/012004

The dielectric tube has radius of 0.23 cm and length of 2.5 cm. The permittivity of dielectric is 10. The surface is 3 cm apart from the tube. The powered electrode is at $r = 0$, the grounded electrode is shown in light blue. In the experiment, the voltage with 4.5 kV amplitude was applied with the frequency of 25 kHz.

From our experimental observation of plasma jet it is seen that the characteristic length of absorption of photons around a streamer head is approximately 0.1 cm. Therefore, in our simplified model of photoionization we assume that electrons due to photoionization are generated within a semi sphere with radius of 0.1-0.15 cm around the streamer head and the excitation rate determines (with some coefficient $k = 0.1 - 1$) the photoionization rate.

The steady-state gas mixture distribution around the helium jet flowing out from the tube inlet with 4 sl/min in atmospheric air (nitrogen with 22% of oxygen) is shown in figure 2. This mixture was calculated with ANSYS Fluent software and used for calculation of electron transport coefficients, the mean electron energy and the ionization rate by the electron impact.

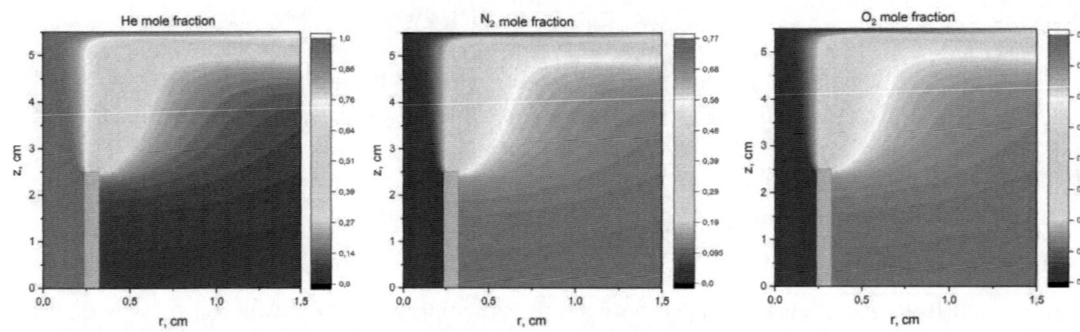

Figure 2 Spatial distributions of fraction of helium (a), nitrogen (b) and oxygen (c) for helium jet, 4sl/min near the tube inlet.

3. Results

At the moment of 'anode phase' of applied sinusoidal voltage, a cathode-directed streamer is created near the powered electrode and after a few nanoseconds it starts to propagate along the dielectric wall. In figures. 1 and 3, first steps of the streamer formation and propagation are shown. The streamer has a toroidal shape as it is indicated in figure 3(b) by dotted line. The measured and calculated velocity of the streamer outside of the tube first is 15 km/s, then rises up to 30 km/s and finally drops in front of the surface.

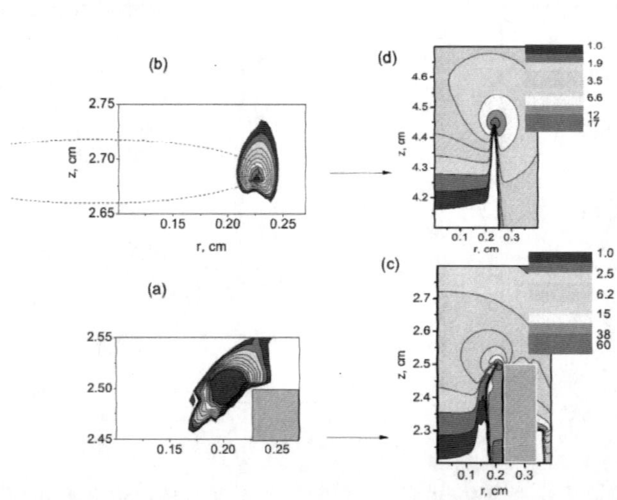

IOP Conf. Series: Journal of Physics: Conf. Series **1112** (2018) 012004 doi:10.1088/1742-6596/1112/1/012004

Figure 3 Ionization rate (a),(b) and electrical field distribution (c),(d). Ionization rate in log scale from 3×10^{18} to 3×10^{20} $cm^{-3}s$-1, the electrical field is in kV/cm.

Measured optical emission distribution around the propagated streamer is shown in figure 4. The intensity of emission is given in log scale. The images show dynamics of the streamer approaching the surface and reflected one for times 2 μs, 2.1 μs and 2.2 μs. These times are calculated from the moment of streamer departure from the tube.

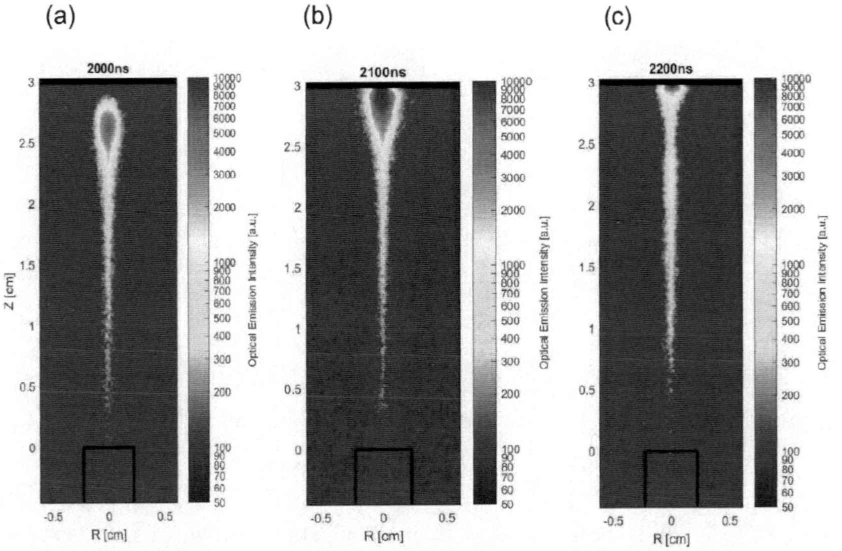

Figure 4 Optical emission intensity for $t = 2$ μs (a), 2.1 μs (b), 2.2 μs(c) for $U = 4.5$ kV. Optical emission intensity for $t = 2$ μs(a), 2.1 μs(b), 2.2 μs(c) for $U = 4.5$ kV.

The calculated ionization rate and the potential distribution for approximately the same times are shown in figure 5 for the streamer approaching the surface (a,d) and for initial stages of reflection of the ionization wave from the surface, (b,e) and (c,f). Both experimental and calculated results demonstrate an increase of ionization rate with approaching to the dielectric surface and the averaged velocity of the streamer is 25 km/s.

The 'seed' electrons arriving from the zone of helium mixture with air appear due to the photoionization. The electrons gain the energy crossing the streamer head where the electrical field is large and provide the ionization avalanche in a vicinity of the streamer head. So, behind the streamer a channel of plasma forms. Reflected ionization wave to the anode direction flows through the channel made by the straightforward streamer. In figure 5 (d),(e),(f), an elongated potential well occurs from the dielectric tube to the surface, tracking the way of the streamer. Note, that the characteristic times of ambipolar diffusion and recombination is much smaller than the characteristic time of a streamer.

In simulations, we performed a test and removed the channel produced by the streamer, artificially increasing the ambipolar diffusion coefficient. In this case, the reflection of the ionization wave does not take place.

In conclusion, in 2D fluid model simulations and in experiments the dynamics of formation and propagation of the cold atmospheric plasma jet have been studied. The propagation of the cathode-directed streamer inside of dielectrically tube takes place over the dielectric wall. Outside of the dielectrically tube the streamer has a toroidal shape. In the experiment and in simulations the reflection of streamer from the metal surface was registered.

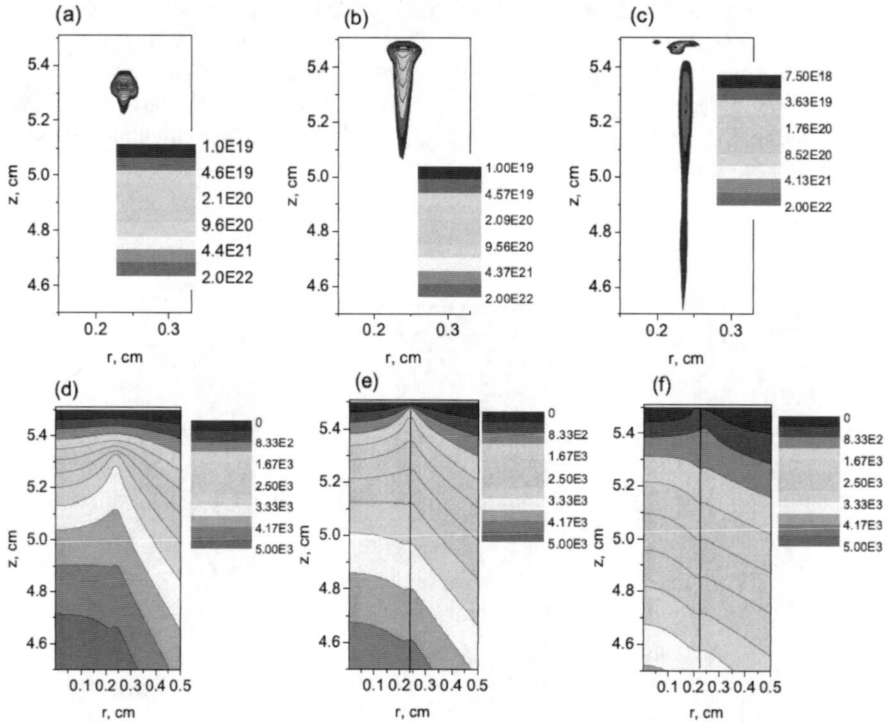

Figure 5 Spatial distributions of ionization rate (a), (b), (c) and potential (d), (e), (f) at different times.

Acknowledgments
Two of authors, IS and SV wish to thank RFBR N 18-08-00510 for financial support.

References

[1] Keidar M, Walk R, Shashurin A, Srinivasan P, Sandler A, Dasgupta S, Ravi R, Guerrero-Preston R and Trink R 2011 *British Journal of Cancer* **105** 12951301
[2] Misra N N, Schluter O and Cullen P J 2016 *Cold Plasma in food and agriculture* (Amsterdam: Elsevie)
[3] Mitsuharu Konuma 1992 *Film Deposition by Plasma Techniques* (Berlin: Springer)
[4] Shneider M N and Pekker M 2018 *J. Appl. Phys.* **123** 204701
[5] Shashurin A, Shneider M N, Dogariu A, Miles R B and Keidar M 2009 *Appl. Phys. Lett.* **94** 231504
[6] Shashurin A and Keidar M 2015 *Phys. Plasmas* **22** 122002
[7] Babaeva N Yu and Naidis G V 1996 *J. Phys.* D: *Appl. Phys.* **29(9)** 2423
[8] Babaeva N Yu and Naidis G V 2016 *Phys. Plasmas* **23** 083527

IOP Conf. Series: Journal of Physics: Conf. Series **1112** (2018) 012005 doi:10.1088/1742-6596/1112/1/012005

Suppression of reflected oblique shock wave by multi-filamentary plasma

S B Leonov[1,2*], **A A Firsov**[2] **and A W Houpt**[1]

[1]University of Notre Dame, Notre Dame, IN 46556, USA
[2]Joint Institute for High Temperature RAS, Moscow, 125412, Russia

*sleonov@nd.edu

Abstract. This paper describes the results of experimental study of stripwise near-surface plasma effect on oblique shock wave (SW) reflection from a solid wall in a $M = 2$ duct-driven shock-dominated airflow. Methods applied include schlieren visualization, plasma characterization with electric probes and optical emission spectroscopy, wall pressure measurements along the test section. The transient filamentary plasma is generated by means of a Quasi-DC electric discharge between flush mounted surface electrodes in a spanwise array. Test arrangement includes a shock wave (SW) generator installed on the bottom wall of the test section and the plasma generator installed on the upper wall. The SW generated from the bottom wall impinges the significantly disturbed boundary layer (BL) generated by the plasma array. The effect of reflected SW mitigation is demonstrated due to the interaction with the plasma-modified BL. The model of interaction is discussed and supported by the results of numerical simulation.

1. Introduction

The control of shock wave interaction with a boundary layer (SWBLI), other SWs, and separated flows in a supersonic/hypersonic airflow is of fundamental and technological interest for aerospace science and industry. Currently airflow conditioning in supersonic inlets is one of the major issues of high-speed airbreathing propulsion [1, 2, 3]. The flow structure in the duct (isolator), at the entrance to the combustor, at the compression surfaces of the inlet, and over other control surfaces is sensitive to the geometrical configuration and main flow parameters, including the state of the boundary layer and the presence of compression/expansion waves previously impinging the BL. In most cases, the control schemes include stationary / movable mechanical elements or gas wall jets. The issue here is in a lack of flexibility, a total pressure loss, and frequently in a long response time, orders of magnitude longer than a characteristic gasdynamic time. Intensive studies of the physical processes at SWBLI are motivated by the need of precise control of shock-dominated flows in ducts, over compression surfaces, or other conditions where BL and shear layers are present. Electrical discharges of different types and in different geometries were tested for such a purpose [4, 5, 6, 7, 8, 9, 10, 11, 12, 13].

The importance of plasma non-uniformity and its transient behavior for SW and BL control has been considered in numerous publications [14, 15, 16, 17, 18, 19, 20, 21, 22]. A filamentary plasmas, to some extent equivalent to heated zones with much higher sonic velocity, significantly changes the SW structure and velocity [21, 22, 20]. The localized heating generated by the plasma produces "hot spots" operating similar to solid obstacles, although this interpretation may be too simplistic. The flow interaction with the hot spots may result in the generation of new flow structures, such as streamwise vortices and wavelets travelling with the flow, while a rapid modulation of discharge energy coupling / body force may lead to effectively tripping the boundary layer. The main premise of this approach was forcing the flow with a high amplitude perturbation. A significant localized heating in the near-surface discharge filaments and repetitive compression wave formation by the plasma were demonstrated, for example in Refs [23, 24, 5]. In addition to this, localized plasma actuators also excite flow instabilities, generate large-scale coherent structures in the flow [7, 6], and result in significant mixing enhancement and modification of shear layers [25, 26, 27].

This work explores the Quasi-Direct-Current (Q-DC) electrical discharge impact on the shock-wave structure and wall pressure distribution in a supersonic duct-driven airflow. A major focus is the effect of a streamwise filamentary plasma array on the reflection pattern of the impinging shock wave at the interaction with the boundary layer.

2. Test setup and instrumentation

The experiments were performed in the supersonic blow-down wind tunnel SBR-50 at the University of Notre Dame. The test section is arranged inline with a Mach 2 Laval nozzle. The cross section at the exit of the nozzle ($x = 0$) is $Y \times Z = 76.2$ mm (width) × 76.2 mm (height), with a 1° expansion angle on the top and bottom walls and a total length of $x = 610$ mm measured to the diffuser, as it is shown in figure 1. The electrode array is flush-mounted on a plane wall as a single unit, indicated in figure1 as "Electrodes". The test section of the SBR-50 high-speed facility is equipped with 2 pairs of quartz windows placed as the side walls of the duct for optical access. In the current experimental series, the conditions were as follows: initial Mach number M=2; total pressure $P_0 = 1 - 3$ bar; stagnation temperature $T_0 = 300$ K; duration of steady-state aerodynamic operation t = 1-2 s. At unit Reynolds number $Re_L = 1.3 - 4 \cdot 10^7$ m^{-1}, a turbulent boundary layer was developed at the plasma location with an initial thickness $\delta = 2 - 3$mm. A trail of reflecting shock waves, related to the plasma generation, is indicated by straight lines. More details are available in [20].

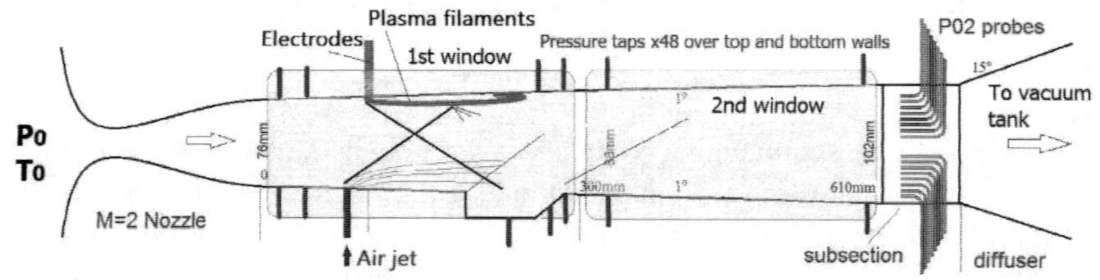

Figure 1 Geometric configuration of the SBR-50 test section with air jet based SW generator.

In this test series a controlled SW generator was installed on the bottom wall: a crossflow array of 5 sonic air jets supplied with a pulsed valve (impulse ratio is approximately $J = 1$). In this scheme a wall cavity was arranged on the bottom wall simulating a cavity-based flameholder. Instrumentation available for the current test include:

- pressure distribution, 48 pressure ports over the top and bottom walls, 16 channels PSI 9116 scanner, response time 2.5ms;
- set of P02 pressure sensors – Pitot probes, installed in a downstream zone between the test section and diffuser;
- fast camera imaging Photron FastCam (black / white);
- high-definition schlieren system, details are described in [20];
- optical emission spectrograph;
- set of electrical probes.

For the plasma generation, the custom-made power supply used in the present experiments is designed to operate with a steep falling voltage-current characteristic and individual control of each output channel [20]. The electrodes were powered by a constant-voltage, $U_{ps} = 5$ kV. The power supply has been connected to the electrodes with the help of a solid-state Behlke™ switch. Typical time sequences of plasma voltage and the discharge power are shown in figure 2 measured at a total pressure $P_0 = 3$ bar; a total plasma power release was $W_{pl} \approx 20.2$ kW in this case. A strong coupling of the plasma to the flowing gas causes fast oscillations of the plasma shape, and consequently, of the plasma voltage and power, as shown in figure 2b. The saw-tooth waveform of the plasma voltage is a result of the movement of individual plasma filaments with the airflow.

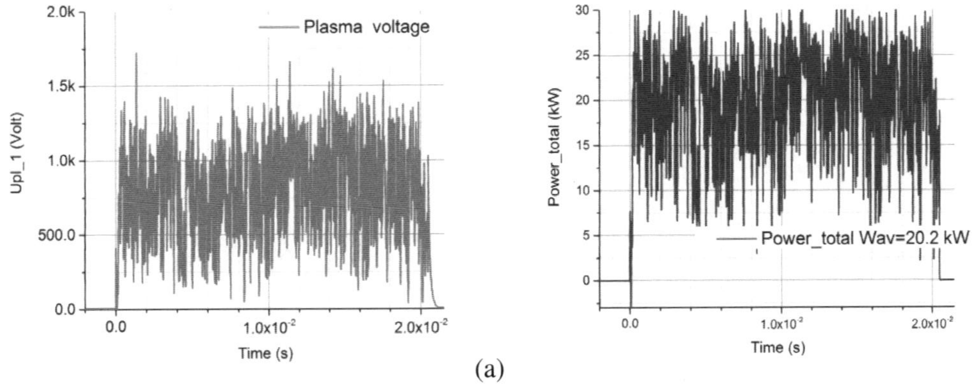

(a) (b)

Figure 2 Typical voltage-power record: (a) voltage on electrode gap U_{pl}; (b) discharge power W_{pl}.

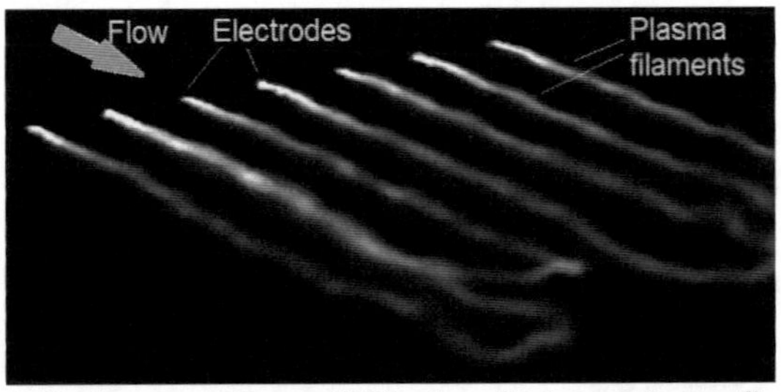

Figure 3 Image of the Q-DC plasma at $P_{st} = 270$mbar, 7 electrode configurations. Exposure 10μs.

The morphology of the Q-DC electric discharge appears in the form of a loop of a constricted (filamentary) plasma connecting the neighboring electrodes, as shown in the camera image in figure 3. The dominant frequency of plasma oscillations is $F = 8 - 15$ kHz, depending on the inter-electrode gap and the gas density / velocity, which govern the length of the plasma loops. Two basic electrode configurations were used (7 and 11 electrodes): the plasma filaments are shorter for the 11 electrode discharge than for the 7 electrode case. At lower pressures the 11 electrode discharge oscillates at a higher frequency, as it is illustrated in figure 4a. As the pressure increases the dominant frequency of the 11 electrode array approaches that of the 7 electrode array. The plasma power is typically in the range of 5-20 kW for the parameters used for this work. Figure 4b shows the nearly linear dependence of the power on the static pressure when the circuit parameters/geometry are held constant. It is important to note that at the same average current the power release for the 11 electrode configuration is significantly lower than for the 7 electrode pattern. The gas temperature in the plasma filaments was measured by optical emission spectroscopy to be as high as $T_g = 3800 - 6000$K depending on conditions [20]. With a high local gas temperature, the multi-filamentary plasma zone presents an array of longitudinal subsonic jets surrounded by a supersonic flow. The volumetrically expanded zone produces a long cone of subsonic flow, where the physical velocity may be close to the gas speed in the supersonic core flow. Such an uncommon gasdynamic structure produces a near-surface stratified zone, which enables a significant redistribution of the gas pressure.

IOP Conf. Series: Journal of Physics: Conf. Series **1112** (2018) 012005 doi:10.1088/1742-6596/1112/1/012005

The effect of the Q-DC plasma generated near the plane wall consists of gas heating in the airflow zone downstream of the electrode system, which causes an increase of the gas pressure and the subsequent propagation of a compression wave. In supersonic airflow, an oblique shock wave (SW) is observed originating from the root part of the plasma filaments [24, 20]. The SW impinges on the opposite (bottom) wall increasing the pressure in the zone of interaction. Then, a series of SW reflections is observed in the test section downstream of the plasma generator, see figure 1. An important feature of the Q-DC generated shock wave should be noted: despite the filamentary and transient nature of the discharge the generated shockwave is nearly planar in the region away from the electrodes [28].

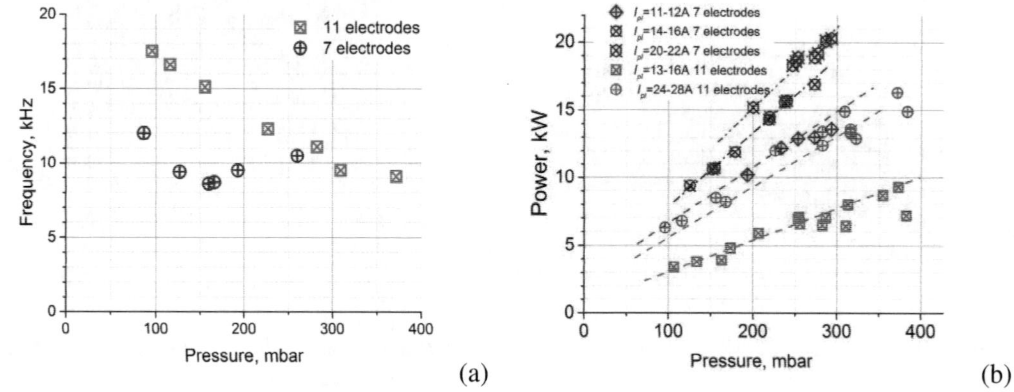

Figure 4 Data for discharge oscillation frequency (a) and plasma power (b) as a function of discharge arrangement, electric current, and static pressure.

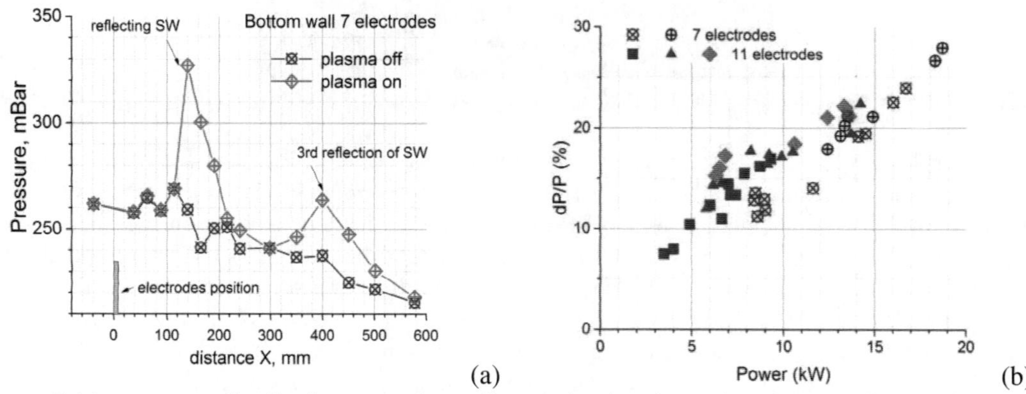

Figure 5 (a) - pressure distributions over the wall opposite the electrodes; (b) - amplitude of relative pressure increase as a function of plasma power.

The 16 channel pressure scanner provides the dataset of pressure distribution throughout the test section. The pressure distribution for the wall opposite the electrodes is presented in figure 5(a) for the 7 electrode discharge. There are two large pressure increases which occur at approximately 140 mm and 400 mm downstream of the electrodes. These pressure peaks are a result of the shock wave-boundary layer interaction between the plasma generated shock wave and the opposite wall. The relative pressure increase (maximum pressure increase normalized by the static pressure at that point) without the discharge for 7 and 11 electrode arrays is shown in figure 5(b). The pressure increase as a function of plasma power follow the linear trend. Note that the 11 electrode system targets a smaller area of influence while achieving the same relative pressure increase.

IOP Conf. Series: Journal of Physics: Conf. Series **1112** (2018) 012005 doi:10.1088/1742-6596/1112/1/012005

3. Test results

The dynamics of the reflected SW was explored when unsteady SW generator was installed on the bottom wall of the test section. The multiple jet configuration allows for a nearly plane SW impinging the upper wall downstream of the electrodes installation, as shown in figure 1. Typically, the plasma was turned on first followed with the impinging SW. The air jets continued operation after the plasma was turned off. During each run of the facility, the data was collected on a baseline, SW structure at the air jets operation, SW structure at plasma operation, and with SW and plasma both. Schlieren images for three cases are shown in figure 6 (a)-(c), correspondingly. The figure 6(c) demonstrates a significant mitigation of the reflected SW. This effect was observed in a wide range of flow parameters, $P_0 = 1 - 3bar$.

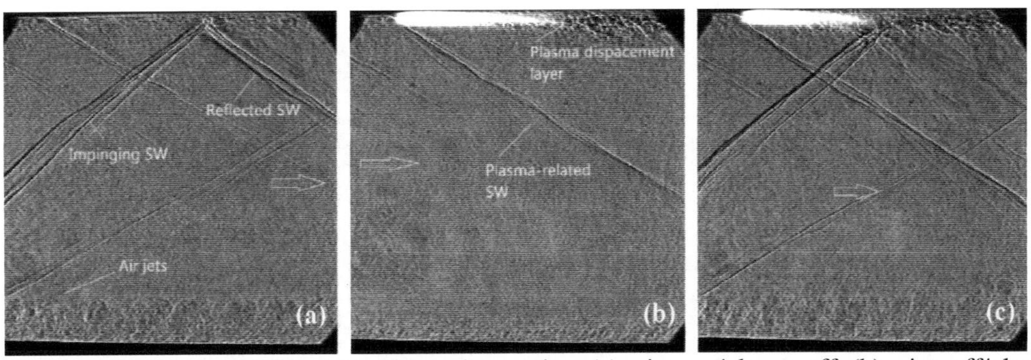

Figure 6 Schlieren images of the SW-plasma interaction. (a) – jets on/plasma off; (b) – jets off/plasma on; (c) – jets on/plasma on.

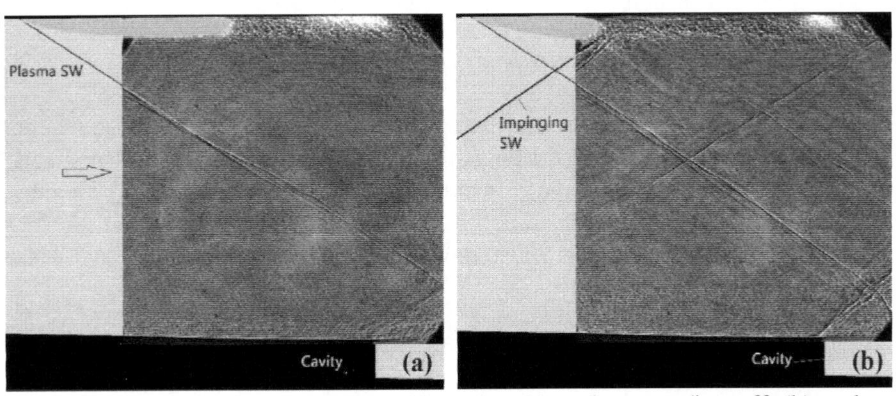

Figure 7. Schlieren image of the interaction with cavity. (a) – plasma on/jets off; (b) – plasma on/jets on.

Presence of the impinging SW modifies the plasma morphology, as shown in figure 7: the filament length is observed to be shorter due to plasma termination in the zone of increased pressure. The zone of the pressure elevation penetrates upstream along with the hot plasma filaments, presenting local subsonic areas. The pressure increases in the plasma zone which enhances the plasma-related SW strength. For the configuration with the cavity on the bottom wall, the plasma-related SW impinges the cavity, increases the cavity pressure, and promotes the SW originated from the cavity rim, as shown in figure 7(b).

The result of the pressure measurements is shown in figure 8 for $P_0 = 2.5$ bar. Figure 8(a) demonstrates the result of the interaction in the vicinity of the electrodes. During air jet operation, the impinging SW interacts with the upper wall around $x = 40 - 60$mm downstream of the electrodes that is represented by a significant pressure rise at the appropriate pressure tap. The reflected SW impinges

IOP Conf. Series: Journal of Physics: Conf. Series **1112** (2018) 012005 doi:10.1088/1742-6596/1112/1/012005

the cavity, leading to some pressure increase there, as shown in figure 8(b). During plasma generation, the pressure increases right behind the plasma zone and then drops down as fast as $x < 30$mm. The plasma-based SW affects the cavity flow similar to the air jet SW. In the case of plasma on /air jets on, the maximal pressure increase on the upper wall is less than the case of SW only. However, the length of the elevated pressure zone appears as long as $x > 60$mm. The pressure effect of the air jet related SW "propagates" forward to the plasma zone through the subsonic zone near the heated plasma filaments. The plasma related SW is amplified which then increases the pressure in the cavity. Note, in this case the pressure in the cavity is measured higher than upstream of the cavity, leading to an enhancement of the gas exchange between the cavity and the core flow and appearance of the SW originating from the cavity rim. Note that, based on the Pitot pressure sensors data, the described manipulation with the air jet and plasma slightly reducing the total pressure in the flow.

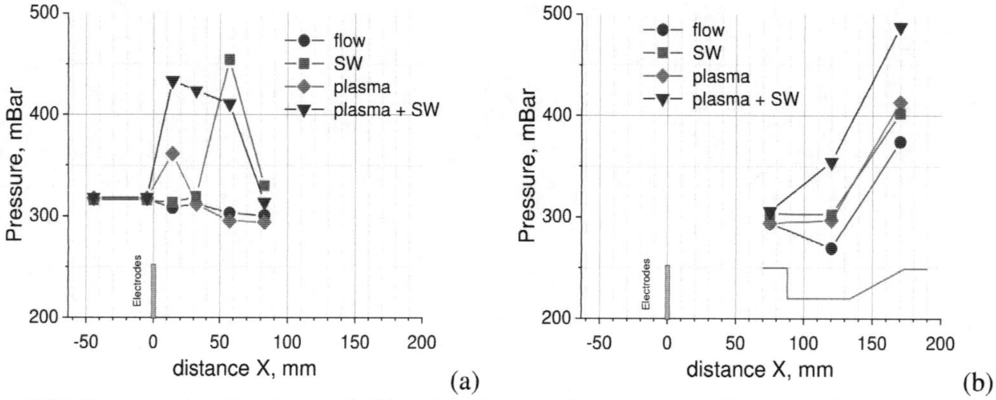

Figure 8 Wall pressure redistribution behind the plasma (a) on upper wall; and in vicinity of the wall cavity (b) on bottom wall.

4. Discussion and summary

In fact, the multi-filamentary plasma zone presents a set of longitudinal subsonic jets surrounded by a supersonic flow, as shown in figure 9(a) for a single plasma filament. The volumetrically expanded zone produces a cone of subsonic flow, where the physical velocity is close to the gas speed in the supersonic core flow. The pressure is increased due to the impact of the impinging SW and affects this cone, increasing the gas pressure in the subsonic zone. The gas expansion induces the conical SW attached to the plasma filament root (electrode). The shape of a "soft" plasma trail complies reducing the reflected SW strength.

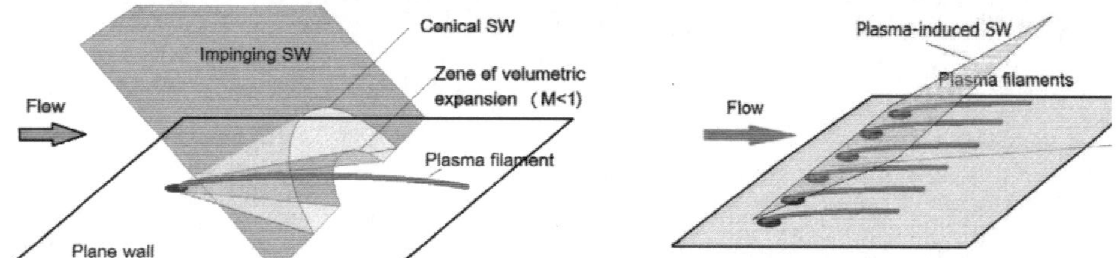

Figure 9 (a) - scheme of a single plasma filament interaction with supersonic flow and impinging SW; (b) plasma array and SW superposition.

In the case of a multi-filament plasma array, figure 9(b), the interference of the conical shock waves produces a combined compression wave attaining the form of a plane shock. A specific structure of the near-surface gas layer, consisting of the intermittent lengthwise zones of supersonic and subsonic flow, possesses a mitigating effect on an external impinging SW. The mechanism of interaction is considered as follows: the pressure, increasing due to the impact of the impinging SW, affects the whole subsonic area, increasing the gas pressure in the upstream zone and reducing the pressure magnitude right after the SW.

The plasma-SW-flow interaction in configuration with a fixed SW generator (analogy of experimentally used jet-based SW generator) was simulated numerically using FlowVision CFD software. The simulation was based on the solution of the three-dimensional unsteady Reynolds averaged Navier–Stokes equations accompanied by the $k-\varepsilon$ turbulence model. The calculation domain includes only a thin layer, which is 4.5mm wide and contains half of one plasma filament located near a side plane of the calculation domain. The symmetry condition was set to both side planes of calculation domain. No-slip conditions and wall functions were used on top and bottom boundaries ($y+$ in range of 10-50). Geometry of the test section corresponds to the experimental one. Influence of the plasma filament produced by Q-DC discharge was simulated by 3D volumetric heat sources. This approach demonstrated good consistency of experimental results and simulation data for plasma flow control in a supersonic inlet [12]. The volumetric heat source was 80mm long and 1.8mm in diameter which approximate the experimental data obtained from image analysis. Power of a single plasma filament was varied in the range of $W_1 = 125 - 1000$ W. Artificial schlieren images (gas density gradient) are shown in figure 10 for two cases: (a) baseline - plasma off; (b) heat source power $W = 8$kW that is close to one of experimental conditions.

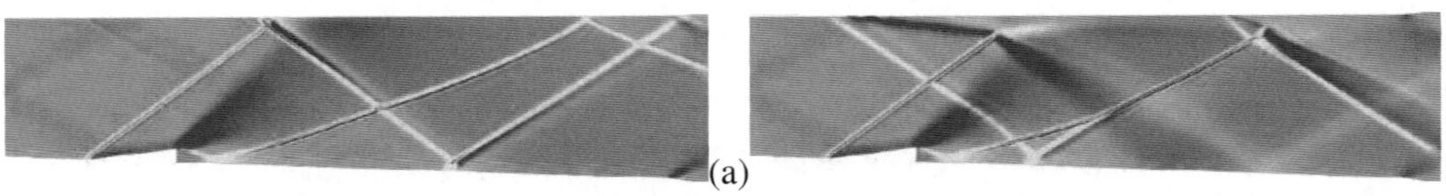

Figure 10 Result of CFD simulations: density gradients. (a)- baseline, plasma off; (b) heat source power $W = 8$kW.

In figure 10(b), there is a significant reduction of the oblique shock caused by the solid wedge. Figure 11 shows the distribution of the Mach number in vicinity of a single heated filament. It is seen that the "plasma" filament produces an extensive subsonic zone of a semi-conical shape. These data support the hypothesis on a stripwise structure producing by the plasma array near the surface with alternating subsonic-supersonic airflow. It was also found that the attenuation of the shock reflection leads to a local increase of total pressure.

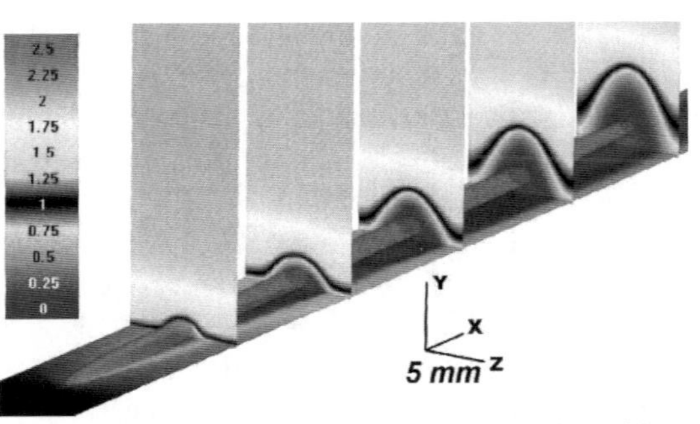

Figure 11 Result of CFD simulations: flow Mach number in vicinity of heated filament, $W_1 = 500$W.

The generation of surface localized discharges in a high-speed flow leads to a substantial change in the structure and parameters of the flow field. The Q-DC electrical discharge affects the flow similar to a soft wedge, whose angle depends on the electrical power deposition. The structure of the plasma-based displacement layer appears as an array of intermittent supersonic and subsonic lengthwise zones. The plasma-based configuration may be applied to mitigate the reflected SW due to the impinging SW interaction with the plasma-related displacement layer. It is beneficial for critical flow configurations because the electrical discharge can be switched on/off electronically at any time and synchronized with all other processes, such as a trajectory change or engine thrust modulation. The active compliant structure may aim in shifting the reflected SW, the mitigation of reflected SW strength, or an elimination of the reflected SW for flow control purposes. This technique demonstrates feasible potential for the control of shock wave – boundary layer interaction.

Acknowledgments
Numerical simulation presented in this work was performed under support of Russian President Grant for young scientists MK 1734.2017.8.

References

[1] VanWie D M 2000 Scramjet Inlets. Scramjet Propulsion *Progress in Astronautics and Aeronautics* vol.189 eds. E.T. Curran and S.N.B. Murthy (Reston, VA, AIAA) p. 447–511.
[2] Surber L E and Tinapple J A 2012 Inlet Flow Control Technology: Learning from History, Reinventing the Future *AIAA Pap.* 2012-0012.
[3] Dolling D S 2001 Fifty Years of Shock-Wave/Boundary-Layer Interaction Research: What Next? *AIAA J.* **39 8** 1517
[4] Greene B R, Clemens N T, Magari P and Micka D 2015 Control of mean separation in shock boundary layer interaction using pulsed plasma jets *Shock Waves* **25 5** 495-505
[5] Hahn C, Kearney-Fischer M and Samimy M 2011 On factors influencing arc filament plasma actuator performance in control of high speed jets *Exp. Fluids* **51** 591-603
[6] Kearney-Fischer M, Kim J-H and Samimy M 2011 A Study of Mach Wave Radiation Using Active Control *J. Fluid Mech.* **681** 261-92
[7] Kim J-H, Nishihara M, Adamovich I, Samimy M, Gorbatov S and Pliavaka F 2010 Development of Localized Arc Filament RF Plasma Actuators for High-Speed and High Reynolds Number Flow Control *Exp. Fluids* **49** 497-511
[8] Macheret S, Shneider M and Miles R 2004 Magnetohydrodynamic and Electrohydrodynamic Control of Hypersonic Flows of Weakly Ionized Plasmas *AIAA J.* **42 11** 1378-87
[9] Narayanaswamy V, Raja L L and Clemens N T 2012 Control of a Shock/Boundary-Layer Interaction by Using a Pulsed-Plasma Jet Actuator *AIAA J.* **50 1** 246-9
[10] Webb N, Clifford C and Samimy M 2013 Control of oblique shock wave/boundary layer interactions using plasma actuators *Exp. Fluids* **54 6** 1545
[11] Wang J, Li Y H, Cheng B Q, Su C B, Song and Y Wu H M 2009 Effects of plasma aerodynamic actuation on oblique shock wave in a cold supersonic flow *J. Phys.* D*: Appl. Phys.* **42** 165503
[12] Falempin F, Firsov A, Yarantsev D, Goldfeld M, Timofeev K and Leonov S 2015 Plasma control of shock wave configuration in off-design mode of M = 2 inlet *Exp. Fluids* **56** 54
[13] Hedlund B, Houpt A, Gordeyev S and Leonov S 2018 Measurement of Flow Perturbation Spectra in Mach 4.5 Corner Separation Zone *AIAA J.* **56 7** 2699
[14] Hanson R E, Lavoie P, Naguib A M and Morrison J F 2010 Transient growth instability cancelation by a plasma actuator array *Exp. Fluids* **49** 1339–48
[15] Rethmel C, Little J, Takashima K, Sinha A, Adamovich I and Samimy M 2011 Flow Separation Control Using Nanosecond Pulse Driven DBD Plasma Actuators *Int. J. Flow Control* **3** 213-32
[16] Kurz A, Grundmann S, Tropea C, Forte M, Seraudie A, Vermeersch O, Arnal D, Goldin N and King R 2013 Boundary Layer Transition Control using DBD Plasma Actuators *Journal AerospaceLab* **6** 2013.
[17] Kopiev V F et. al. 2014 Instability wave control in turbulent jet by plasma actuarors *J Phys* D*: Appl Phys* **47** 505201
[18] Moralev I, Boytsov S, Kazansky P and Bityurin V 2014 Gas-dynamic disturbances created by surface dielectric barrier discharge in the constricted mode *Exp. Fluids.* **55 5** 1747

IOP Conf. Series: Journal of Physics: Conf. Series **1112** (2018) 012005 doi:10.1088/1742-6596/1112/1/012005

[19] Choi K-S., Jukes T and Whalley R 2011 Turbulent boundary-layer control with plasma actuators *Phil. Trans. R. Soc.,* A **369** 1443–58

[20] Houpt A, Hedlund B, Leonov S, Ombrello T and Carter C 2017 Quasi-DC Electrical Discharge Characterization in a Supersonic Flow *Exp. Fluids* **58 4** 25

[21] Bityurin V, Klimov A, Leonov S, Wie D V, Brovkin V, Kolesnichenko Y and Lutsky 1999 A Effect of heterogeneous discharge plasma on shock wave structure and propagation *AIAA Pap.* 1999-4940.

[22] Soloviev V R, Krivtsov V M, Konchakov A M and Malmuth N D 2003 Drag Reduction by Plasma Filaments over Supersonic Forebodies *AIAA J.* **41 12** 2403-09

[23] Utkin Y G, Keshav J, Kim J-H, Kastner J, Adamovich I V and Samimy M 2007 Development and Use of Localized Arc Filament Plasma Actuators For High-speed Flow Control *J. Phys.* D*: Appl. Phys.* **40** 685-94.

[24] Leonov S B and Yarantsev D A 2008 Near Surface Electrical Discharge in Supersonic Airflow: Properties and Flow Control *J. Propul. Power* **24 6** 1168-81

[25] Nishihara M, Frederickson K and Lempert W R 2016 Dual-Pump CARS Measurements in a Vibrationally Nonequilibrium Supersonic Mixing Layer *AIAA Pap.* 2016-1762.

[26] Leonov S, Firsov A, Savelkin K and Yarantsev D 2015 Plasma-enhanced mixing and flameholding in supersonic flow *Philos. Trans. R. Soc.,* A **373** 20140337

[27] Leonov S B, Isaenkov Y I, Yarantsev D A, Kochetov I V, Napartovich A P and Shneider M N 2009 Unstable Pulse Discharge in Mixing Layer of Gaseous Reactants *AIAA Pap.* 2009-0820

[28] Hayashi T, Houpt A, Hedlund B E, Leonov S and Sakaue H 2017 Pressure-Sensitive Paint Measurement under Transient Plasma in M=2 Airflow *AIAA Pap.* 2017-0704

[29] Shneider M N, Macheret S O, Zaidi S H, Girgis I and Miles R B 2008 Virtual Shapes in Supersonic Flow Control with Energy Addition *J. Propul. Power* **24 5** 900-15.

IOP Conf. Series: Journal of Physics: Conf. Series **1112** (2018) 012006 doi:10.1088/1742-6596/1112/1/012006

Discontinuities dynamics after the interaction of a plane shock wave with pulse volume discharge

I Znamenskaya[1], A Kuznetsov[1], I Mursenkova[1*] and I Doroschenko[1]

[1]Faculty of Physics, Moscow State University, 119991 Leninskie gory, Moscow, Russia

*murs_i@physics.msu.ru

Abstract. The dynamics of the nanosecond combined volume discharge radiation in the air flow with the plane shock wave was studied. Shock wave with Mach number 1.9-4.5 was passing through the discharge chamber and coming out of it while the electric pulse was switched on. The radiation spectra and the discharge currents were registered in the air pressure range 10-150 Torr, a pulsed voltage of 25 kV, and an electric current ~1 kA. It is shown that the electric current duration depends on shock wave position relative to the discharge gap and does not exceed 500 ns at various conditions. As a result of discharge energy input, the shock breakdown occurs with the formation of shock waves and contact surfaces. The high-speed shadow imaging was used to study the flow evolution after the discharge pulse. The experimental data on shock waves and contact surfaces positions is compared with numerically calculated positions, making it possible to determine the amount of thermal energy release during the discharge electric current time.

1. Introduction

Low-temperature non-equilibrium plasma of gas discharges is widely used in many technological processes. In plasma aerodynamics, the generation of controlled plasma areas can be used to reduce dynamic and thermal loads on the surface of an aircraft, to correct the flow regime and to ignite the fuel in the engines [1-4]. Pulse discharges are characterized by a change in the state of a gas and fast heating of the gas as a result of pulsed energy input during the short time of the discharge current. In the initiation of pulsed discharges, the gas heating [1, 3] and shock-wave mechanism [4, 5] determine the action on the flow. Shock waves generated as a result of a sharp change in the state of the medium can noticeably affect the gasdynamic flow. The intensity of these shock waves and their dynamics depend on the processes occurring at the stage of the discharge current.

The combined volume discharge with plasma electrodes providing uniform energy input into the gas due to the diffusive form was studied in stationary air and in the gasdynamic flow with a plane shock wave [6]. The presence of shock waves in the discharge volume may change the discharge current regime, the spatial distribution of charged particles, and the radiation structure and duration [6, 7]. When a pulsed discharge is initiated in a channel gas flow with a shock wave, the discontinuity breakdown conditions are created at the shock front [8].

To use the discharge effect on high-speed flows, it is necessary to know the mechanism of interaction of the shock wave with the plasma region. The goals of the work were registration of discharge current and radiation under interaction of discharge with a plane shock wave inside the discharge volume and outside it and analysis of the flow field after the discharge on the base of high-speed digital shadow imaging.

2. Experimental Setup

The investigations were carried out in discharge test chamber mounted into the gas-dynamic shock tube [4-6]. A pulsed combined volume discharge with preionization by ultraviolet radiation from plasma sheets was initiated in a discharge chamber. Discharges of this type are used in gas-discharge and excimer lasers [9]. The discharge was realized at operating voltage of 25 kV and switched by a controller. The discharge current was ~ 1 kA, duration was ~ 300 ns. The sliding surface discharges with an area of 3×10 cm^2 were created on the upper and lower walls of the chamber at a distance of 24 mm from each other. The sliding surface discharges provided the preionization and operated as the electrodes of volume discharge. After reaching a sufficient concentration of electrons, a volume discharge was formed. Two opposite walls of the discharge chamber were the quartz glasses within a length of 17 cm without changing the cross section of the channel (figure 1).

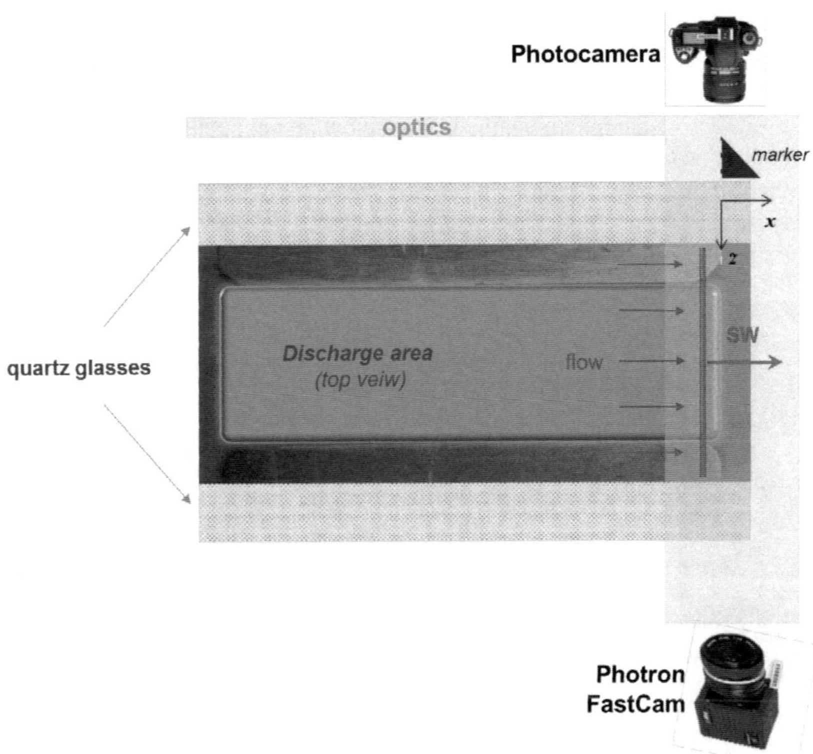

Figure 1 Scheme of discharge chamber channel with a shock wave in discharge volume.

The high-speed shadow imaging was used to study the flow pattern evolution after the discharge. Shadow images of the gas-dynamic flow field after the discharge were recorded with a high-speed Photron Fastcam camera with a frame rate of up to 500,000 fps and exposure time 1 μs. The laser beam (532 nm) was formed and directed perpendicular to the glasses of the discharge chamber by the optical system of shadow visualization (figure 1). An important advantage of high-speed imaging was the ability to trace the evolution of the shock-wave structure in the case of each particular experiment, since the discharge structure, as a rule, differs in each case of discharge initiation, and accordingly the evolution of the shock wave configuration is different.

Using the signals of piezoelectric pressure sensors in the shock tube channel, the velocity of the shock waves was measured and synchronization of the processes in the experiments was provided. The

IOP Conf. Series: Journal of Physics: Conf. Series **1112** (2018) 012006 doi:10.1088/1742-6596/1112/1/012006

radiation of the discharge were recorded by photo cameras. Multi-frame images of the discharge radiation were obtained using ICCD camera BIFO K011 during 20 μs after discharge. The spectra of radiation are recorded by a spectrometer AvaSpec-2048FT (174-1100 nm). The discharge current is measured by means of a calibrated low-inductance shunt.

The combined volume discharge was investigated in quiescent air and in the airflows behind the plane shock wave at different positions of the shock front in the discharge volume. The length of the discharge region along the direction of motion of the shock wave was 10 cm; the end of the electrodes was taken as the zero value of the coordinate x (figure 1). Thus, the position of the shock wave front inside the discharge volume corresponded to -30 mm $\leq x_0 \leq 0$ (first series of experiments), and beyond the discharge volume $0 \leq x_0 \leq 20$ mm (second series of experiments). Initial pressure was 10-50 Torr, Mach numbers 1.9-4.5. The shock wave front shifted less than 0.5 mm during the discharge current in the shock tube channel. The triangular marker in figure 1 indicates the discharge gap edge. Arrows indicate the direction of propagation of the shock wave (SW) and concurrent flow.

3. Experimental Results

The uniform radiation of two plasma sheets provides spatially uniform flow volume preionization. As a result, the volume discharge in motionless air (figure 2 a) and in homogeneous flow (figure 2 d) is characterized by a high spatial uniformity. In the presence of a shock wave that is in the discharge volume at the moment of a discharge pulse, the plasma glow is localized in front of the shock wave in the area of low pressure [6, 7]. In this part of the volume the value of the reduced electric field E/N and, correspondingly, the electron concentration are greater in the moment of the discharge phase (E is the electric field strength and N is the concentration of molecules). The photo image of the discharge during interaction with the planar shock wave front inside the discharge volume is shown in figure 2 b, and outside it in figure 2 c, figure 3 a.

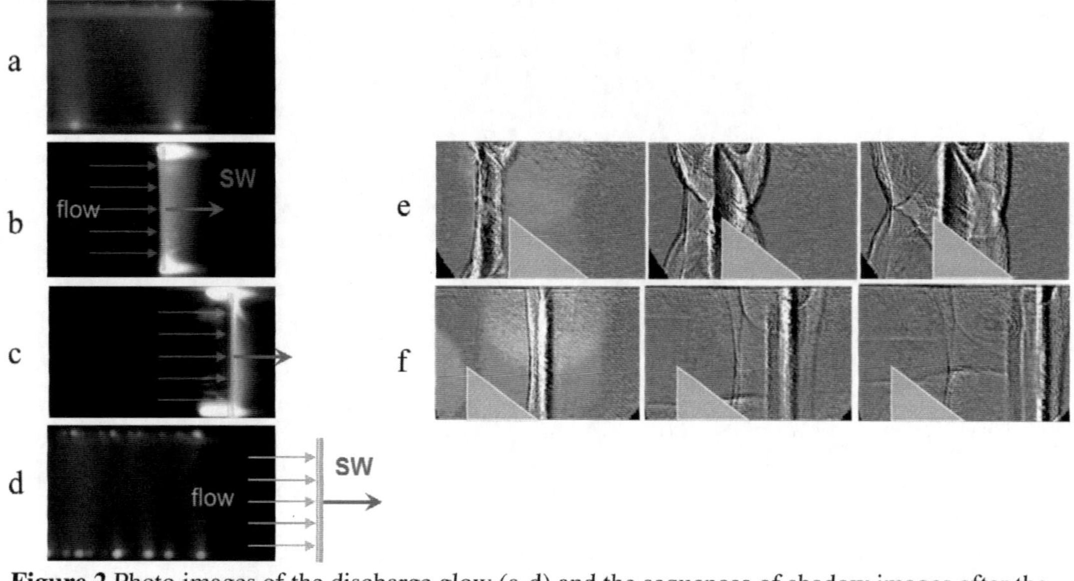

Figure 2 Photo images of the discharge glow (a-d) and the sequences of shadow images after the discharge (e, f). The optical axis of the photo camera is perpendicular to the quartz glasses.

Depending on the plane shock wave position front relative to the discharge volume, the discharge plasma can occur both ahead of the shock wave front in the region of low density and behind the front of the shock wave in the flow [5], as photo images of the discharge show. The surface discharges, in addition, can run along U-shaped channels, which include the line of surface intersection with front of the initial shock wave (see figure 3). The volume energy input into the gas is realized in front of the shock wave in the first series of experiments. The volume energy input realizes in a narrow region in front of the shock wave front and/ or in the flow behind it in the second series of experiments. The waveforms of discharge electric current in the experiments 2 a, b, c, d are rather different.

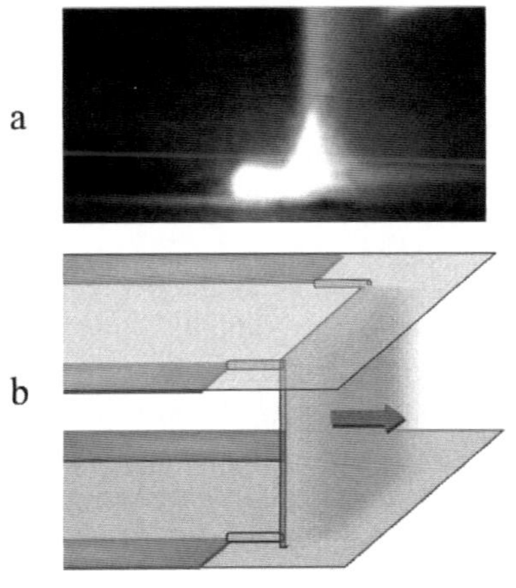

Figure 3. The photo image of the discharge during interaction with the planar shock wave front outside the discharge volume (a); and the scheme of arrangement of U-shaped channels of the surface discharges (b). The optical axis of the photo camera is inclined to the surface discharge plane at a small angle. The arrow indicates the direction of movement of the plane shock wave.

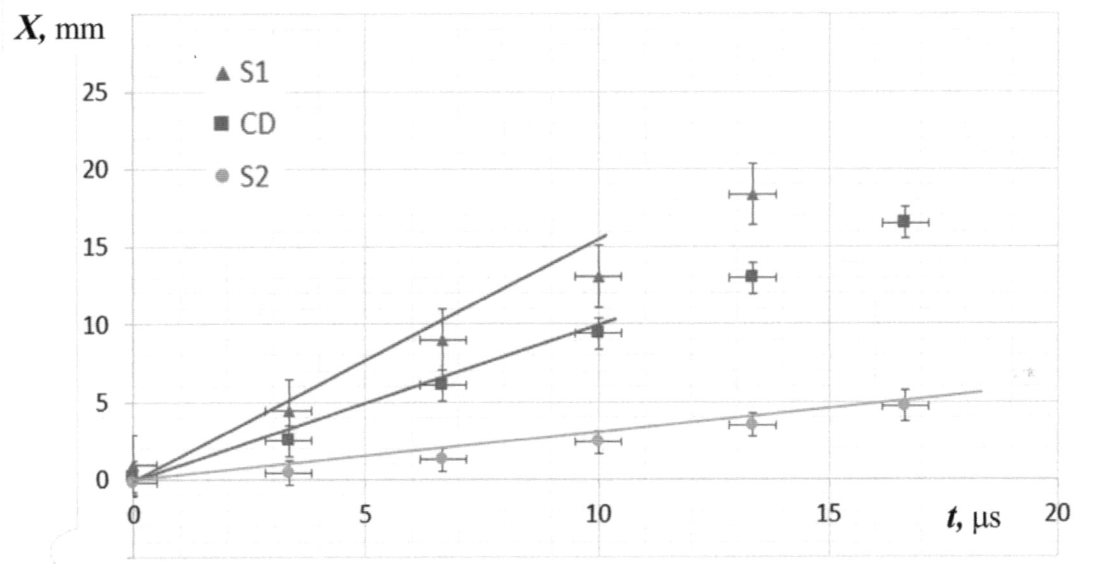

Figure 4 *X-t* diagram of the motion of discontinuities after discharge.

High-speed shadowgraphy showed the peculiarities of the motion of shock-wave configurations after discharge. In the first series of experiments, gas-dynamic discontinuities break down on the left boundary of the plasma volume containing the shock wave front (airflow - discharge plasma), and on the right boundary (discharge plasma - quiescent air) (fig. 2 b). When the initial plane shock wave breaks down, two shock waves and a contact surface are formed. Shock wave, a contact surface and a rarefaction wave generate on the plasma - quiescent air boundary. One of the formed shock waves moves to the right along the region of the relaxing plasma, compressing it and leading to long afterglow [7]. The narrow area near the shock wave front continues to radiate after the termination of the discharge current, and the afterglow duration depends on the experimental conditions and may exceed 2 μs. Surface discharges initiate shock waves moving in the direction transverse to the front of the initial shock wave and affecting the structure of the flow. The shock-wave configuration changes with time because of the motion and interaction of all waves with each other. The sequence of shadow images after the discharge are shown in figure 2 e (Mach number of shock wave 2.2, $x_0 = -7$ mm). In this case, the objects formed after the discontinuities breakdowns are concentrated in a narrow region and form the shock-wave configuration described in [8]. For the conditions of this experiment, the average velocity of the right boundary of the shock-wave configuration (shock wave S1) is about 800 m/s in the first 20 microseconds. It is higher than the velocity of the initial shock wave. The velocity of the contact surface (CD) is ~ 420 m/s. The left boundary of the shock-wave configuration (shock wave S2) moves in the direction opposite to the flow, at a velocity of ~ 110 m/s. Figure 4 shows the $X - t$ diagram of the motion of discontinuities after discharge at Mach number of initial shock wave 4.2, $x_0 = -13$ mm ($X = x - x_0$).

Points denote the experimental values obtained by processing of images of high-speed shadow visualization. The lines indicate the results of the numerical calculation (see paragraph 4).

In the second series of experiments, when the shock wave front is beyond the discharge gap volume, the structure of the discharge glow and the structure of the shock-wave configuration after discharge are different in the cases: 1) $0 < x_0 \leq 7$ mm, 2) $7 < x_0 \leq 15$ mm, 3) x_0>15 mm. In the first case, the discharge current occurs along the shock wave and in front of shock wave; the shock-wave configuration is close to that realized in the first series of experiments. In the second case the volume energy input zone is narrow (practically the plane), and the surface discharges are stretched into U-shaped channels, enveloping the initial shock wave front (figure 2 c, figure 3). Two shock waves moving in opposite directions and two contact surfaces bounding the gas region heated by the discharge are formed in the volume (figure 2 f, Mach number of shock wave 2.2, $x_0 = 8$ mm). Surface discharges generate shock waves in the flow and near the front of the shock wave. In the third case, discharges are initiated only in the flow, without interacting with the front of the shock wave (figure 2 d).

4. Numerical Simulation

The dynamics of the gas flow discontinuities after pulse ionization of a space in front of a flat shock wave was studied numerically. The fast energy release in front of the shock wave changes the gas parameters in this area and breaks the Rankine-Hugoniot relations at the initial shock wave. As a result, the initial shock wave breaks down into three discontinuities: two shock waves and a contact discontinuity between them. This structure corresponds to the classical Riemann problem solution [8]. At the other boundary of the energy input, a shock wave, a contact surface and a rarefaction wave are formed.

The numerical simulation of the decay of the gas-dynamic discontinuity in the one-dimensional calculation was realized. The initial temperature was 300 K, and the initial density of air corresponded to the experimental conditions. The system of Euler equations was solved using the second-order Godunov method [10]. At the initial instant, three regions of the gas with different values of gas dynamic parameters were specified at the boundaries of these regions. The conditions of discontinuity decay were created including the energy input region in front of the shock wave, stationary air on the right of it at the initial experimental conditions, and the region on the left of the shock front with parameters of shock-compressed air in the airflow behind the shock wave. The energy input varied depending on the size of the energy input region. The adiabatic exponent for air was assumed to be constant 1.4. The Mach number of the shock wave varied in the limits 1.9-4.5.

Figure 4 shows the comparison results of the numerical simulation, which corresponds to 40% part of the discharge's energy converting to the instant heating (Mach number of shock wave 4.2, $x_0 = -13$ mm) with the experimental data (positions of discontinuities at different time moments). The $X - t$ diagram of the discontinuities motion shows that the results of 1D numerical simulation coincides with the experimental results within 7-10 μs after the discharge.

5. Conclusions

Digital films of the high speed flow with combined discharge of nanosecond duration were obtained and analyzed when interacting it with a plane shock wave in the channel of discharge chamber. High-speed shadow recording of the flow field evolution after the discharge initiating was carried out with 1 microsecond exposition. The dependence of the type of gas dynamic discontinuity breakdown on the initial position of the shock front is recognized.

A one-dimensional numerical simulation of the discontinuity breakdown at the front of a plane shock interacting with volume discharge is carried out. The experimental data of the discontinuity motion taken from digital films are compared with the calculated data; the discharge energy rate converted into the internal energy of the gas at the stage of the discharge electric current is determined in different experimental conditions.

The realization of various types of energy input for modification of flow patterns can be used in technologies of high-speed flow control.

Acknowledgments
The work was conducted under the support of RSF Grant № 18-19-00672.

References

[1] Moreau E 2007 Airflow control by non-thermal plasma actuators *J. Phys.* D **40** 605
[2] Starikovskiy A and Aleksandrov N 2013 Plasma-assisted ignition and combustion *Prog. Energy Combust. Sci.* **39** 61–110
[3] Russell A, Zare-Behtash H and Kontis K 2016 Joule heating flow control methods for high-speed flows *J. Electrost.* **80** 34–68
[4] Znamenskaya I A, Latfullin D F, Lutskii A E and Mursenkova I V 2010 Energy deposition in boundary gas layer during initiation of nanosecond sliding surface discharge *Tech. Phys. Lett.* **36** 795
[5] Mursenkova I V, Znamenskaya I A and Lutsky A E 2018 Influence of shock waves from plasma actuators on transonic and supersonic airflow *J. Phys.* D*: Appl. Phys.* **51 5** 105201
[6] Arkhipov N O, Znamenskaya I A, Mursenkova I V, Ostapenko I Yu and Sysoev N N 2014 Development of nanosecond combined volume discharge with plasma electrodes in an air flow *Moscow Univ. Phys. Bull.* **69** 96
[7] Kuznetsov A, Mursenkova I and Znamenskaya I 2017 Decay of radiation of the sliding surface discharge and the combined volume discharge *Proc. of the XXXIII Int. Conf. on Phenomena in Ionized Gases (ICPIG), Estoril, Portugal. 9-14 July 2017* p. 139
[8] Doroshchenko I, Znamenskaya I, Koroteev D and Kuli-zade T 2017 When shock is shocked: Riemann problem dynamics at pulse ionization of a shock wave *Phys. Fluids* **29 10** 101701
[9] Karlov N V, Kisletsov A V, Kovalev I O, Kuz'min G P, Movshev V G, Nesterenko A A, Prokhorov A M and Toker G R 1986 High-pressure CO_2 laser with plasma cathode *Sov. Tech. Phys. Lett.* **12 5** 254–55
[10] Znamenskaya I A, Ivanov I E, Koroteeva E Yu and Orlov D M 2012 Investigation of the interaction of a shock wave with the zone of a pulsed surface discharge in a rectangular channel *Moscow Univ. Phys. Bull.* **67 6** 541

IOP Conf. Series: Journal of Physics: Conf. Series **1112** (2018) 012007 doi:10.1088/1742-6596/1112/1/012007

Peculiarities of the charged particles distribution near the front of a shock wave in a glow discharge plasma

A S Baryshnikov[1], I V Basargin[1], N O Bezverhniy[1,2], S V Bobashev[1,2], N A Monakhov *[1], P A Popov[1], V A Sakharov[1] and M V Chistyakova[1]

[1]Ioffe Institute, Politechnicheskaya 26, Saint-Petersburg, 194021, Russia
[2]Peter the Great Saint-Petersburg Polytechnic University, Polytechnicheskaya 29, Saint-Petersburg, 195251, Russia

*nm1988@mail.ru

Abstract. The charged particles distribution in a glow discharge plasma near the front of shock wave moving perpendicularly to the discharge axis has been investigated using a double electrical probe. It has been shown that the interaction of the shock wave with the glow-discharge plasma is accompanied by a decrease of charged particles density before the shock wave arrival to the measurement point. This decrease occurs simultaneously through the entire plasma volume. The short-term increase of charged particles density has been revealed. Its appearance caused by interaction of a shock wave with the cathode layer of a glow discharge. These peculiarities of charged particles distribution compared with the case when shock wave propagates through decaying plasma. A significant difference between the distributions of the charged particles in these two cases was revealed.

1. Introduction

The possibilities of plasma influence on supersonic flows in order to aerodynamic characteristics modification are actively investigated [1-4]. The problem of interaction with a shock wave with plasma is part of a fundamental problem directly related to development of modern aeronautics and cosmic technique [5-7], because it reveals a number of unexplained phenomena so far. In particular, the mechanism of decreasing of the charged particles concentration in plasma ahead of the shock wave front in the longitudinal glow discharge is not clear [8-9]. Attempts to explain this effect are made in theoretical works [10-12]. A similar phenomenon is observed in a transverse gas discharge [13-16]. A distinctive feature of the experiments [13-16] is the small linear size of the gas discharge in comparison with the size of the working chamber, which makes it possible to consider the influence of the chamber walls on the discharge to be negligible. This circumstance is important, since in this case self-organization of the discharge occurs in accordance with the conditions in the external gaseous medium.

In this paper, one more feature of the distribution of charged particles near the shock wave front was found, namely, a brief increase in this concentration during the motion of the shock wave through the discharge plasma.

2. Experimental setup and diagnostics

This investigation was carried out on the setup shown at figure 1. It consisted of working chamber of 350 mm in diameter and 400 mm in height and two conical electrodes (2 and 3), which were placed vertically at a distance of 100 mm from each other. Air or argon at a pressure of 4 kPa were used in the working chamber. An external power supply provides the discharge current of 1.1 A with a constant voltage between the electrodes of 650 V. The anode was placed near the bottom of working chamber, and cathode was placed near the top of working chamber and was grounded. This arrangement provided the special stability of axisymmetrical glow discharge with the cross section increasing toward the cathode. The gas temperature in the center of the discharge is ~ 1300 K, electrons ~ 12000 K, the degree of ionization does not exceed 10^{-6}.

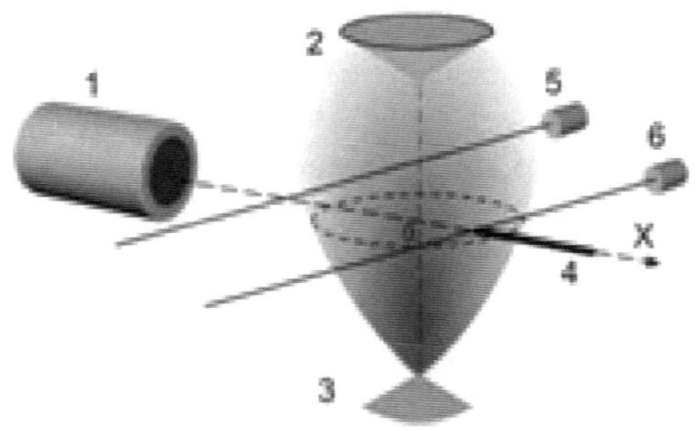

Figure 1 Experimental setup. 1 – electromagnetic shock tube, 2 – cathode, 3 – anode, 4 – double electric probe, 5,6 – laser schlieren systems.

Gas discharge was localized far from the lateral chamber walls, therefore, in the central part of discharge the gas temperature is three times higher than the temperature of ambient gas. It leads to the strong gradients of gasdynamic and electrodynamic parameters and intense convection. The characteristic feature of regarded processes is the shock wave propagation through spatially non-homogeneous media. Gasdynamics peculiarities of this interaction were regarded at the paper [6].

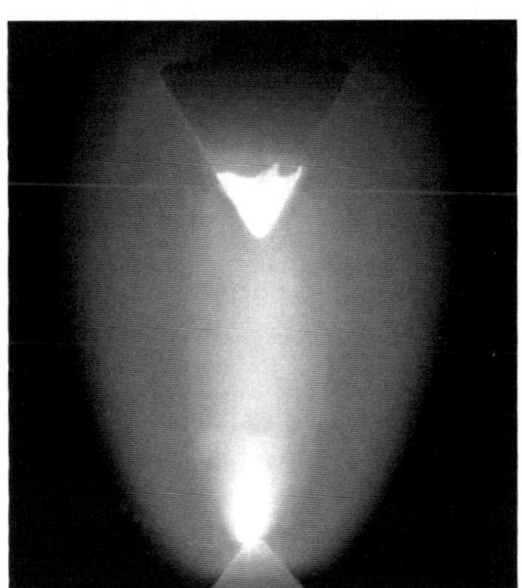

Figure 2 Visible structure of the discharge.

The glow discharge self-organization appears with visually distinguishable areas of the discharge (figure 2): a bright "core" with a diameter of about 20 mm in the center and an outer part with a diameter of the order of 60 mm. The transverse dimension of the cathode luminescence coincides with the diameter of the "core", which allows us to assume that a significant part of the discharge current is concentrated in this region.

The shock wave (SW) is generated in the electrical shock tube (1) and propagates perpendicularly to the axis of the discharge. Outlet cross-section of the shock tube is located at a distance of 100 mm from the discharge axis. The shock wave velocity at the shock tube outlet is 1,000 m/s, which corresponds to Mach number M=3.

The shock wave was registered by an optical schlieren system consisting of a laser source, a photodiode, and a Foucault knife. When the shock wave crosses the laser beam, the latter deviates from the Foucault knife and thus changes the signal from the photodiode. Optical beam 5 remained fixed and it was placed between electrodes and electrical shock tube outlet at the distance of 60 mm from discharge axis; beam 6 passed through the location point of the electrical probe. The accuracy of shock wave arrival to the measurement point was about 3 μs.

In this study the main facility for discharge plasma diagnostics is the double electric probe. The probe is fixed to the end of the rod 4, which moves in a horizontal plane coaxially with the shock tube. The position of the probe is determined by its distance from the discharge axis. The probe consists of two parallel platinum electrodes with a diameter of 0.5 mm and a length of 10 mm, spaced 8 mm apart.

The probe current meter is an operational amplifier with an input resistance of 1 kOhm. One of the electrodes of the double probe is connected to the common bus of the amplifier. The amplifier input is connected to a circuit consisting of a serial connection of a 38 V source of constant voltage and a second probe electrode. This ensures the operation of the probe in the ion current saturation regime, in which the probe current is determined by the electron temperature and the ion concentration in the vicinity of the negative electrode of the probe [17].

In addition, a double probe was used as a voltmeter. For this, the parameters of the input circuit of the amplifier varied, and, in particular, the 38 V voltage source was eliminated and the input impedance of the operational amplifier was increased. This made it possible to measure the potential difference between the electrodes of a double probe. With this method, the vertical component of the field was measured. It was equal to 30 V/cm at the discharge axis and decreasing by an order of magnitude at a distance of 50 mm from the discharge axis. Thus, the distribution of the vertical component of the electric field in the interelectrode gap and the position of the visible regions of the discharge are in correspondence with each other.

3. Charged particles distribution near the shock wave front in glow discharge plasma

Fig. 3 shows the double probe signals at three different points located at different distances to the axis of the discharge gap: 1 - 0 mm, 2 - 10 mm, 3 - 20 mm. The moments of arrival of the shock wave in the measurement currents are shown by vertical lines. It can be seen from the figure that, firstly, before the shock wave, a decrease in the probe current corresponding to a decrease in the concentration of charged particles is observed, and, secondly, these changes begin simultaneously at all points of measurement. The latter fact suggests that the decrease in local charged particles density is caused not by the shock wave as such, but by a physical mechanism that alters the electrical characteristics of the discharge, for example, as a result of the deformation of the current channel of the gas discharge. In such a case, the propagation of the shock wave can influence on the integral parameters of the gas discharge, such as current and voltage between the electrodes.

IOP Conf. Series: Journal of Physics: Conf. Series **1112** (2018) 012007 doi:10.1088/1742-6596/1112/1/012007

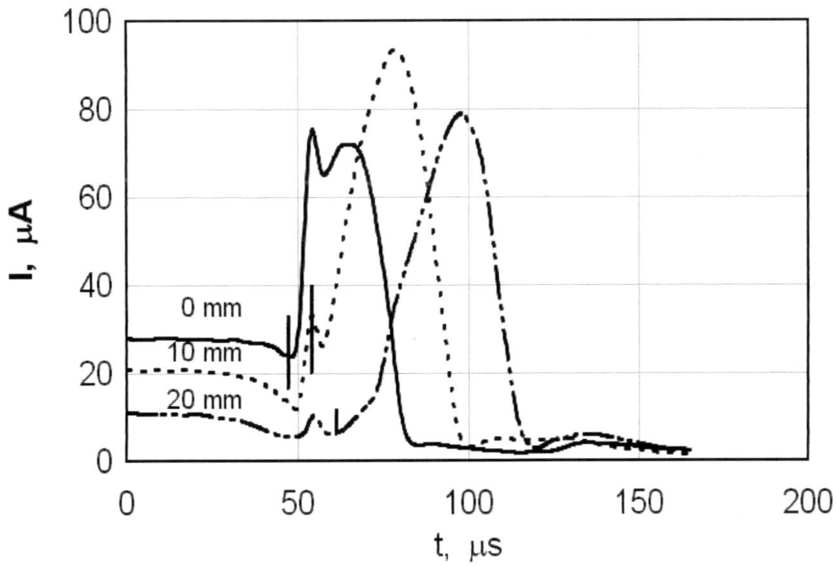

Figure 3 Double probe current variation at the shock wave propagation through glow discharge.

Figure 4 shows the variation of the current I and the voltage U between the electrodes of the discharge gap as the shock wave passes through the discharge. Time is counted from the moment when the shock wave is at a distance of 60 mm to the axis of the discharge, which corresponds to the signal of the first schlieren system on the figure 1. The graph shows that the change in the discharge current and voltage at the electrodes of the discharge gap begins at t = 0 μs, i.e. almost immediately at the moment when shock wave enters the discharge plasma. It can be seen from the figure that by the time the shock wave arrives to the discharge axis (t~50 μs), the discharge current is reduced by 20%, and the voltage is increased by 10% of the initial level.

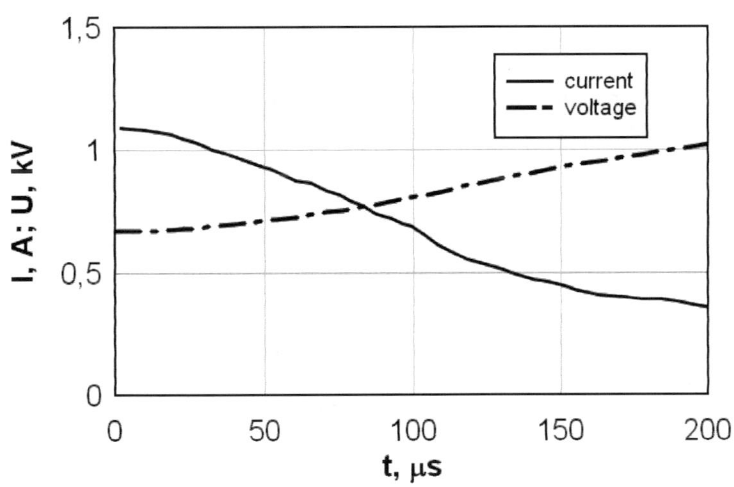

Figure 4 Discharge current and interelectrode voltage variation at a shock wave propagation through glow discharge.

This analysis isn't a direct evidence of a fact that charged particles decrease is caused by the change of integral discharge parameters. Nevertheless, their influence on plasma parameters is doubtless.

Another feature of the charged particles distributions near the shock wave front the positive current pulse, the temporal position of which remains practically constant irrespective of the double probe position. In other words, this effect is observed simultaneously in the entire cross-section of the discharge plasma. This suggests that the origin of the current pulse is caused by the interaction of the shock wave with either the elements of the internal structure of the discharge (as in the case of a decrease in the current before the front of the shock wave) or with fixed elements of the working chamber, for example, with the cathode of the discharge gap.

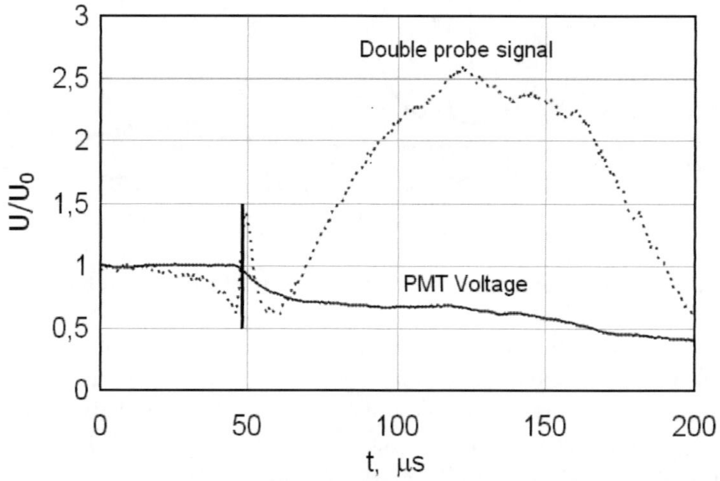

Figure 5 Comparison of the double probe signal and the cathode emission intensity at the shock wave propagation through the glow discharge.

To verify the last assumption, the time signal of the probe and the radiation intensity of the cathode emission region were compared. A fragment of the image of the luminous cathode was projected onto a photosensitive element whose signal was recorded simultaneously with the probe signal. The axis of the optical Schlieren system was located near the surface of the cathode from the approach side of the shock wave. The results of these measurements are shown in figure 5.

The dashed curve is the double probe signal, the solid curve is the photomultiplier signal, the vertical line marks the moment when the shock wave reaches the cathode. Thus, the moment the shock wave arrives at the cathode coincides with the onset of a change in the intensity of the cathode luminescence and the current of the probe. This allows one to unambiguously associate the appearance of a positive pulse on the probe current signal with the impact of a shock wave on the cathode glow region of a glow discharge.

The size and position of the current pulse caused by the shock wave impact to the cathode layer were studied in dependence on gas pressure in the chamber. Since the pressure in the chamber changes, the shock wave intensity and the size of the glow discharge changes, therefore the measurements were carried out on the discharge axis since this point is common to all regimes.

Experiments have shown that a pressure decrease in the working chamber causes the increase of discharge current. The pressures in the chamber equal to 2; 4; 7 kPa correspond to the following sequence of values of the discharge current - 1.3, 1.1, 1.0 A. When discharge current increases, the region of cathode glow and also diameter of the visible part of the glow discharge increases.

The results of double probe measurements at the discharge axis are shown at figure 6. The time of the shock wave arrival to the point of measurement (the axis of the discharge) is indicated by a vertical line. The double probe current related to the initial current value at a pressure of 4 kPa. The figure

IOP Conf. Series: Journal of Physics: Conf. Series **1112** (2018) 012007 doi:10.1088/1742-6596/1112/1/012007

shows that a decrease in pressure in the working chamber causes the shock wave speed and the initial double probe current level increase. Thus, at a pressure of 2 kPa, the double probe current increased by a quarter in comparison to the other two stationary glow regimes.

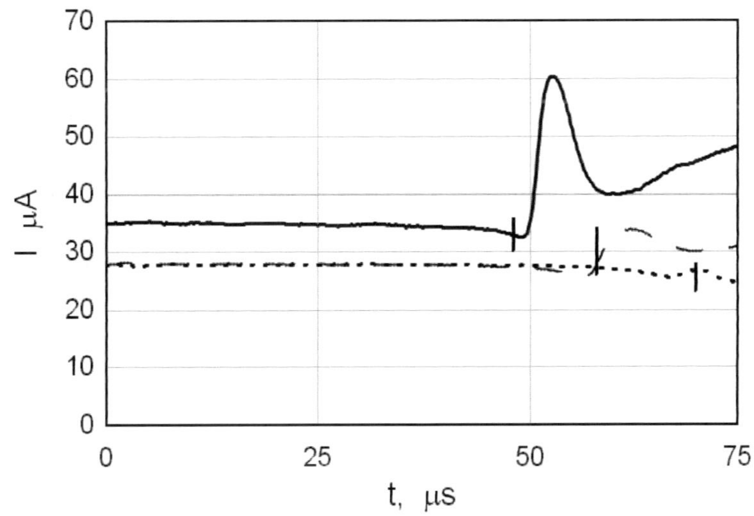

Figure 6 Relative double probe current on the axis of the discharge at different pressures in the working chamber: 2 kPa – solid, 4 kPa – dashed, 7 kPa – dotted.

The goal of this experiment is to show how pulse form depends on shock wave parameters. Times of current pulses caused by the shock wave interaction with the cathode are different because of the different shock wave speed. The magnitudes of the current pulse also significantly differ. Decreasing the pressure in the chamber causes the increase of the pulse magnitude.

4. Charged particles distribution near the shock wave front in decaying plasma

The simultaneity of charged particles density decrease at shock wave entrance to the discharge leads to a conclusion that observed effect may be caused by electric field between the electrodes. This conclusion has determined the further experiments which were carried out with plasma without electric field.

Turning off the discharge was carried out by shunting discharge gap, the voltage across the electrodes and the discharge current decreased to zero within 1 microsecond. One part of this investigation was the study of decaying plasma properties itself. Figure 7 shows the probe currents at the different distances from the discharge axis. Time was counted from the moment when the discharge was turned off. The values of probe current in the area $t<0$ corresponded to the stationary discharge plasma parameters. After turning off the discharge, the decay of plasma begins and probe current monotonically decreases. Measurements have shown that the conductivity of decaying plasma remains non-zero for 30-40 ms after turning off the discharge. The characteristic time of air plasma decay is about 1 ms.

The interaction between shock wave and decaying plasma was realized at the time about 1 ms after turning off the discharge when the recombination rate of charged particles became small and their concentration remained nearly constant during the observation time. The level of that concentration is about 5 times smaller in comparison with stationary discharge plasma. Figure 8 shows the probe currents at the discharge axis for two different cases: dotted curve– shock wave interaction with the decaying plasma, solid curve – shock wave interaction with the discharge plasma. Time was counted from the moment of the shock wave arrival to point X=-60 mm. The probe measurements have shown that there is no charged particles density decrease before the shock wave front in the decaying plasma.

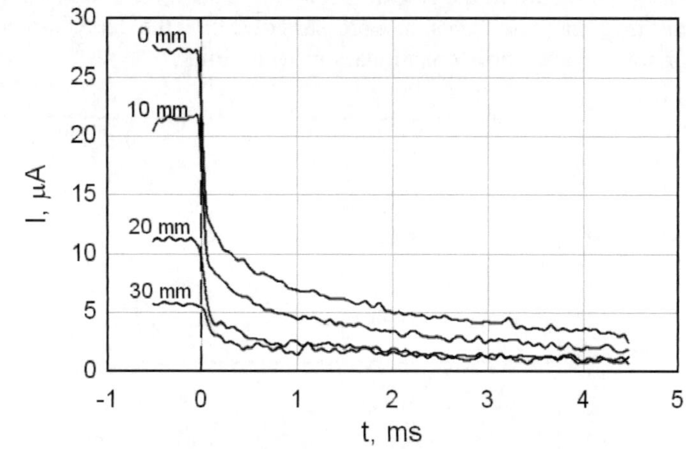

Figure 7 Temporal evolution of decaying plasma at different distances from discharge axis.

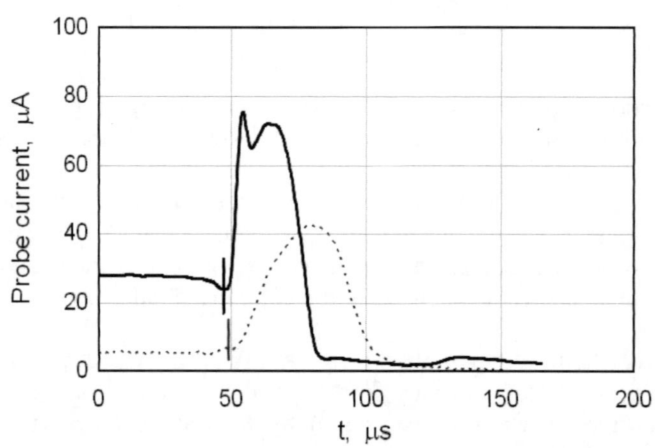

Figure 8 Probe signals at the shock wave interaction with the decaying plasma (dotted curve) and with the discharge (solid curve)

5. Charged particles distribution behind the shock wave front in glow discharge and decaying plasma

As it was shown at figures 3 and 8, the probe current, which corresponds to charged particles concentration, increases monotonically at shock wave front. This peculiarity of probe current oscillogram allows us to raise a question about shock wave existence in the discharge plasma. Per contra, in our previous works [15-16] we measured the pressure distribution at a shock wave propagation through the discharge plasma and that results had shown the existence of shock wave at the discharge plasma. An increase in the gas density at the shock wave front, calculated according to the measured shock wave velocity, did not exceed 2. The relative increase in the concentration of the charged component compared with the unperturbed value differs for different measurement points, while the maximum concentration of the charged component behind the shock front remains approximately constant (figure 3). The time delay between the shock wave arrival to the measurement point and the attainment of the maximum concentration of charged particles increases when the distance between discharge axis and measurement point increases. Such a character of variation of the charged particles distribution variation behind the shock wave front is determined by the inhomogeneity of the charged particles distribution in the unperturbed plasma, and the instant of reaching the maximum on the oscillogram of the probe current corresponds to the arrival of a compressed plasma from the central region of the discharge to the double probe.

6. Conclusion

We have experimentally confirmed the effect of local conductivity decrease before the shock wave arrival. We manifest that this decrease simultaneously at the entire volume of gas discharge plasma. The probable reason of this phenomenon is the decrease of electrical power which was put in the discharge region. In a decaying plasma, the concentration of charged particles remains constant until the shock wave arrival.

A short-term increase in the current of the double probe observed before the front of the shock wave is apparently caused by the interaction of the shock wave with the cathode of the discharge gap. Thus, the shock wave interaction with the cathode glow region of a glow discharge affects the plasma parameters of the positive column of the glow discharge. This influence is determined, in particular, by the gas-dynamic interaction of the shock wave with the cathode surface, i.e. depends on the shape of the cathode, the state of the medium near its surface, the intensity of the shock wave and the angle of its approach to the cathode.

At the shock front, the charged particles concentration varies continuously, in contrast to the stepwise variation of the neutral component concentration. The charged particles concentration maximum arises at the noticeable distance from the front. The complex nature of the variation of the charged particles concentration behind the shock wave front is determined by the inhomogeneity of the distribution of charged particles in the unperturbed plasma.

References

[1] Georgievsky P Yu, Levin V A 2003 *Fluid Dynamics* **38** 794
[2] Zheltovodov A A, Pimonov E A, Knight D D 2007 *AIAA Pap.* 2007–1230
[3] Adamovich I V 2010 *Encyclopedia of Aerospace Engineering* (John Wiley & Sons Ltd)
[4] Starikovskiy A, Aleksandrov N 2011 Nonequilibrium Plasma Aerodynamics *Aeronautics and Astronautics,* Ed. By M. Mulder (InTech)
[5] Bletzinger P, Ganguly B N, Van Wie D, and Garscadden A. 2005 *J. Phys.* D*: Appl. Phys.* **38** R33
[6] Baryshnikov A S, Basargin I V, Bobashev S V, Monakhov N A, Popov P A, Sakharov V A and Chistyakova M. V. 2015 *Tech. Phys. Lett.* **41** 753
[7] Lapushkina T A and Erofeev A V 2017 *Tech. Phys. Lett.* **43** 17
[8] Gorshkov N, Klimov A, Mishin G, Fedotov A, Yavor I. 1987 *Sov. Phys. —Tech. Phys.* **57** 1893
[9] Ershov P, Klishin S V, Kuzovnikov A A, Ponomareva S E, and Pyt'ev Y P 1989 *Sov. Phys. —Tech. Phys.* **34** 936
[10] Avramenko R F, Ruhadze A and Teselkin S F 1981 *JETP Lett.* **34** 463
[11] Naidis G V 1991 *High Temp.* **29** 15
[12] Teselkin S F 1991 *Sov. Phys. —Tech. Phys.Lett.* **17** 50
[13] Baryshnikov A S, Basargin I V, Bobashev S V, Monakhov N A, Popov P A, Sakharov V A and Chistyakova M V 2016 *Journ. Eng. Phys.Thermophys.* **89** 565
[14] Bobashev S V, Basargin I V, Baryshnikov A S, Monahov N A, Popov P A, Sakharov V A and Chistyakova M V 2016 *Abstracts of 15th International Workshop on Magneto-Plasma Aerodynamics, Moscow, April 19-21, 2016* 112
[15] Baryshnikov A S, Basargin I V, Bobashev S V, Monakhov N A, Popov P A, Sakharov V A and Chistyakova M V 2017 *Tech. Phys. Lett.* **43** 511
[16] Baryshnikov A S, Basargin I V, Bezverkhnii N O, Bobashev S V, Monakhov N A, Popov P A, Sakharov V A and Chistyakova M V 2018 *Tech. Phys.* **63** 180
[17] Kotelnikov V A and Kotelnikov M V 2017 *High Temp.* **55** 477

IOP Conf. Series: Journal of Physics: Conf. Series **1112** (2018) 012008 doi:10.1088/1742-6596/1112/1/012008

High Frequency Feature of Pinhole-to-Plate Sinusoidal micro-DBD

VA Bityurin and A N Bocharov

Joint Institute for High Temperatures RAS, Moscow, Russia

E-mail: valentin.bitryurin@gmail.com

Abstract. The high (~ GHz range) frequency features of the sinusoidal axi-symmetrical micro-DBD discharge are studied with 2Dt numerical model. It was found that both in the negative and positive driving electrode polarities the electrode current reveals the rather regular "pulse" mode similar to the well-known Trichet pulse structure typically observed in the negative corona discharges in air. The preliminary analyses shown the reason of that is probably the strong nonlinear drift-diffusion-ionization interaction near the electrode surface.

1. Introduction

The surface DBD configuration is widely used in many plasma aerodynamics applications to control flow field or/and temperature distribution near the vehicle surface. The most popular configuration of sDBD is a 2D structure formed by the strip high-voltage electrode mounted above the grounded electrode covered by the thin insulator film (see, for example [1-3] and references there). The discharge is typically initiated by the voltage from couple to several tens of kV. In the case of the long duration experiment the sinusoidal shape of voltage applied can be considered as a good example. The applied voltage frequency is in a range of 1 – 100 kHz. In analysis, the sDBD configuration is usually assumed as 2D [1-4]. At the same time the experiments at high applied voltage and long duration have also shown that the discharge becomes constricted along the strip high voltage electrode edge and 2D consideration is now questionable. Nevertheless, the 2D based approach is still very popular because the 3Dt numerical simulation is enormously hard for everyday practice and, from the other hand, can provide acceptable accuracy to treat experimental characteristics.

In frame of such an approach, we performed earlier the analysis of the heat and dynamics effects of surface DBD actuators with the in-house created code PlasmAero [5,6] based on full self consisted solution of 2Dt Navier-Stokes equations, the drift-diffusion model for real gases, the physical-chemical kinetics and the Poisson equation. Used model has demonstrated the acceptable operability and flexibility to describe the in-time evolution of the electrophysical and thermo- and gas-dynamics fields in good correspondence with experimental observations under sinusoidal load in range from tens to thousands kHz.

The recent studies of long duration operation of the «classical» sDBD actuators proposed for practical applications [7] have revealed another non-2D effects - the formation of the micro structure on the oxidized aluminum electrode surface. This structure is characterized by the thin holes penetrating the oxide film normally to electrode surface to provide the contact of the discharge plasma with the metal electrode surface. This configuration is similar to the micro discharge panels used in some applications [8]. One can expect the rather important role of the transfer process on these small scales. To focus on the discharge processes taking place in close vicinity of the effective "pinhole" electrode the simplified problem formulation is considered in this work.

2. Problem formulation

For well detailed description of the fast ionization and drift- diffusion processes in the electrode vicinity the rectangular computational region of 0.15 × 0.075 mm² is used (Fig. 1). We consider the first sinusoidal period of voltage applied for several frequencies from 0.1 to 10 MHz for positive and negative polarities.

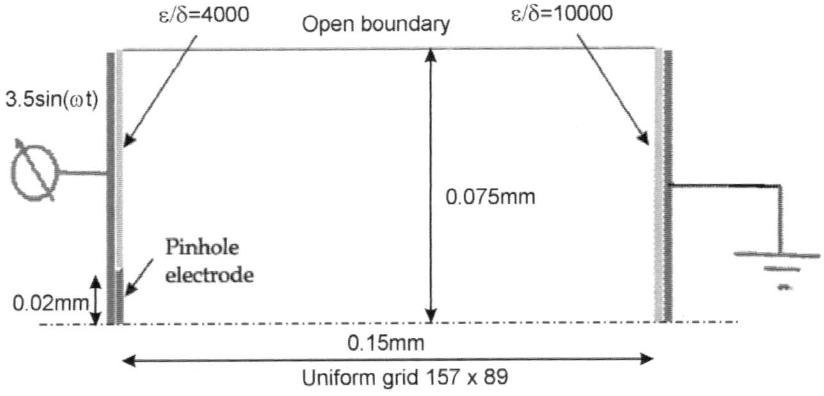

Figure 1 Computational domain.

The computational model is basically same as those described in paper [4,5]. We assume that the flow and electric discharge characteristics can be described by the set of equations reflecting conservation of mass, momentum and total energy for the whole fluid. Transport equations for charged particles, electrons and ions, as well as the Poisson equation for the electric field strength are solved assuming the drift-diffusion approach is valid. The set of governing equations is given below:

$$\frac{\partial \rho}{\partial t} + \nabla(\rho \mathbf{U}) = 0, \quad (1)$$

$$\frac{\partial \rho \mathbf{U}}{\partial t} + \nabla(\rho \mathbf{U} \otimes \mathbf{U}) + \nabla \boldsymbol{\tau} = \frac{\partial P}{\partial \mathbf{r}} + \mathbf{F}_e, \quad (2)$$

$$\frac{\partial \rho e^0}{\partial t} + \nabla(\rho \mathbf{U} h^0) + \nabla(\mathbf{U}\boldsymbol{\tau}) + \nabla \mathbf{q} = Q_e. \quad (3)$$

Here, ρ is the density, $\mathbf{U} = (U_x, U_y)$ is the velocity, P is the thermodynamic pressure, e^0 is the specific total energy, h_0 is the specific total enthalpy. Terms $\mathbf{F}_e$ and Q_e reflect the influence of plasma on the flow field and will be defined later. Equations of state are read as follows

$$h^0 = e^0 + \frac{P}{\rho}, e^0 = e + \frac{U^2}{2}, (\gamma - 1)e = P/\rho, P = \rho R. \quad (4)$$

Here, $\gamma = 1.4$ is the ratio of specific heats, and T is the temperature.
Viscous stress tensor components and heat fluxes are specified as usually:

$$\tau_{ij} = \frac{2}{3}\eta \delta_{ij} \nabla \mathbf{U} - \eta\left(\frac{\partial U_i}{\partial x_j} + \frac{\partial U_j}{\partial x_i}\right), \mathbf{q} = -\lambda \frac{\partial T}{\partial \mathbf{r}}, \quad (5)$$

where δ_{ij} is Kronecker symbol, η is viscosity, λ is heat conductivity. Gradient and divergence operators are defined as follows:

$$\frac{\partial}{\partial \mathbf{r}} = \mathbf{e}_x \frac{\partial}{\partial x} + \mathbf{e}_y \frac{\partial}{\partial x}, \nabla = \frac{\partial}{\partial x} + \frac{1}{y^\xi} \frac{\partial}{\partial x} y^\xi, \quad (6)$$

where $\mathbf{e}_x$ and $\mathbf{e}_y$ are unit base-vectors, $\xi = 0$ for Cartesian coordinate system, and $\xi = 1$ for axi-symmetric one.

The transport equations for concentrations and Poisson equation for electric field are read as follows:

IOP Conf. Series: Journal of Physics: Conf. Series **1112** (2018) 012008 doi:10.1088/1742-6596/1112/1/012008

$$\frac{\partial n_i}{\partial t} + \nabla \mathbf{\Gamma}_i = \dot{Q}, \tag{7}$$
$$\frac{\partial n_e}{\partial t} + \nabla \mathbf{\Gamma}_e = \dot{Q}, \tag{8}$$
$$\varepsilon \nabla \mathbf{E} = \rho_e, \rho_e = q(n_i - n_e). \tag{9}$$

In equations (7) through (9) n_i is ion number density, n_e is electron number density, ρ_e is the electric charge density, **E** is electric field, $\mathbf{E} = -\frac{\partial \varphi}{\partial \mathbf{r}}$, φ is the electric potential, q is electron charge, ε is permittivity of vacuum, $\dot{Q}$ is the source term defined later. Strictly speaking the Poisson equation with zero right-hand-side should also be solved in the bulk of dielectric. In this paper, simplified approach is used. Namely, the dielectric surface is considered via special boundary conditions on the plasma-dielectric boundary.

Ion and electron fluxes, $\mathbf{\Gamma}_i$ and $\mathbf{\Gamma}_e$, are defined as follows

$$\mathbf{\Gamma}_i = n_i(\mathbf{U} + \mathbf{V}_{di}) - D_i \frac{\partial n_i}{\partial r},$$
$$\mathbf{\Gamma}_e = n_e(\mathbf{U} + \mathbf{V}_{de}) - D_e \frac{\partial n_e}{\partial r}. \tag{10}$$

Here, drift velocities $\boldsymbol{V}_{di}$ and $\boldsymbol{V}_{de}$ are defined as

$$V_{di} = \mu_i E, V_{de} = -\mu_e E, \tag{11}$$

where μ_i and μ_e are ion and electron mobilities, respectively:

$$\mu_{i,e} = \frac{q/m_{i,e}}{\upsilon_{i,e}}, \upsilon_{i,e} = \frac{2\sqrt{2}}{3}\pi d_{i,e}^2 \left(\frac{8kT_{i,e}}{\pi m_{i,e}}\right)^{1/2}. \tag{12}$$

In equation (12) $d_{i,e}$ are mean collision cross-section diameters for ions and electrons, respectively. T_i and T_e are ion and electron temperature, respectively, m_i and m_i are ion and electron mass, $v_{i,e}$ are mean collision frequences for ion-neutral and electron-neutral collisions.

Diffusion coefficients, D_i and D_e in (10) are defined as

$$D_i = \frac{\mu_i k_B T_i}{q}, D_e = \frac{\mu_e k_B T_e}{q} \tag{13}$$

In (13) k_B is the Boltzman's constant. In paper [2] we used an approximation for electron temperature as function of the reduced electric field, $E_r = E/n$ (Td), where $n = P/k_B T$ is the total number density. Here, we simply take $T_e = 2\text{eV}$, and $T_i = T$.

Electric current density is determined as

$$\boldsymbol{j} = \boldsymbol{j}_i + \boldsymbol{j}_e = q(\boldsymbol{\Gamma}_i - \boldsymbol{\Gamma}_e).$$

Source-terms in equations (2), (3), (7) and (8) are determined as follows

$$\boldsymbol{F}_e = \rho_e \boldsymbol{E},$$
$$Q_e = \eta_e j_e E + \eta_i j_i E, \tag{14}$$
$$\dot{Q} = \alpha n_e - \beta n_e n_i. \tag{15}$$

The approximations used for ionization coefficient, α, and recombination coefficient, β, follow recommendations of the paper [5]:

At the dielectric surface, the jump condition for electric field strength projection is specified:

$$\varepsilon_P(\mathbf{E}_P\mathbf{n}) - \varepsilon_D(\mathbf{E}_D\mathbf{n}) + \left(1/_q \int \mathbf{jn}\, dt\right) = 0. \tag{16}$$

Here, ε_P is electric permittivity of plasma, $\varepsilon_P = \varepsilon_0$, ε_D is electric permittivity of dielectric material. In the paper $\varepsilon_D = 3\varepsilon_0(\mathbf{E}_P\mathbf{n})$ is projection of electric field strength onto normal-to-surface from the side of plasma, and $(\mathbf{E}_D\mathbf{n})$ is projection of electric field from the side of dielectric. The latter term in equation (16) represents the accumulation of the charge on the surface due to the flux of charged particles onto the surface.

At the electrode surface potential is specified as: $\varphi_{HV} = E_0 \cdot sin(2\pi\nu t)$, where E_0=-3.5kV, ν=1-10 MHz. Cathode-type boundary conditions are set as follows. Zero normal gradient is set for ion number density. Relation $(\Gamma_e\, \mathbf{n}) = -\gamma_e(\Gamma_i\, \mathbf{n})$ is used for electrons. In this relation, γ_e is the coefficient of secondary electron emission. Calculations were carried out with $\gamma_e = 0.02$.

The uniform grid with steps $10^{-6}\times10^{-6}$ m is used in the simulation presented below. The air at the atmospheric pressure and the room temperature is specified as the initial condition, the background ionization as $n_e = n_i = 10^6$ m^{-3} is specified in full domain, in thin sheet of 5 grid steps over pinhole electrode the initial ionization is increased up to 10^{16} m^{-3} to provide smooth start of calculations. It was found that there is no significant influence of the initial level of ionization degree on the studied phenomena. The time step integration is limited from above by 10^{-12} sec and automatically changed during the calculation to provide stability and accuracy and typically varied within the interval 10^{-12} – 10^{-14} sec, lower time steps results in unacceptable total calculation time for desktop computing.

3. Results

The results of the simulation for two values of the power supply frequencies of 1 MHz and 10 MHz are presented below. These values are chosen to reduce the componential time requirement defined mostly by the physical time interval duration.

In figure 2 the evolution of the main integral characteristics: Voltage of driving (pinhole) electrode, V_{ph}, the total pinhole current I_{ph} , the total electrical charge extracted from the domain Q_{ext}, the average potentials of barrier V_b and the isolator cover of the main electrode V_s. The important value of the gap voltage $V_g = V_{ph} - V_b$ will be also used in the analysis.

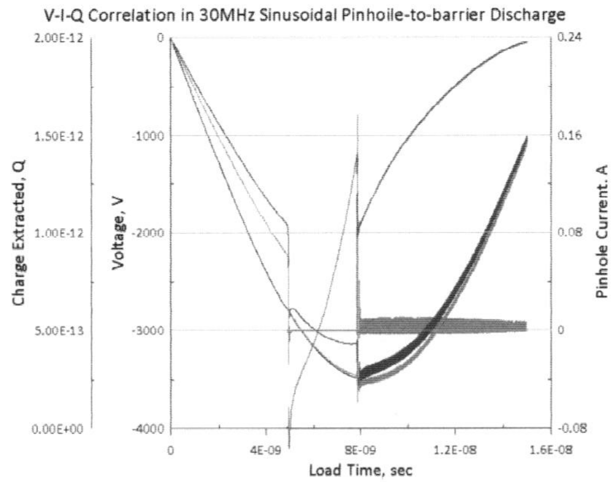

Figure 2 V-I Evolution in 10 MHz discharge

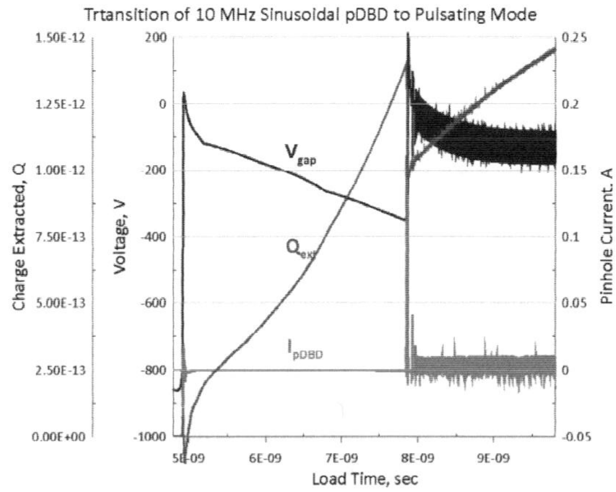

Figure 3 The fine structure of the burst dicharge mode

The first regular mode corresponds to electric field formation through the hole domain with minor currents provided by initially introduced charged particles. In the system under consideration this linear mode exists until electric field will reach the value to start the fast ionization. In this model the only mechanism of ionization is included – electron impact ionization, and the maximal electric field is located at the pinhole outer edge. After the first breakdown newly produced electrons can support rather gradually developing current. At this quasi stable "glow" mode the electric field is reduced significantly due to charge deposition on the barrier and, consequently, the ionization production rate remains at low level. As it will be seen from the next pictures in figure 4 the glow mode is terminated first by the next three breakdowns producing high enough amount of electrons to provide the required level of charge deposition on the barrier for regular operation with rather stable level of the gap voltage V_g. The latter is more than an order of magnitude lower in compare with the applied voltage V_d. Further, the regular pulsating mode establishing as it is seen in figure 5.

The similar features are observed and in the case of the lower power supply frequency. Some examples of data for $f = 1$ MHz are presented in figure 6 and figure 7. In addition to the evolution of the integral characteristics the behavior of the local parameters looks also very informative. The maximal value of the electric field strength is located as rule at the singularity of the "triple" point connecting plasma, insulator and electrode at the outer edge of pinhole electrode. In the considered configuration this point initiated as practically all 'fast' processes. This 'hot spot' is the location of maximal power dissipation, ionization production rate, temperatures and other local parameters.

IOP Conf. Series: Journal of Physics: Conf. Series **1112** (2018) 012008 doi:10.1088/1742-6596/1112/1/012008

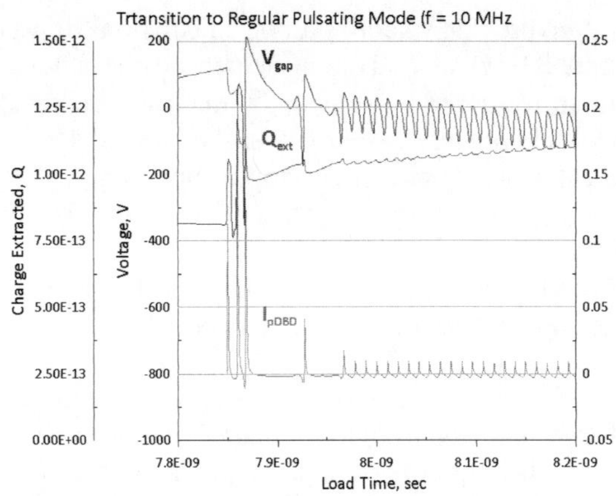

Figure 4 Details of the transition to the regular pulsating mode (f = 10 MHz)

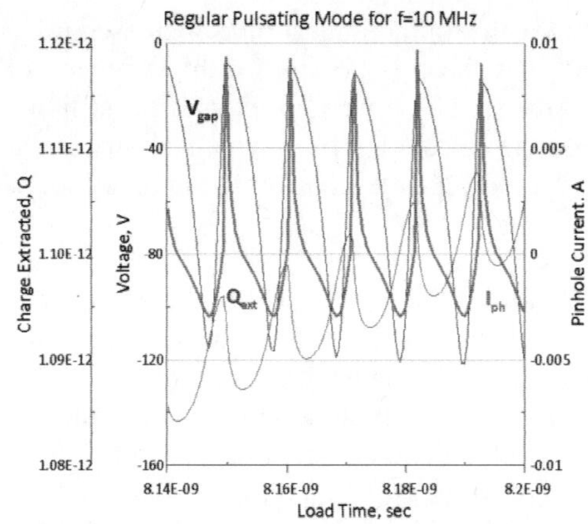

Figure 5 The fine structure of the regular pulsating mode (f = 10 MHz)

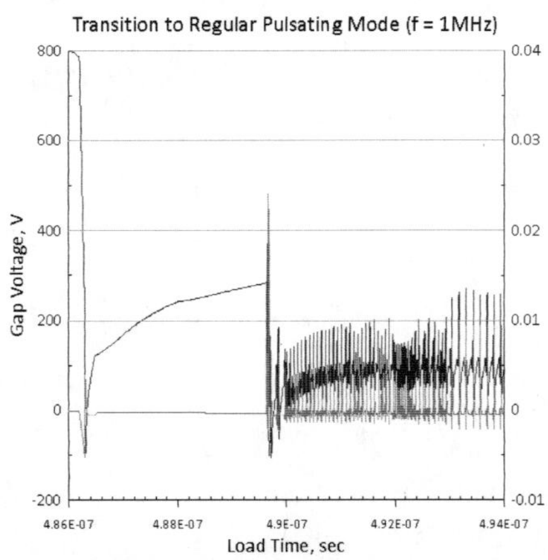

Figure 6 Details of the transition to the regular pulsating mode (f = 1 MHz), dark blue for gap voltage, red – pinhole current.

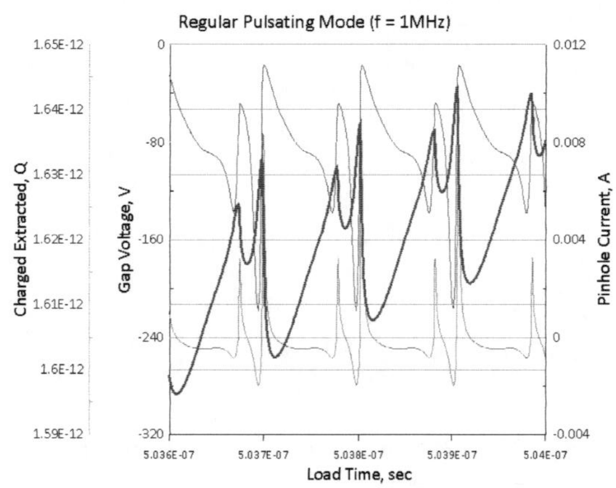

Figure 7 The fine structure of the regular pulsating mode (f = 1 MHz), dark blue for gap voltage, red – pinhole current, green – charge.

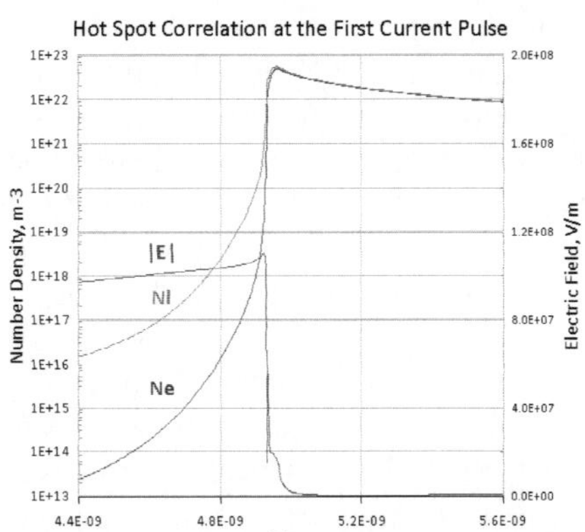

Figure 8 The first pulse formation at the hot spot (f = 10 MHz)

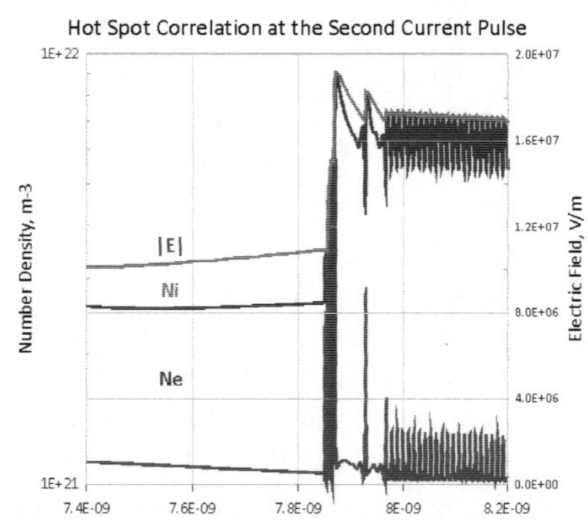

Figure 9 Details of the transition from glow to the regular pulsating mode at the hot spot (f = 10 MHz)

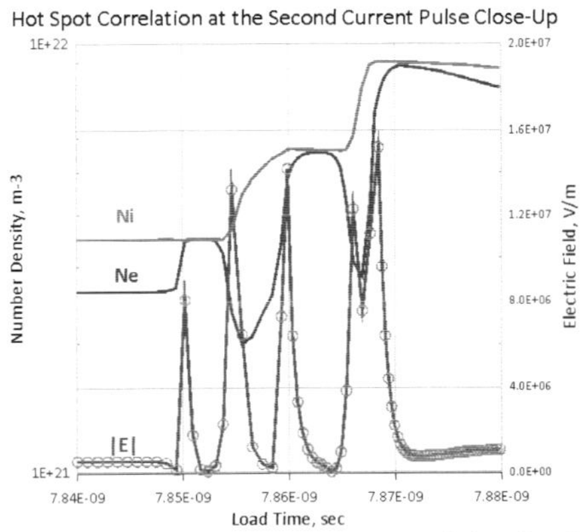

Figure 10 The fine structure of the final preparation of the pulsating mode at the hot spot (f = 10 MHz)

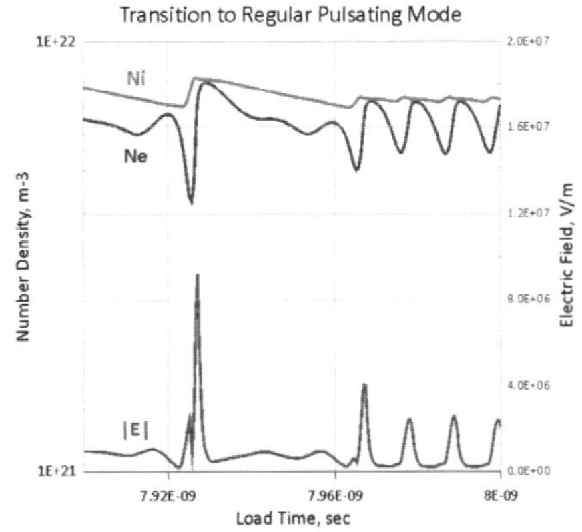

Figure 11 The fine structure of the pulsating mode at the hot spot (f = 10 MHz)

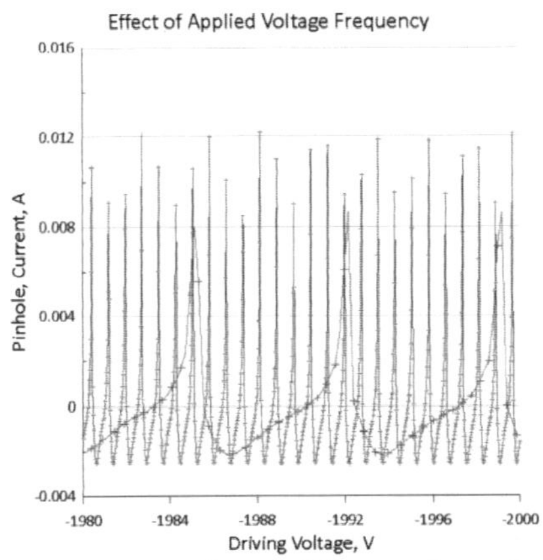

Figure 12 Comparison of the regular pulsating modes for 1 (red) and 10 (blue) MHz power supply frequencies

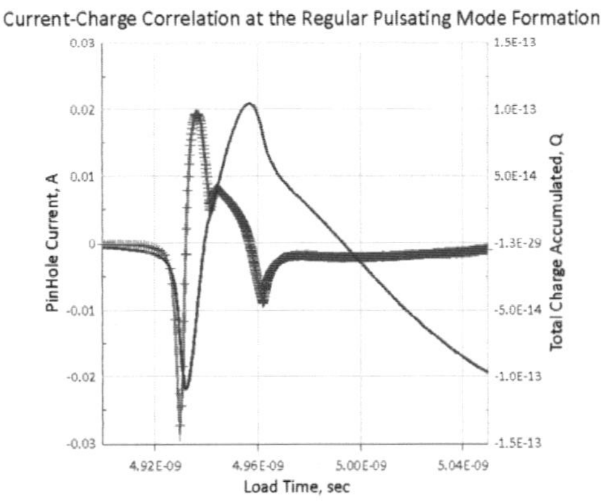

Figure 13 The total domain charge establishing at the first pulse

In figures 8-11 the transition process for the parameters at the hot spot from the initial stage to the fully developed pulsating mode are presented. In the figure12 the effect of the power supply frequency is shown with comparison of the pulsating mode structure. Figure 13 shows the charge formation at the very beginning of the discharge inception. Finally, the framing of the selected current pulse development in terms of electron number density, current density and electric field vectors distribution in the region are presented.

4. Discussion

The particular goal of this publication to report the interesting features of the axisymmetric micro-DBD found during our study of the discharge characteristics for another application.

In the numerical simulation of the sinusoidal DBD with accurate and fine time resolution (time step from 10^{-12} to less than 10^{-14}) we observed the well resolved pulsation of the driving electrode current. Pulsation is observed in both negative and positive polarity. At the negative polarity the pulses look very similar although not identical to the well-known Trichel pulses (see, for example, the work [3] and the good reference list there). The main difference is the frequency which is in two-three order of magnitude higher in compare with “classical“ cases. As the preliminary explanation based on the results obtained in this study (only couple of cases are cited above because of the limited available space) is that the ionization process in this case governed by two physical factors – the local electric field defined by the global charge distribution and the local electron number density defined by ionization rate and electron drift-diffusion transfer. Micro scale configuration is considered: calculating domain linear scale is

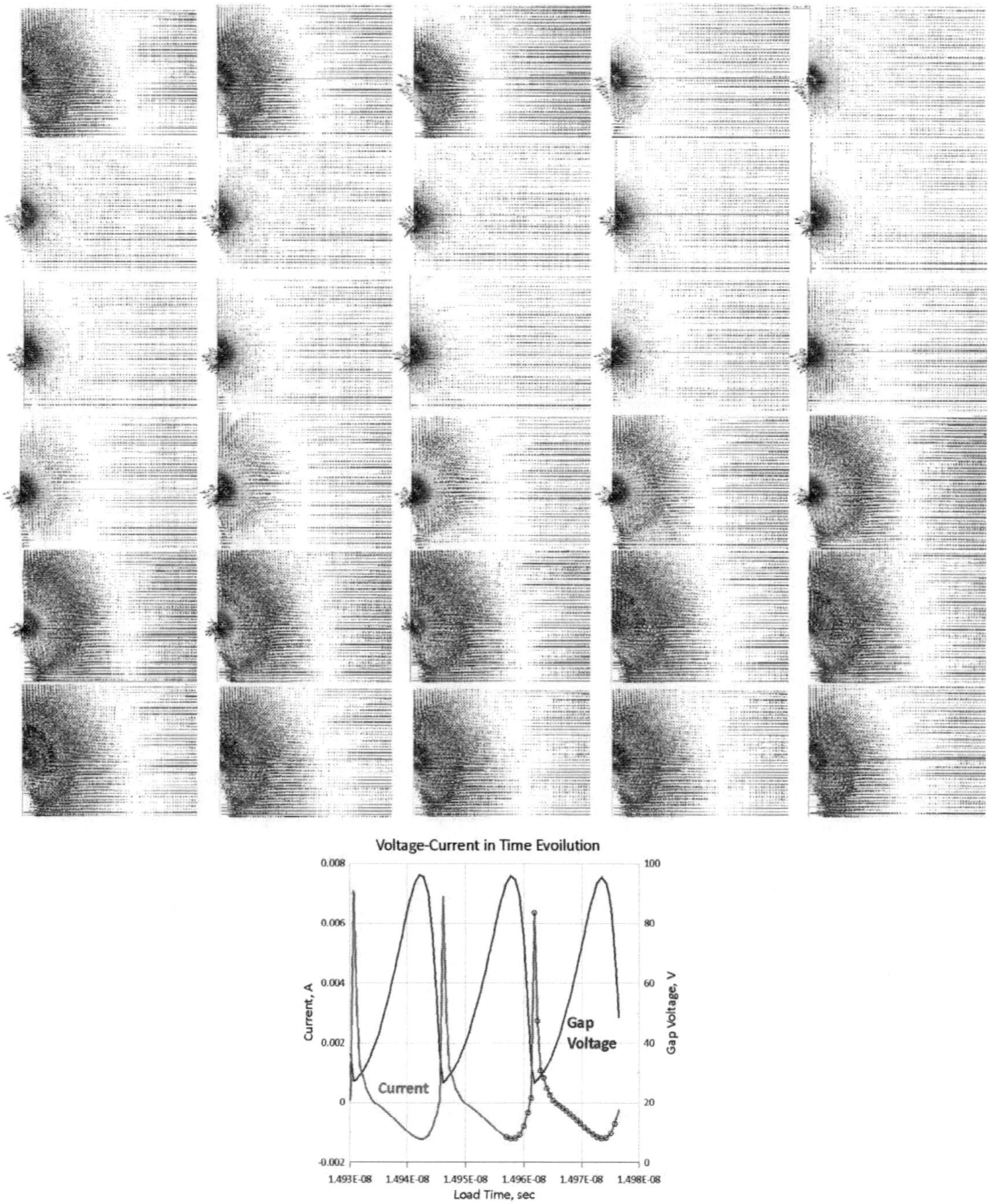

Figure 14 The discharge domain flashing at the regular pulsating mode (f = 10 MHz). Frame consequence from left to right and from top to down corresponds to the symbol order on the plot. Electron number density is represented by level line in blue, current density and electric field vectors in blue and red.

two-three order magnitude larger than the free path length and characteristics times are smaller than the electron collisional time. The framing of the (one of thousands) selected pulse in figure 14 may help us to preliminary suggest the rough model of the regular 10 GHz-pulsing as concurrent chain mechanisms: (a) ionization, (b) electron drift-diffusion, resulting in polarization of the strongly ionized area, (c) decrease of the electric field due to polarization (or screening effect), and (d) ionization rate drops catastrophically (practically stops), drift current becomes less than the diffusion flux (current can change the direction!), all process are stabilized , but (f) external voltage (slowly) recovers the electric field up to breakdown level – pulse start with the fast ionization, and so on. We found in the considered range of parameters that pulsating frequency is rather proportional to the frequency of the extremal applied voltage.

5. Concluding Remarks

1. The new very high frequency (1 – 10 GHz) pulsating electrode current is observed in the numerical simulation of the sinusoidal (0.1 – 10 MHz) micro-DBD in room air environment.
2. The regular (stable pulse shape and frequency) pulsating modes are observed both in the negative and positive polarities but they can differ significantly from each other.
3. The preliminary proposed mechanism can probable include several main processes: electric field induced ionization, drift/diffusion polarization, external circuit slow recovery action, capacity effects of surface and free charge reconfiguration.
4. The more detailed analysis using the wider simulation data based are in progress.

Acknowledgments
The study was carried out at the expense of a grant from the Russian Science Foundation (project No. 14-50-00124).

References

[1] Boeuf U P, Lagmich Y, Unfer Th, Callegari Th, Pitchford L C 2007 Electrohydrodynamic force in dielectric barrier discharge plasma actuators *J. Phys.* D: *Appl. Phys.* **40 3** 652–62 DOI 10.1088/0022-3727/40/3/S03

[2] Benard N, Morea E 2014 Electrical and mechanical characteristics of surface AC dielectric barrier discharge plasma actuators applied to airflow control *Exp. Fluids* **55** 1846 DOI 10.1007/s00348-014-1846-x

[3] Leonov S B, Adamovich I V, Soloviev V R 2016 Dynamics of near-surface electric discharges and mechanisms of their interaction with the airflow, *Plasma Sources Sci. Technol.* **25** 063001, DOI 10.1088/0963-0252/25/6/063001

[4] Bityurin V A, Bocharov A N 2017 Dynamics and Heat Effects in Pulse Periodic Surface DBD, *IOP Conf. Ser. Mater Sci Eng.* **249** 012008 DOI:10.1088/1757-899X/249/1/012008

[5] Bityurin V, Bocharov A, and Popov N 2007 Numerical Simulation of the Discharge in Supersonic Flow Around a Sphere *Proc. 45th AIAA Aerospace Sciences Meeting & Exhibit, 5-8 January 2007, Reno, NV, AIAA Pap.* 2007-0223

[6] Bityurin V A, Bocharov A N and Popov N A 2011 A High Frequency Discharge on a Dielectric Surface *High Temp.* **49 5** 758–61.

[7] Selivonin I, Lazukin A V, Moralev I A, Krivov S 2018 Effect of the electrode degradation on the electrical characteristics of the surface dielectric barrier discharge *Plasma Sources Sci. Technol.* **27** 085003 DOI: 10.1088/1361-6595/aacbf5

[8] Eden J G, Park S-J, et al 2013 Plasma Science and Technology in the Limit of the Small: Microcavity Plasmas and Emerging Applications *IEEE Trans. Plasma Sci.* **41** 661

[9] Trichel G W 1938 The mechanism of the negative point to plane corona near onset *Phys. Rev.* **54 12** 1078

IOP Conf. Series: Journal of Physics: Conf. Series **1112** (2018) 012008 doi:10.1088/1742-6596/1112/1/012008

[10] Akishev Y, Grushin M, Kochetov I, Karal'nik V, Napartovich A, & Trushkin N. (2005). Negative corona, glow and spark discharges in ambient air and transitions between them. Plasma Sources Science and Technology, 14(2), S18–S25. doi:10.1088/0963-0252/14/2/s03

Properties of the coaxial DBD forming plasma jet in the air flow at atmospheric pressure

Yu S Akishev[1,2*], G I Aponin[1], A A Balakirev[1], V B Karalnik[1], A V Petryakov[1] and N I Trushkin[1]

[1] SRC RF TRINITI, 108840, Moscow, Troitsk, Pushkovykh street, 12, Russia
[2] NRNU MEPhI, 115409, Moscow, Kashirskoe shosse, 31, Russia

*akishev@triniti.ru

Abstract. The non-thermal plasma source based on the high-frequency (100 kHz) sinusoidal dielectric barrier discharge (DBD) generating the plasma jet in air at atmospheric pressure is described. This source uses the electrode system of a coaxial geometry and able to work at the low- and high-power regimes. The features of DBD in these regimes are discussed. A narrow plasma jet was generated in airflow at the velocities in the range of 20-80 m/s. The presented experimental results give insight into the spatial-temporal behaviour of the gas discharge plasma inside the coaxial DBD and the spatial characteristics of plasma jet outside the discharge zone as well.

1. Introduction

Non-thermal plasma (NTP) generated by discharges in gas glow at atmospheric pressure is widely used for many applications [1-13]. Along with gas discharges, electron beams of moderate and high energy are also used for the NTP generation [14-16]. In most cases, plasma jets are being formed by discharges in the flow of the inert gases (He, Ar, N_2 and their mixtures) as work gases. The ambient air contains the electronegative components (oxygen and water molecules) which are able to efficiently quench electronically excited species and, besides, they diminish the concentration of electrons in plasma due to strong attachment processes. This circumstance leads to diminishing of the concentration of reactive species and shortening of the action range of these species in air plasma jet. Therefore there was always a strong doubt - whether the plasma jet in ambient air can provide the concentration of reactive species which is sufficient for effective plasma processing. On the other hand, atmospheric pressure air plasma jets are of great interest because the replacement of the expensive inert gases by the ambient air promises a high profit for the plasma-based technologies.

In this paper, we present the newly developed gas discharge system generating plasma jet using the ambient air as working gas at atmospheric pressure. This system is based on the usage of the coaxial high-frequency (100 kHz) sinusoidal DBD in fast flow velocity (20-80 m/s) of the ambient air with a high volume flow rate of 4-24 L/min. High flow velocity provides faster delivering of reactive species from discharge zone to the surface that reduces their loss due to the quenching or other plasma-chemical processes in the jet and increases the total amount of the delivered reactive species onto the surface to be treated. The latter leads to the intensification of the plasma surface processing. We used a high-voltage sinusoidal power supply with a frequency of 100 kHz and output power about 1 kW. Note that high-frequency DBD is able to provide a high specific energy deposition into the working gas. Due to fast quenching processes at atmospheric pressure, it leads to strong air heating that, in its

IOP Conf. Series: Journal of Physics: Conf. Series **1112** (2018) 012009 doi:10.1088/1742-6596/1112/1/012009

turn, can lead to a positive synergy effect of the simultaneous action of hot air and reactive plasma species at the surface treatment.

2. Experimental setup

General scheme of the single-tube reactor is shown in figure 1. Note the coaxial electrode system was never applied before for plasma jet generation in the ambient air as working gas. The main feature of this system is that the inner HV pin electrode was wholly located outside the zone enclasped by the outer cylindrical electrode. This feature allowed us to increase both the volume of discharge zone in the tube and the residence time of gas being activated by the discharge.

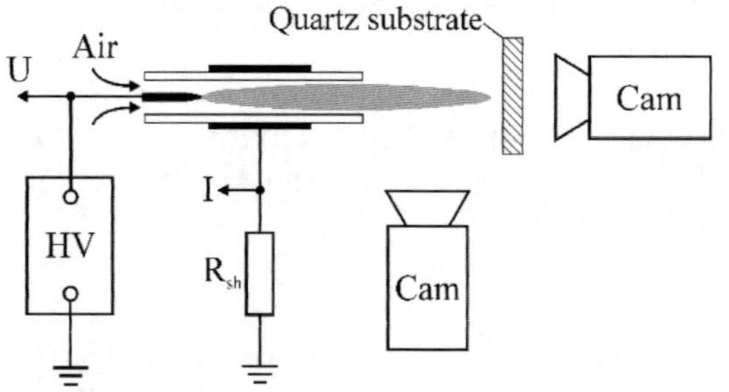

Figure 1 Setup of the coaxial DBD in fast air flow at atmospheric pressure. The inner and outer diameters of the dielectric tube are 2 and 4.5 mm. The length of discharge zone is 20 mm.

The applied high voltage was measured by HV divider PINTEK HVP-39 (1000:1, 40 kV, 200 MHz). The discharge current was measured by a low-inductive shunt with resistance of 50 Ohm. All electrical signals were recorded by the digital oscilloscopes such as Tektronix TDS 520, Tektronix DPO2024 and Tektronix TDS 2012 with bandwidth of 500, 200 and 100 MHz. The current and voltage waveforms were used to calculate the average discharge power $< W >$:

$$< W > = \frac{1}{nT}\int_0^{nT} I(t)U(t)dt,$$

where $I(t)$ and $U(t)$ are the current and voltage waveforms of the sinusoidal DBD recorded in the experiment by the digital oscilloscope, n is the number of periods T taken into consideration.

The discharge and plasma jet images were taken by digital camera Canon EOS 550 with the exposure time down to 10^{-4} s and multi-frame fast camera (equipped with the intensifier) with the exposure time down to 50 ns. The concrete exposure time has been determined by both the light intensity of the registered discharge and plasma jet and the sensitivity of the used optical device.

The barrier discharge was generated in the airflow inside thin quartz or ceramic tube. The sizes of tubes were chosen close to each other as much as possible. The inner and outer diameters of the quartz and ceramic tubes are of 2.0 and 4.5 mm and 2.5 and 4.5 mm, respectively. At given inner diameters of tubes and gas flow velocities (20-80 m/s) Reynolds number of flows inside tubes exceeds the critical number $Re \approx 2320$. It means that flow of the ambient air inside tubes is turbulent one.

3. Experimental results and discussion

The images in figure 2 show the evolution of the plasma structure inside the discharge zone at the increasing the amplitude U of the applied sinusoidal voltage. These images are averaged over many half-cycles. The current-voltage waveforms of the airflow DBD in the regimes corresponding to the discharge photos at $U = 4.5$ kV and $U = 7.6$ kV is shown in figure 3.

One may see in figure 2 that the discharge strongly changes its structure at the increase in the amplitude U, i.e. the discharge occupies the larger volume at higher voltage. These changes in the spatial structureare accompanied by the essential changes in the discharge current waveform. Figure 3 presents these changes by the example of the low-power (LP) and high-power (HP) regimes. In the LP regime, the discharge current consists only of the high amplitude sharp peaks. In the HP regime, the current consists of two components: major component is the smoothly changing current and minor

IOP Conf. Series: Journal of Physics: Conf. Series **1112** (2018) 012009 doi:10.1088/1742-6596/1112/1/012009

component is the sharply changing current with peaks of low amplitude. The existence of the smoothly changing background current we attribute to the existence in each half-cycle of the survived plasma of high concentration (about 10^{12} cm^{-3}) generated in the previous half-cycle.

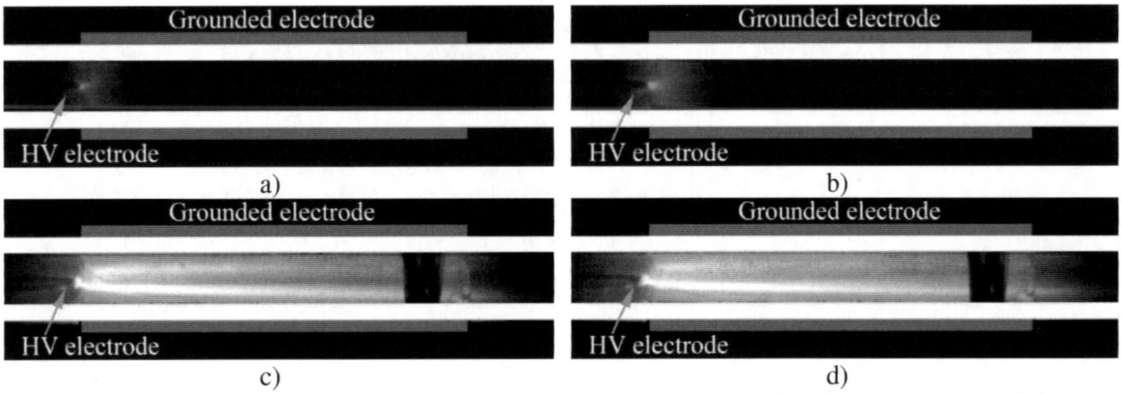

Figure 2 The averaged in time images (side view) of DBD structure inside the quartz tube at different applied voltages. The length of discharge zone is 20 mm. Exposure time is 1 ms. V = 60 m/s. a) U = 4.5 kV, $< W >=$ 3.7 W, b) 5.3 kV, $< W >=$ 6.4 W, c) 7 kV, $< W >=$ 68.5 W, d) 7.6 kV, $< W >=$ 86.3 W.

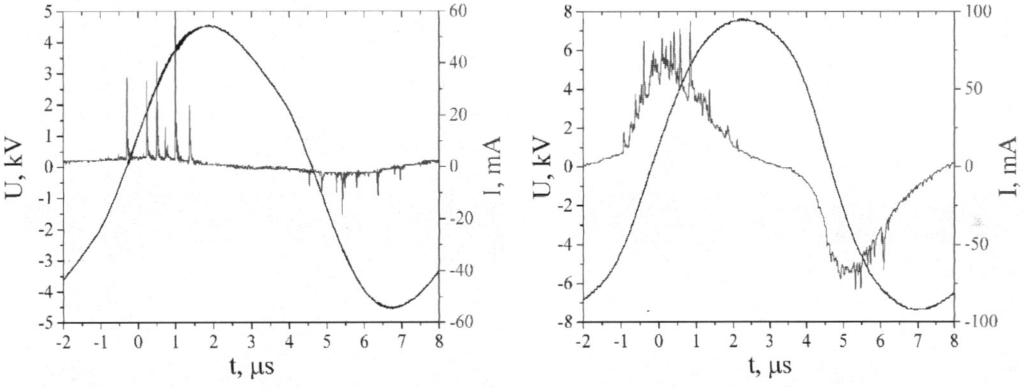

Figure 3 The current and voltage waveforms of the airflow DBD in the regimes corresponding to photos at U = 4.5 kV (a) and U = 7.6 kV (b) in figure 2 respectively.

Evolution of the visual length of the air plasma jet with changes of the airflow velocity V is shown in figure 4. The visual information is complimented with data in figure 5. The dependence of the average discharge power $< W >$ on the applied voltage amplitude U at the gas flow velocity V = 60 m/s is given in figure 5a. The dependence of visual length of plasma jet on $< W >$ and V is depicted in figures 5b and 5c. The dotted arrows in figures 5a and 5b show a sharp DBD transition (like jump) from the low-power regime into high-power regime. The arrow in figure 5c shows the termination of DBD if the velocity exceeds the critical magnitude about of 75 m/s. It was observed also that airflow DBD in the low-power regime generates ozone but transition into the high-power regime terminates the ozone production.

IOP Conf. Series: Journal of Physics: Conf. Series **1112** (2018) 012009 doi:10.1088/1742-6596/1112/1/012009

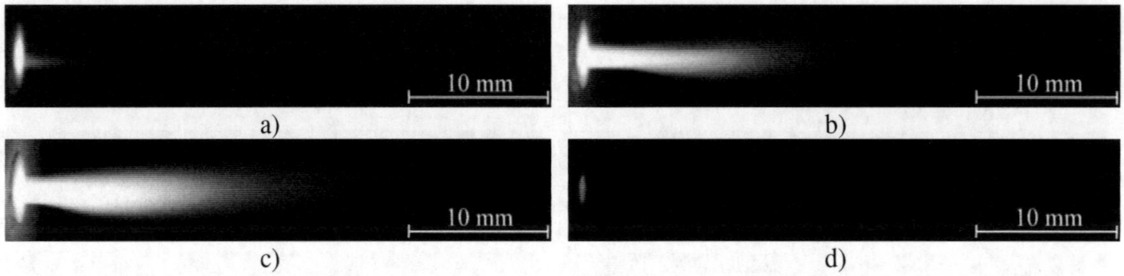

Figure 4 The visual length of the air plasma jet vs the airflow velocity *V*. Exposure time of each photo is 3 s, $U = 5.5$ kV. a) 20 m/s, 8W; b) 30 m/s, 65.2W; c) 50 m/s, 53 W; d) 75 m/s, 10.7 W

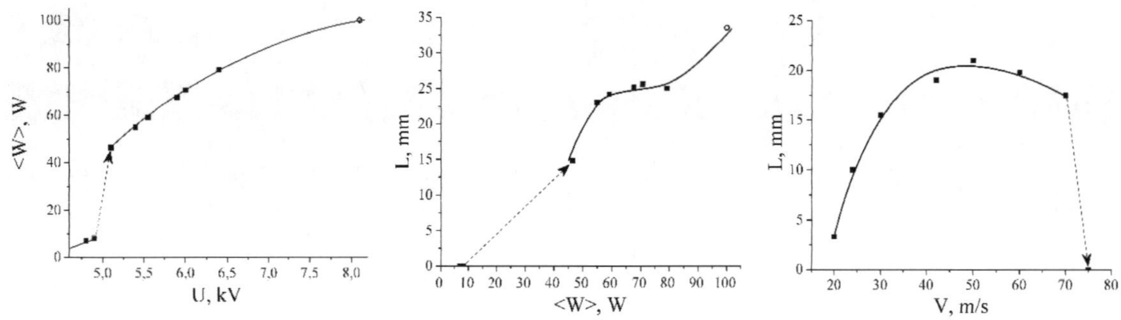

a) b) c)

Figure 5 (a) The dependence of $< W >$ vs U at the gas flow velocity $V = 60$ m/s. The visual length of the air plasma jet vs the average discharge power <*W*> (b) and vs gas velocity *V* at $U = 5.5$ kV (c). The squares and circle correspond to ceramic and quartz tube. The arrows in (a) and (b) show the sharp transition of DBD from the low- to high-power regime; arrow in (c) shows the termination of DBD at higher *V*.

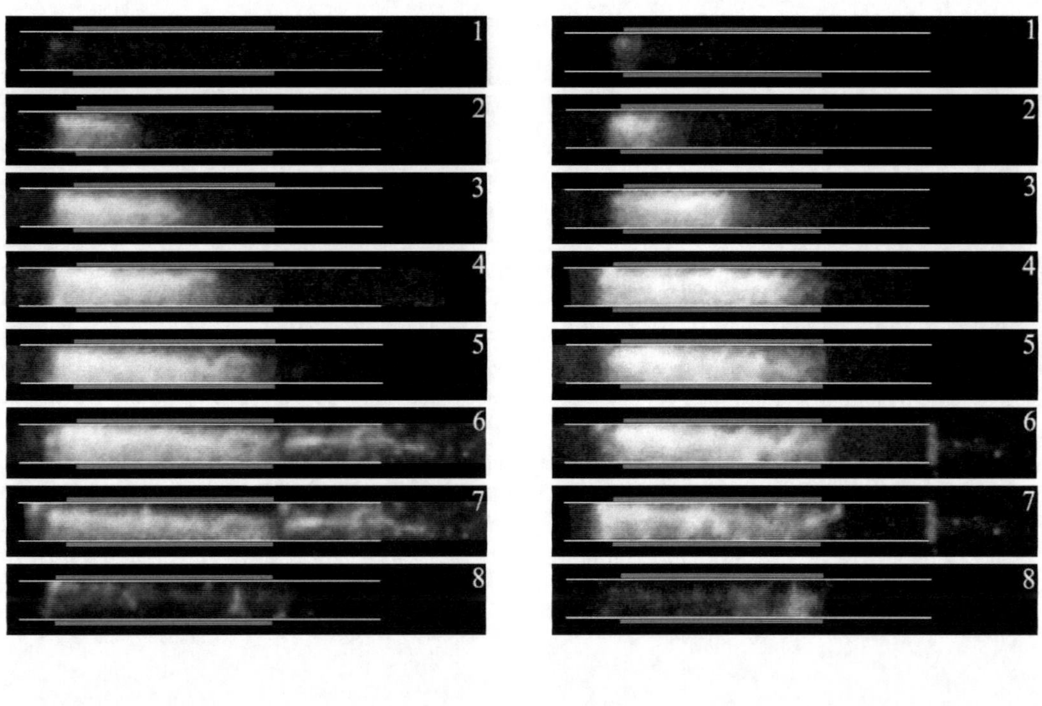

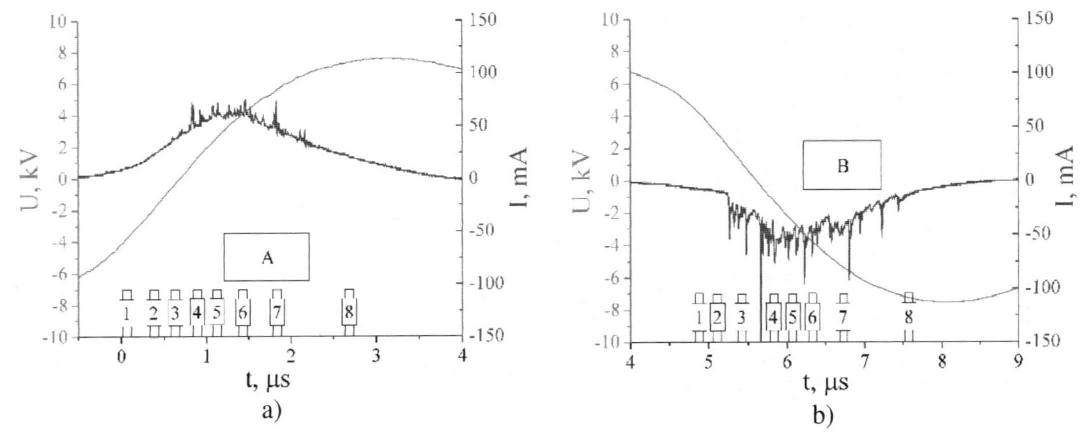

Figure 6. Images of the discharge zone and plasma jet taken with a short exposure time. The instant images are correlated with the proper *I-U* waveforms placed below the pictures. High voltage was applied to outer electrode. $V = 60$ m/s, high-power regime $< W >\approx 75$ W. The exposure time of the discharge zone is 100 ns and plasma jet 1.0 μs (window A) and 1.0 μs (window B). The figures show the position of the shots on the time axis.

It is interesting to trace in detail the evolution of plasma structure with high time resolution. Such information is presented in figure 6. Note that the brightness of the discharge zone inside tube strongly exceeds the brightness of the plasma jet. In fact, the dim air plasma jet is hardly visible with the naked eye. Therefore the "bright" plasma jet depicted in figure 4 is the trick which was obtained due to the chosen exposure time for plasma jet. This exposure time was much longer compared to that for the discharge zone. Besides, figure 4 shows the images averaged over long time but figures 6a and 6b show the instant images of the discharge zone at positive and negative phases of the discharge current. Note that the propagation of so-called "plasma bullets" along hot air plasma jet outside the tube was not observed.

4. Conclusion

We have developed novel plasma source of the "efflux" type that is based on the usage of the coaxial high-frequency (100 kHz) sinusoidal DBD in fast flow velocity (20-80 m/s) of the ambient air with a high volume flow rate of 4-24 L/min per tube. The configuration electrode and gas-dynamic and electrical regimes were never been used before for generation of plasma jet in the ambient air as working gas at atmospheric pressure. The usage of high-frequency voltage allows one to increase appreciably the specific energy deposition in the discharge and to heat the airflow up to the high temperature of 500-600 C. We have revealed two regimes (low-power and high-power) of airflow DBD and determined the existence of upper limit of gas flow velocity that terminates the work of DBD. Our findings promote more insight into physics of the airflow DBD at atmospheric pressure.

Acknowledgement

This work was supported by the RFBR. Grants No 17-02-00234, No 18-02-00539.

References

[1] Kim M C, Yang S H, Boob J H and Han J G 2003 *Surf. Coat. Technol.* **174-175** 839
[2] Laroussi M and Lu X 2005 *Appl. Phys. Lett.* **87** 113902
[3] Temmerman E, Akishev Yu, Trushkin N, Leys C and Verschuren J 2005 *J. Phys. D: Appl. Phys.* **38** 505
[4] Winter J, Brandenburg R, and Weltmann K D 2014 *Plasma Sources Sci. Technol.* **24** 064001
[5] Korolev Y 2015 *Russ. J. Gen. Chem.* **85** 1311

[6] Akishev Yu, Grushin M, Dyatko N, Kochetov I, Napartovich A, Trushkin N, Duc T and Descours S 2008 *J. Phys. D: Appl. Phys.* **41** 23520
[7] Babaeva N Y and Kushner M J 2014 *Plasma Sources Sci. Technol.* **23** 015207
[8] Sosnin E A, Panarin V A, Skakun V S, Baksht E K and Tarasenko V F 2017 *Eur. Phys. J. D* **74** 25
[9] Lu X, Naidis G V, Laroussi M, Reuter S, Graves D B and Ostrikov K 2016 *Phys. Rep.* **630** 1
[10] Akishev Yu S, Grushin M, Karalnik V, Petryakov A and Trushkin N 2012 *IEEE Trans. Plasma Sci.* **40** 2806
[11] Bartis E A, Graves D B, Seog J and Oehrlein G S 2013 *J. Phys.* D*: Appl. Phys.* **46** 312002
[12] Akishev Yu, Balakirev A, Grushin M, Karalnik V, Kochetov I, Napartovich A, Petryakov A and Trushkin N 2015 *IEEE Trans. Plasma Sci.* **43** 745
[13] Korolev Y D, Frants O B, Geyman V G, Landl N V and Kasyanov V S 2011 *IEEE Trans. Plasma Sci.* **39** 3319
[14] Akishev Yu S, Karal'nik V B, Petryakov A V, Starostin A N, Trushkin N I and Filippov A V 2016 *Plasma Phys. Rep.* **42** 14
[15] Akishev Yu S, Karal'nik V B, Petryakov A V, Starostin A N, Trushkin N I and Filippov A V 2017 *J. Exp. Theor. Phys.* **124** 231
[16] Levko D, Krasik Y, Tarasenko V, Rybka D and Burachenko A 2013 *J. Appl. Phys.* **113** 196101

IOP Conf. Series: Journal of Physics: Conf. Series **1112** (2018) 012010 doi:10.1088/1742-6596/1112/1/012010

Optical spectrum of the coaxial DBD forming the plasma jet in fast air flow at atmospheric pressure

Yu S Akishev[1,2*], V B Karalnik[1], A V Petryakov[1], K B Popov[1], O N Sattarov[1] and N I Trushkin[1]

[1] SRC RF TRINITI, 108840, Moscow, Troitsk, Pushkovykh street, 12, Russia
[2] NRNU MEPhI, 115409, Moscow, Kashirskoe shosse, 31, Russia

*akishev@triniti.ru

Abstract. The non-thermal plasma source based on the high-frequency (100 kHz) sinusoidal dielectric barrier discharge (DBD) generating the plasma jet in air at atmospheric pressure is described. This source uses the electrode system of a coaxial geometry and able to work at the low- and high-power regimes. A narrow plasma jet was generated in airflow at the velocities in the range of 20-80 m/s. The emissive UV-Visual spectra of the discharge zone and plasma jet are presented. The change of the emissive spectrum along the length of plasma jet and on the surface of the quartz substrate to be treated is studied as well. The spectrum reflects the composition of reactive species generated by DBD. The presented experimental results give insight into the properties of the gas discharge plasma inside the coaxial DBD and the spatial characteristics of plasma jet outside the discharge zone as well.

1. Introduction

Non-thermal plasma jets at atmospheric pressure are widely used for many applications [1-11]. In most cases, plasma jets being formed by discharges using the inert gases (He, Ar, N_2 and their mixtures) as work gases. The ambient air contains the electronegative components (oxygen and water molecules) which are the efficient quenchers of electronically excited species and, besides, they diminish the concentration of electrons in plasma due to strong attachment processes. This circumstance leads to diminishing the concentration of reactive species and shortening the action range of these species in air plasma jet. Therefore, there was always a strong doubt - whether the plasma jet in the ambient air as work gas can provide the concentration of reactive species which is sufficient for effective plasma processing. On the other hand, atmospheric pressure air plasma jets are of great interest because the replacement of the expensive inert gases by the ambient air promises a high profit for the plasma-based technologies.

In this paper, we present the newly developed gas discharge system generating the plasma jet using the ambient air as the working gas at atmospheric pressure. This system is based on the usage of the coaxial high-frequency (100 kHz) sinusoidal DBD in fast flow velocity (20-80 m/s) of the ambient air with a high-volume flow rate of 4-24 L/min per tube. High flow velocity provides faster delivering of reactive species from discharge zone to the surface that reduces their loss due to the quenching or other plasma-chemical processes in the jet and increases the total amount of the delivered reactive species onto the surface to be treated. The latter leads to the intensification of the plasma surface processing. We used the high-voltage sinusoidal power supply with a frequency of 100 kHz and output

IOP Conf. Series: Journal of Physics: Conf. Series **1112** (2018) 012010 doi:10.1088/1742-6596/1112/1/012010

power about 1 kW. Note that high-frequency DBD is able to provide a high specific energy deposition into the working gas. Due to fast quenching processes at atmospheric pressure, it leads to strong air heating that, in its turn, can lead to a positive synergy effect of the simultaneous action of hot air and reactive plasma species at the surface treatment.

Note that the effectiveness of the wanted plasma-chemical processes happening at the interaction of plasma jet with the surface depends on the composition of the reactive species delivered by plasma jet onto the surface. However, the reactive species interacting with the substrate are not the same as those generated by the gas discharge itself. A reason is that there is a plasma-chemical transformation of the primary reactive species into the secondary (and so on) ones in the course of their transportation by the plasma jet from the gas discharge zone onto the substrate to be treated. So, to evaluate correctly the efficiency of the surface treatment by plasma jet, it is necessary to take into account the properties of the used gas discharge generating the primary reactive species and also the transformation of reactive species in the course of their transportation by plasma jet on the surface. This task was implemented by us with use of the spectroscopic method. Information on the composition of the reactive species generated by the DBD plasma has been obtained from their emissive spectra in UV-Visual region. Of course, such information cannot satisfy this question in full because some species do not emit in this spectral region or the intensity of their emission either is too weak (for instance, metastable) or overlaps with the strong lines (bands) of other species. Nevertheless, some conclusions can be done.

2. Experimental setup

General scheme of the single-tube reactor is shown in figure 1. The barrier discharge was generated in the airflow inside thin quartz or ceramic tube. The sizes of tubes were chosen close to each other as much as possible. The inner and outer diameters of the quartz and ceramic tubes are of 2.0 and 4.5 mm and 2.5 and 4.5 mm, respectively. At given inner diameters of tubes and gas flow velocities (20-80 m/s) Reynolds number of flows inside tubes exceeds the critical number $Re \approx 2320$. It means that flow of the ambient air inside tubes is turbulent one.

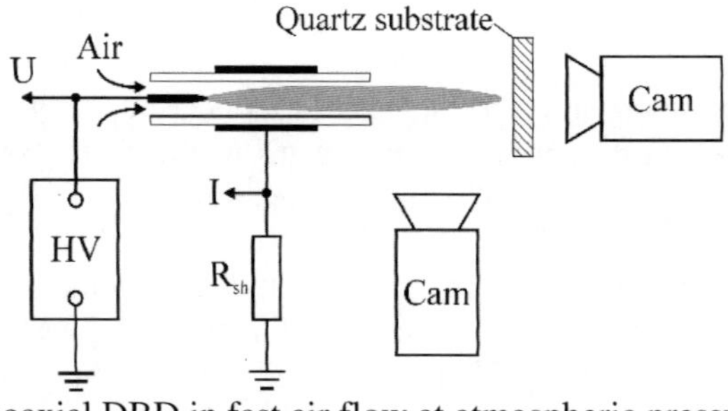

Figure 1 Setup of the coaxial DBD in fast air flow at atmospheric pressure. The inner and outer diameters of the dielectric tube are 2 and 5 mm. The discharge zone length is 20 mm.

The applied high voltage was measured by HV divider PINTEK HVP-39 (1000:1, 40 kV, 200 MHz). The discharge current was measured by a low-inductive shunt with resistance of 50 Ohm. All electrical signals were recorded by the digital oscilloscopes such as Tektronix TDS 520, Tektronix DPO2024 and Tektronix TDS 2012 with the bandwidth of 500, 200 and 100 MHz. The current and voltage waveforms were used to calculate the average discharge power $< W >$:

$$< W > = \frac{1}{nT}\int_0^{nT} I(t)U(t)dt,$$

where $I(t)$ and $U(t)$ are the current and voltage waveforms of the sinusoidal DBD recorded in the experiment by the digital oscilloscope, n is the number of periods T taken into consideration.

To register the UV and visual spectrum of the discharge and plasma jet, the spectrometers AvaSpec-2048FT-6-RM and AvaSpec-2048-L were used. The first of them has a low sensitivity to the registered light but high spectral resolution $\Delta\lambda = 0.1$ nm in the region $\lambda = 240 - 400$ nm and $\Delta\lambda =$

IOP Conf. Series: Journal of Physics: Conf. Series **1112** (2018) 012010 doi:10.1088/1742-6596/1112/1/012010

0.2 nm in the region $\lambda = 400 - 800$ nm. The second one has a high sensitivity to the registered light but low spectral resolution $\Delta\lambda = 1$ nm in the region $\lambda = 200 - 1000$ nm. The optical system was focused in such a way to collect the light either from the middle of discharge zone or from different parts of the plasma jet.

The data capture time was different and has been determined by the intensity of the registered light and sensitivity of the spectral device. The emissive spectra were used to determine the sorts of reactive species in the gas discharge zone and different parts of plasma jet. The second positive system of N_2 was used to determine the rotation and vibration temperature in plasma using the SpecAir code [12]. The average translation gas temperature T inside the non-equilibrium plasma jet was measured by the thermocouple located 5 mm away from the tube exit.

3. Experimental results and discussion

The coaxial DBD in fast airflow exhibits two regimes: the low-power (LP) and high-power (HP) regimes which being happened at low and high amplitudes of the applied voltage. The images (a) and (b) in figure 2 show the side view of the plasma structure inside the discharge zone of DBD in quartz tube at the LP and HP regimes respectively. These images are averaged over many half-cycles. The UV-Visual spectra of the discharge zone of the airflow DBD in the LP and HP regimes are shown in figures 2c and 2d. The $I - U$ waveforms of DBD in LP and HP regimes are given in figures 2e and 2f.

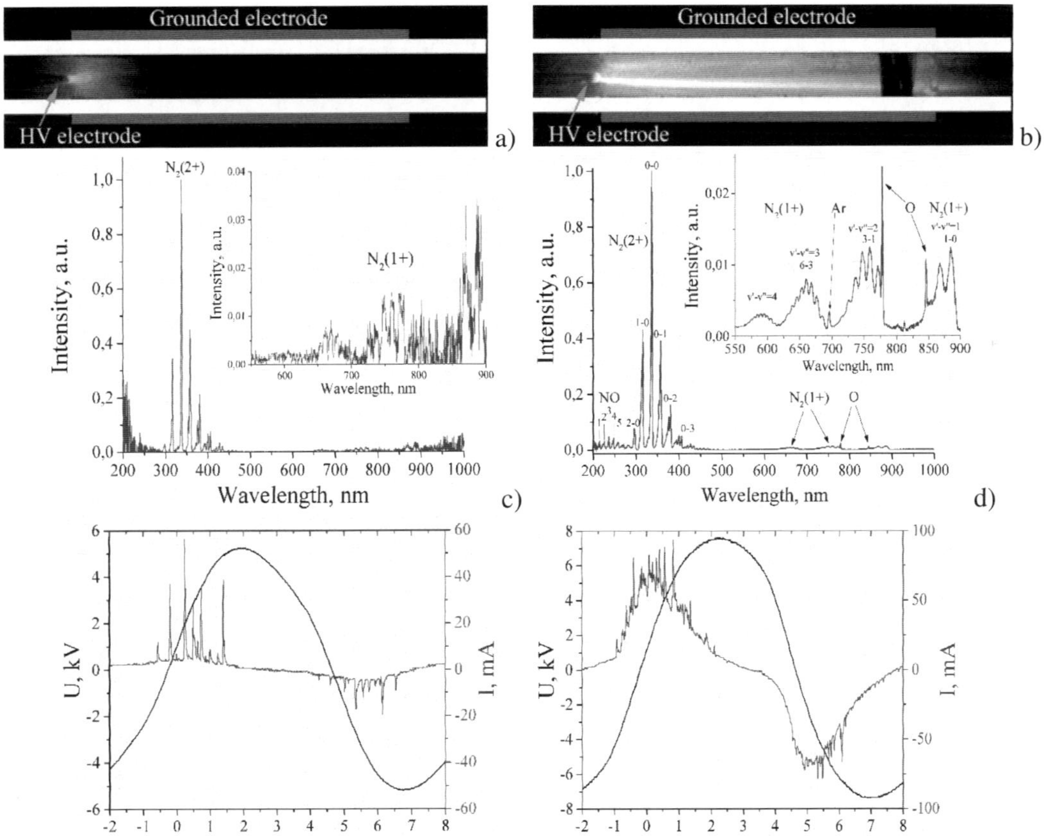

Figure 2 Images of plasma structure inside discharge zone of DBD in quartz tube at low-power (a) and high-power regime (b); airflow is directed from left to right. The UV-Visual spectra of the discharge zone at low-power (c) and high-power regime (d). Images and spectra were taken under the exposure time of 1 ms (a, b) and 20ms (c) and 10 ms (d). The current-voltage waveforms of DBD in

IOP Conf. Series: Journal of Physics: Conf. Series **1112** (2018) 012010 doi:10.1088/1742-6596/1112/1/012010

LP (e) and HP (f) regimes. Parameters: (a, c, e) $V = 60$ m/s, $U = 5.3$ kV, $< W >= 6.4$ W; (b, d, f) $V = 60$ m/s, $U = 7.6$ kV, $< W >= 86$ W. Quartz tube. The length of discharge zone is 20 mm.

One may see in figure 2 that the emissive spectrum of the discharge zone of DBD in HP regime is formed by such excited species as N_2, Ar, O, NO, OH. It is surprising that argon contributes to the plasma chemical kinetics in the discharge zone as well. The reason is that the concentration of Ar in ambient air is high and equal to 0.93% (vol.) or $2.3 \cdot 10^{17}$ cm^{-3} at normal atmospheric condition.

Figure 3 presents information about the image of air plasma jet in HP regime, its emissive spectrum in different parts of jet and the intensity distribution of N_2(2+) system ($C^3\Pi_u \rightarrow B^3\Pi_g, \Delta v = 0\text{-}0, \lambda =$ 337 nm) along the plasma jet.

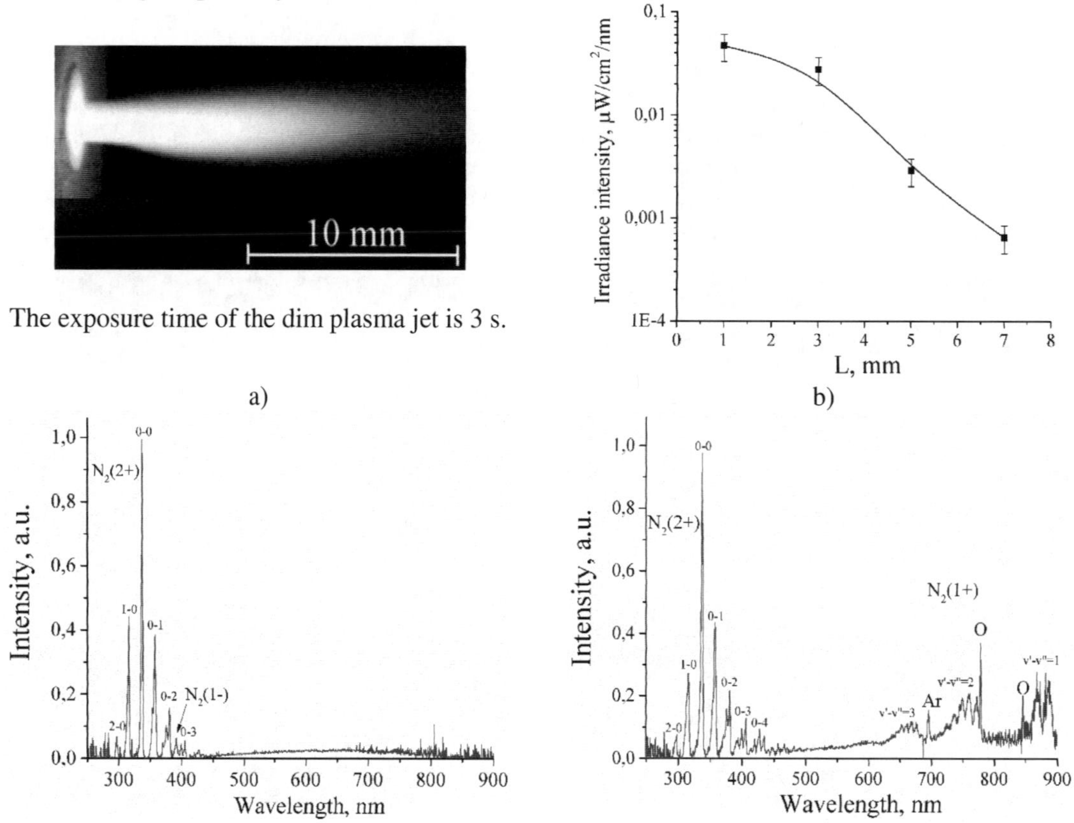

Figure 3 Image of free air plasma jet (a). The intensity distribution of $N_2(C^3\Pi_u \rightarrow B^3\Pi_g, \Delta v = 0\text{-}0, \lambda = 337$ nm) along the plasma jet. Two examples of the UV-Visual spectra taken at $L = 1$ mm (c) and L=5 mm(d) away from the tube exit. Parameters of DBD: $V \approx 60$ m/s, $U = 5.3$ kV, $< W >\approx$ 60W. Exposure time for all spectra is 40 s.

Figures 4 and 5 show the radial distribution of both the UV-Visual spectrum emitted by plasma sheet being formed at striking the substrate by plasma jet and the intensity of emission of $N_2(C^3\Pi_u \rightarrow B^3\Pi_g, \Delta v = 0\text{-}0, \lambda = 337$ nm), NO(γ-band, $A^3\Sigma^+ \rightarrow X^2$), O(^{5}P $\rightarrow 5S, \lambda = 777$ nm).

IOP Conf. Series: Journal of Physics: Conf. Series **1112** (2018) 012010 doi:10.1088/1742-6596/1112/1/012010

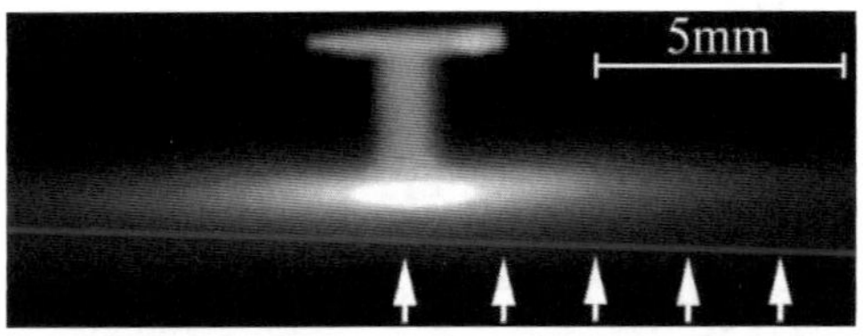

The arrows show several points at the plasma sheet from which the spectra were observed.

a)

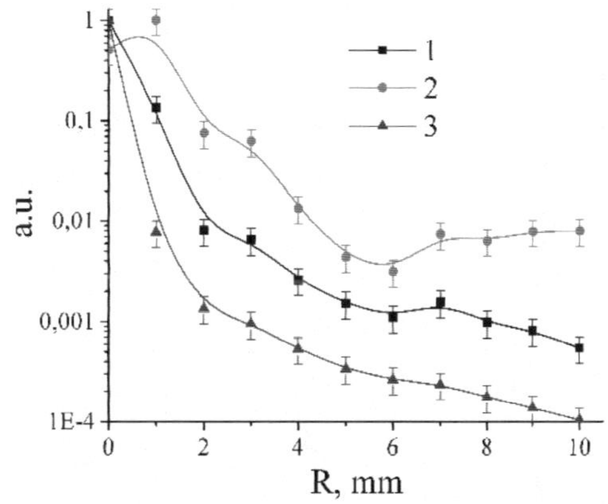

b)

Figure 4 The image of plasma sheet being formed by the air plasma jet on the quartz substrate (a). Radial distribution along plasma sheet of the emission intensity for the excited species (b): $N_2(C^3\Pi_u \rightarrow B^3\Pi_g, \lambda = 337$ nm) (1), NO(γ-band, $A^3\Sigma^+ \rightarrow X^2\Pi, \lambda = 236$nm) (2), O(^{5}P $\rightarrow 5S, \lambda =$ 777 nm) (3).

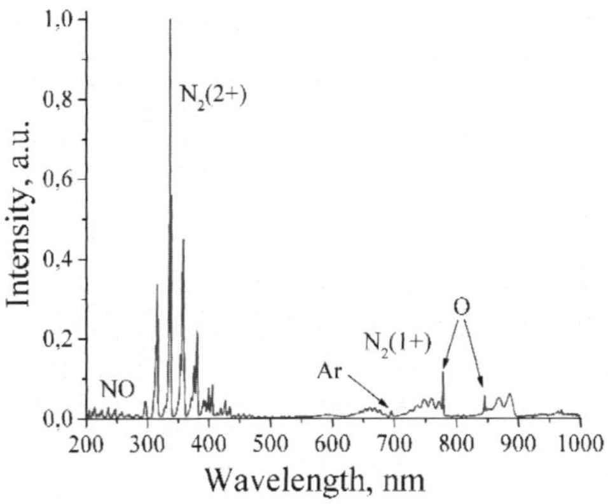

a)

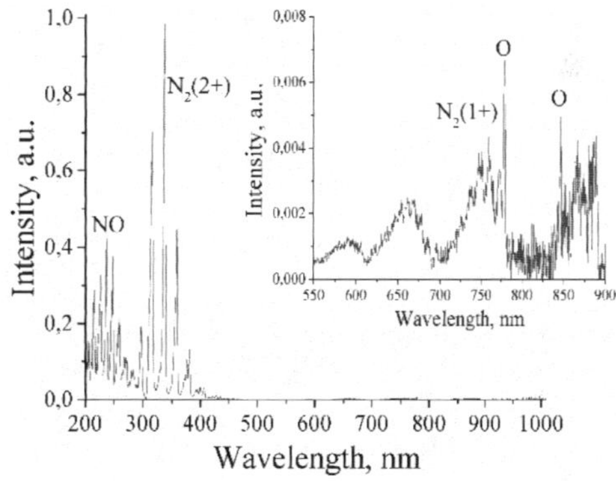

b)

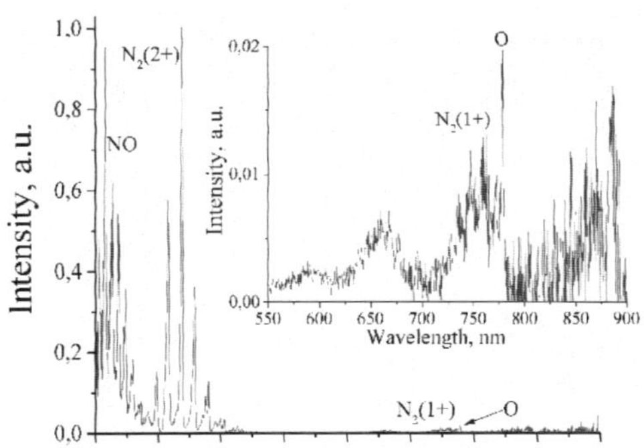

c)

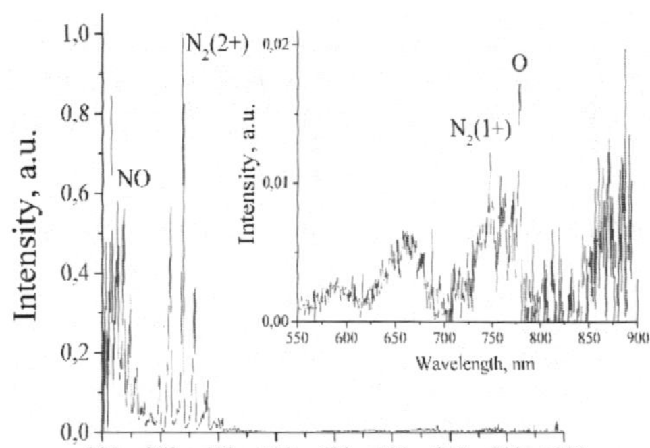

d)

IOP Conf. Series: Journal of Physics: Conf. Series **1112** (2018) 012010 doi:10.1088/1742-6596/1112/1/012010

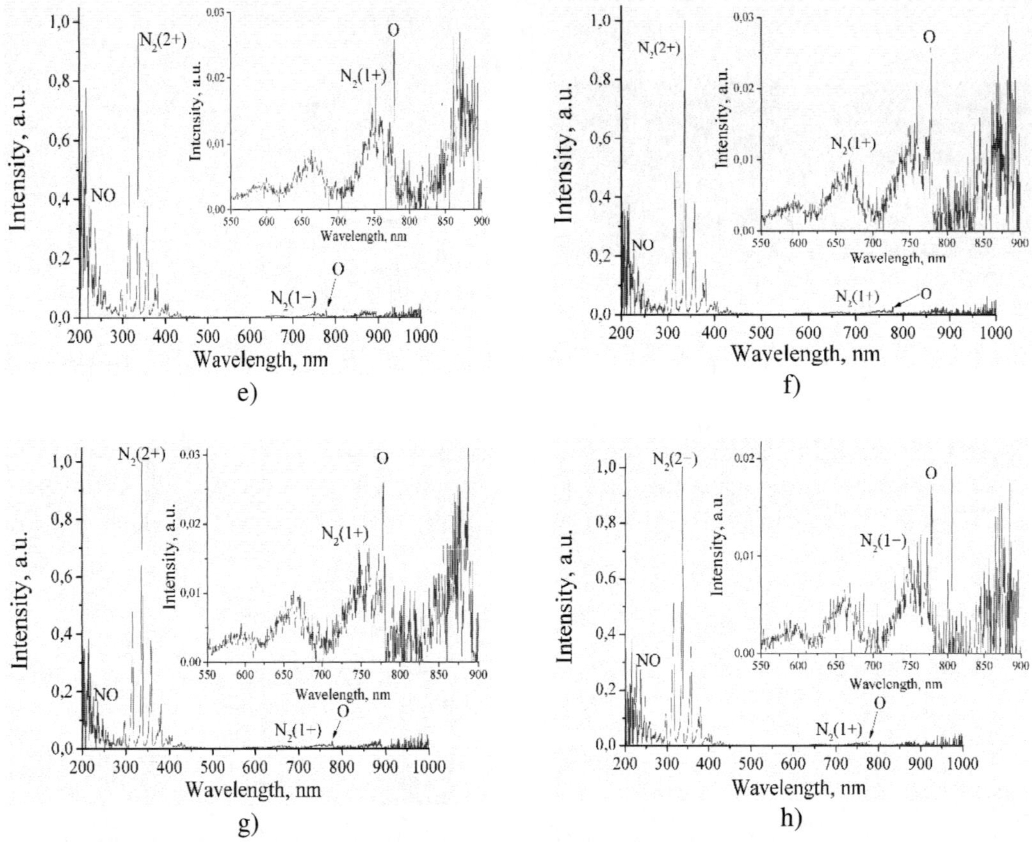

Figure 5 Evolution of the spectrum along radius R of the plasma sheet on the quartz substrate placed at 5 mm away from the tube exit. $V \approx 60$ m/s, $U = 5.3$ kV, $< W > \approx 60$ W. a) $R = 0$mm, $\tau = 40$ ms; b) $R = 1$ mm, $\tau = 500$ ms; c) $R = 2$ mm, $\tau = 500$ ms; d) $R = 3$ mm, $\tau = 3$ s; e) $R = 4$ mm, $\tau = 3$ s; f) $R = 5$ mm, $\tau = 10$ s; g) $R = 6$ mm, $\tau = 20$ s; h) $R = 8$ mm, $\tau = 20$ s.

The probe measurements revealed that the electric field strength in plasma jet is very low ($E/N \approx$ 1 Td). The features of the probe measurements at an elevated gas pressure are described in detail in [13,14]. Based on the analysis of the possible reactions in the plasma jet with a low electric field, one may state that the plasma chemical reactions in the jet are induced not by electrons but metastable $O_2(a^1\Delta)$, $O(^1D)$, $N_2(A^3\Sigma^+)$ and $Ar(^3P)$ being formed in the discharge zone and blown out by flow. Besides, the non-excited species such as $O(^3P), OH, O_3, NO$ present also in the air plasma jet. The close examination of the possible reactions in the air plasma jet leads to the following list:

$$Ar(^3P) + N_2(X^1\Sigma^+) \rightarrow N_2(A^3\Sigma^+) \quad (1)$$
$$N_2(A^3\Sigma^+) + O_2 \rightarrow N_2(X^1\Sigma^+) + O(^3P) + O(^3P) \quad (2)$$
$$N_2(A^3\Sigma^+) + H_2O \rightarrow N_2(X^1\Sigma^+) + OH(X^2\Pi) + H \quad (3)$$
$$N_2(A^3\Sigma^+) + O(^3P) \rightarrow NO(X^2\Pi) + N \quad (4)$$
$$N + O_2 \rightarrow NO(X^2\Pi) + O(^3P) \quad (5)$$
$$O(^3P) + O_2 + N_2(X^1\Sigma^+) \rightarrow O_3 + N_2(X^1\Sigma^+) \quad (6)$$
$$OH + OH \rightarrow O(^3P) + H_2O \quad (7)$$
$$O(^1D) + H_2O \rightarrow OH + OH \quad (8)$$
$$O(^1D) + O_2 \rightarrow O_2(a^1\Delta) + O(^3P) \quad (9)$$
$$O(^1D) + O_2 \rightarrow O_2 + O(^3P) \quad (10)$$

IOP Conf. Series: Journal of Physics: Conf. Series **1112** (2018) 012010 doi:10.1088/1742-6596/1112/1/012010

$$O(^1D) + N_2(X^1\Sigma^+) \rightarrow N_2(X^1\Sigma^+) + O(^3P) \quad (11)$$
$$O_3 + O(^1D) \rightarrow O(^3P) + O_3,\ O_2 + O_2 \quad (12)$$
$$O_3 + O_2(a^1\Delta) \rightarrow O(^3P) + O_2 + O_2 \quad (13)$$
$$O_3 + NO(X^2\Pi) \rightarrow NO_2 + O_2 \quad (14)$$

One may see in the above list that primary reactions in the jet are induced by metastable $O_2(a^1\Delta)$, $O(^1D)$, $Ar(^3P)$ and $N_2(A^3\Sigma^+)$. According to the above list, the excitation of these species happens due collisions with metastable $N_2(A^3\Sigma^+)$. However, there is a question: how these metastable themselves are being generated in the plasma jet and the plasma sheet. We assume this mechanism is based on bimolecular reactions between the highly excited $N_2(X,v>12)$ molecules having the rate coefficient $k_{v1,v2} \approx 10^{-16}$ cm^{-3} [15]:

$$N_2(X,v_1) + N_2(X,v_2) \xrightarrow[v_1,v_2>12]{} N_2(X) + N_2(A^3\Sigma^+) \quad (15)$$

To prove the existence of highly vibration excited nitrogen molecules, we have performed the spectroscopic measurements of the average vibration T_{vib} and rotation T_{rot} temperatures of the electron-exited state $N_2(C)$ in the discharge zone and the translation gas temperature T in the plasma jet as well. The gas temperature was measured by the thermocouple located 5 mm away from the tube exit. The obtained results are presented in figure 6. The behaviour of the averaged DBD power $\langle W \rangle$ in dependence of gas flow velocity is shown in this figure as well. The dotted arrows denote the back transition of DBD from the high-power regime into the low-power regime. This transition is induced by high velocity of airflow inside the discharge zone.

In air at the atmospheric pressure there is a fast exchange between rotational and translation degrees of freedom, therefore, the rotation temperature is equal to the gas temperature: $T_{rot} = T$. Also, due to fast exchange between vibration degrees of freedom of $N_2(X,v)$ and $N_2(C,v)$, vibration temperatures of these states can be equal to each other. The spectroscopic measurements showed that, despite the high gas temperature ($T \approx 900$ K), the air inside the discharge zone is non-equilibrium, $T_{vib} > T_{rot}$, and in some cases the measured vibration temperature reaches up to 4600-4800 K. High gas temperature leads to that the rate of electron detachment from negative ions practically overcomes the rate of electron attachment to oxygen and the role of negative ions in the charged particles kinetics is negligible. In other words, the main role in the generation and excitation of reactive species in the discharge zone belongs to electrons. However, because of very low electric field in the plasma jet, the role of electrons in the kinetics of reactive species in this zone is negligible.

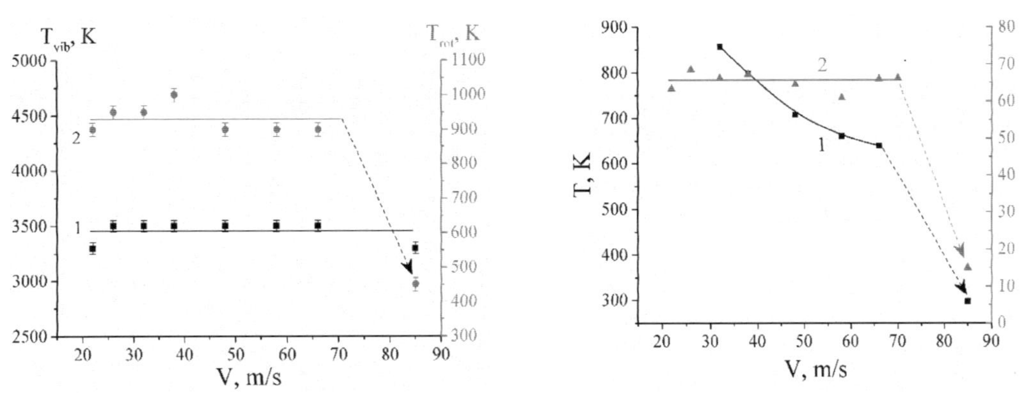

a) b)

Figure 6. a) The dependence of the spectroscopically determined T_{vib} (curve 1) and T_{rot} (curve 2) of the electron-exited state in the discharge zone inside the ceramic tube on the air flow velocity V. b) The dependence of gas temperature T inside the plasma jet (curve 1) measured by the thermocouple located 5 mm away from the tube exit vs air flow velocity V, $\langle W \rangle$=65 W; the behaviour of the averaged DBD power $\langle W \rangle$ in dependence of gas flow velocity (curve 2).

4. Conclusion

We have developed novel plasma source of the "efflux" type that is based on the usage of the coaxial high-frequency (100 kHz) sinusoidal DBD in fast flow velocity (20-80 m/s) of the ambient air at atmospheric pressure with a high volume flow rate of 4-24 L/min per tube. This discharge exhibits two regimes (low-power and high-power ones). The information on composition of the reactive species generated by airflow DBD has been obtained from the emissive spectra of the discharge zone plasma jet and plasma sheet in the UV-Visual region. The obtained spectral data allowed us to trace the composition of reactive species being generated inside the discharge zone and the change of this composition along both the plasma jet and plasma sheet on the surface formed by plasma jet on the substrate. The basic set of the elementary plasma-chemical processes responsible for the generation of reactive species in the discharge zone and their further changes in plasma jet is offered. Our findings promote more insight into physics of the airflow DBD at atmospheric pressure.

Acknowledgement
This work was supported by the RFBR. Grants No 17-02-00234, No 18-02-00539.

References

[1] Laroussi M and Lu X 2005 *Appl. Phys. Lett.* **87** 1139
[2] Akishev Yu, Grushin M, Dyatko N, Kochetov I, Napartovich A and Trushkin N 2010 In: *Plasma Technology for Hyperfunctional Surface. Food, Biomedical and Textile Applications* Ed. Rausher H, Perucca M, Buyle G (Weinheim: WILEY-VCH, Verlag) 295
[3] Bárdos L and Báranková H 2010 *Thin Solid Films* **518** 6705
[4] Lee H W, Nam S H, Mohamed A-A H, Kim G C and Lee J K 2010 *Plasma Processes Polym.* **72** 74
[5] Winter J, Brandenburg R, and Weltmann K.-D 2014 *Plasma Sources Sci. Technol.* **24** 064001
[6] Korolev Yu 2015 *Russian Journal of General Chemistry* **85** 1311
[7] Akishev Yu, Balakirev A, Grushin M, Karalnik V, Kochetov I, Napartovich A, Petryakov A and Trushkin N 2015 *IEEE Trans. Plasma Sci.* **43** 745
[8] Lu X, Naidis G V, Laroussi M, Reuter S, Graves D B and Ostrikov K 2016 *Phys. Rep.* **630** 1
[9] Babaeva N Y and Kushner M J 2014 *Plasma Sources Sci. Technol.* **23** 015207
[10] Sosnin E A, Panarin V A, Skakun V S, Baksht E K and Tarasenko V F 2017 *Eur. Phys. J.* D **74** 25
[11] Akishev Yu, Grushin M, Dyatko N, Kochetov I, Napartovich A, Trushkin N, Duc T and Descours S 2008 *J. Phys.* D: *Appl. Phys.* **41** 23520
[12] Laux C 2002 *Radiation and nonequilibrium collisional-radiative models* (*Physico-Chemical Modeling of High Enthalpy and Plasma Flows)*, D. Fletcher, J.-M. Charbonnier, G. S. R. Sarma, and T. Magin, Eds.Rhode-Saint-Gen'ese, (Belgium: von Karman Institute Special Course)
[13] Akishev Yu S, Medvedev M A, Napartovich A P, Petryakov A V, Trushkin N I and Shafikov A G 2017 *Plasma Phys. Rep.* **43** 472
[14] Akishev Yu S, Karal'nik V B, Medvedev M A, Petryakov A V, Trushkin N I and Shafikov A G 2017 *J. Phys.: Conf. Ser.* **789** 012002
[15] Guerra V and Loureiro J 1997 *Plasma Sources Sci. Technol.* **6** 373

IOP Conf. Series: Journal of Physics: Conf. Series **1112** (2018) 012011 doi:10.1088/1742-6596/1112/1/012011

Effect of DC-discharge geometry on ignition efficiency in supersonic flow

A A Firsov[1,*], E V Dolgov[1] and S B Leonov[1,2]

1 - Joint Institute for High Temperatures (JIHT RAS), 125412, Moscow, Russia

2 - University of Notre Dame, Notre Dame, IN 46556, USA

*af@jiht.org

Abstract. The plasma-assisted combustion is considered as a prospective approach for ignition of a hydrocarbon fuel and flameholding in a supersonic airflow. Stable flameholding of gaseous and liquid hydrocarbon fuel was achieved by means of surface Q-DC discharge without employing mechanical flameholders in a supersonic combustion chamber. However, a high level of electric power, typically required to realize this method, may limit its application in a real apparatus. The current experimental and computational efforts continue the study of a distributed plasma system with the aim of reduction the total energy consumption and extending the life cycle of the electrode system. In the work described in this paper, the interaction of two separated plasma modules was investigated. The fuel ignition caused by individually controlled plasma modules was explored showing a significant effect of the plasma filament length on ignition efficiency.

1. Introduction

The plasma-assisted ignition and flameholding based on surface DC discharge is known as a realizable method to stimulate combustion in a supersonic flow [1-2]. In work [3], the feasibility of plasma actuators was investigated for ignition of hydrocarbon fuel in high-speed airflows. From the viewpoint of entire system design, it was demonstrated that the plasma torch employing air or oxygen for plasma generation is effective for ignition and flameholding in a scramjet. The work [4] considers the mechanism of ignition due to generation of hot zones of highly excited gas with peak temperature up to 5000 K at a low total input power about 2kW. Results from these studies suggest that the DC torch is more effective in producing a flame than the AC torch with a similar levels of input peak power for the configurations tested. In work [5], various configurations of fuel mixtures used for injection into the plasma torch were considered. It was found that stability of plasma-stimulated combustion significantly depends on the chemical composition of injected mixture: for example, stability of the plasma jet drastically improved with the addition of a small amount of O2 to the feedstock. In Ref [6], plasma jets were used for ignition of the liquid fuel (kerosene). Results of this study indicate that the critical factor affecting the ignition of liquid hydrocarbon fuel in the scramjet is the mixing of the fuel with oxidizer. One of the possible ways of mixing improvement, ignition, and flame-holding can be use the plasma actuator based on Q-DC discharge [7]. It was applied for ignition of gaseous hydrocarbon fuel in a wide range of flow conditions without employing mechanical flameholders in a supersonic combustion chamber. The complication in employment of plasma-based actuators for ignition and combustion support in high-speed airflow is typically associated with high power consumption. In order to develop the technology, an intensive work is carried out to reduce the total power deposited in the discharge keeping the ignition to be still feasible. The reduction in power intended for plasma generation not only reduces the power requirements for entire system, but also extends the lifetime of the electrodes, which is especially important at long-term use without an option for emergency replacement of the units. Reduced power budget also positively affects the volume and weight associated with the power supply of the combustion support system in a flying vehicle.

The typical discharge power in a series of previous experiments was about 12-18 kW for model combustor with 60x70mm cross-section and Mach number M=2 flow [7], which is a reference point for the current research described in this paper. Previously it was shown that distributed plasma system, consisting of two plasma modules, accelerates the combustion process comparing to a single plasma module [8]. These test series, to some extent, verify the two-stage mechanism of plasma-assisted ignition and combustion [9]. The current experimental and computational work continues to investigate the plasma-based system based on distributed electrodes arrangement, which beneficial in terms of total energy consumption and potentially allows the extending the life cycle of the electrode system.

2. Experimental setup

The experiments were performed in supersonic wind tunnel PWT-50 of JIHT RAS. Flow parameters were as follows: Mach number M=2, static pressure P_{st}=170 torr, and stagnation temperature T_0=300K. Detailed description of experimental facility could be found in work [3]. Fuel injectors were flush-mounted on the wall in the test section and provided fuel supply perpendicular to the flow. In our experiments, ethylene injection was used in a range of mass flow rates G_{fuel}=0.5-4 g/s. Two ceramic inserts with ∅1.5 mm electrodes are located downstream of the fuel injectors to create a quasi-direct current (Q-DC) discharge near the wall. Each insert has 8 electrodes placed in pairs (anode-cathode), with a distance between the centers of the electrodes in one pair of 5mm and the distance between the electrodes from the adjacent pairs is 7mm. Both electrode systems were powered independently of each other through two fast high-voltage solid-state switches. The main idea of testing the described system was to organize two low-power electric discharges that cannot ignite the fuel separately, but, when working together, they provide stable ignition and combustion. This approach is based on the concept of a two-stage combustion mechanism in a supersonic flow [5]: first, a cold flame is realized that promotes the production of active particles, and after a while (in supersonic flow conditions, we can talk about the distance traveled by the elementary volume of the mixture) the hot flame arises, which is characterized by active heat release and ensuring the growth of pressure. The reduction in the power of one of the plasma actuators or even its shutdown can be the next step, since it is expected that the combustion mode requires a lower discharge power in comparison with moment of fuel ignition.

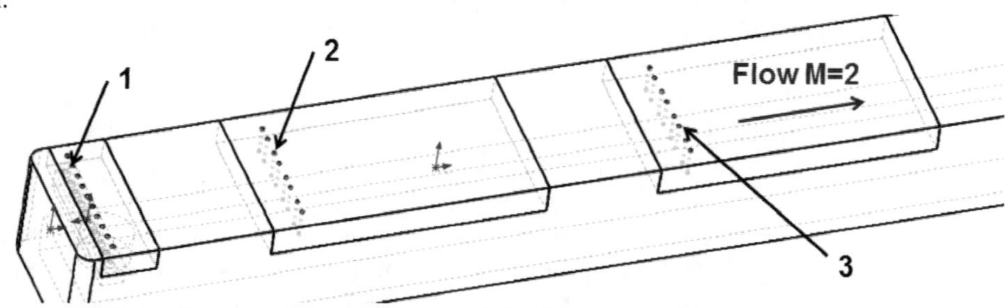

Figure 1. Wall of test section: 1 – ∅1,5 mm fuel injectors with 0.5-4g/s ethylene flow rate, 2 – electrodes of first actuator, 3 – electrodes of second actuator

3. Experimental results for simultaneous operation of two actuators

3.1. Interaction of two actuators

The first series of experiments was performed without fuel injection to test the operation of the actuators - each individually, as well as their joint work to explore the effect of plasma operation of different modules on each other. The following feature was observed during these tests: activation of the first actuator leads to a decrease in the length of the plasma filaments and a voltage reduction on the second plasma module. This can be clearly seen at following test sequence: initially the second actuator (which is downstream) was turned on, after a while, the first one was turned on (which is located upstream). After activation of first actuator the discharge voltage on the second actuator was reduced from 1200V to 700V, and discharge length was decreased in 1.8-2 times. For this experiment, voltage time series for both actuators and images of plasma filament of second actuator at different period of time are presented in figure 2.

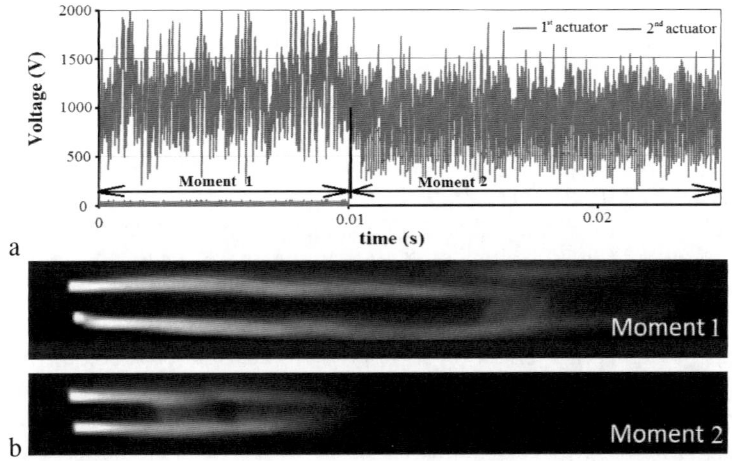

Figure 2. Typical voltage dependencies on time for first and second actuator (a) and typical discharge images of one loop for second actuator for different time moments (b)

In order to explain this effect, numerical modeling of supersonic flow in a rectangular channel was performed, whose geometric dimensions and flow parameters correspond to the experimental ones. The simulation was performed in the software package FlowVision 3.10.02. The simulation is based on the solution of a three-dimensional nonstationary system of Navier-Stokes equations supplemented by the k-ε model of turbulence using wall functions for modeling the boundary layer. The simulation took into account the work of only the first discharge module, while the discharge was modeled by a three-dimensional heat source inside the imported object with a total heat release close to the experimental one (1кW for one plasma filament or 2kW for one plasma loop). The imported object was located in the zone of the first actuator, and its shape was similar in shape to the discharge filament, 40 mm in length and 1.3 mm in diameter. The calculation was carried out for a thin three-dimensional layer with one half of the one discharge filament to decrease the number of cells and simulation time. The ceramic insertions from figure 1 (zones in which the actuators are located) are marked with purple rectangles in Figure 3. Analyzing the obtained results it is clearly seen (see figure 3) that the temperature of the gas flow downstream the 1st actuator significantly increases in the area of second actuator. But it is important that the temperature significantly depends on distance from actuator and at the distance more than 300mm it is restored close to initial state because of flow turbulence. Temperature increase is accompanied by decrease of the density and the voltage of breakdown for second actuator. This results in a decrease in the discharge length and a drop in power.

IOP Conf. Series: Journal of Physics: Conf. Series **1112** (2018) 012011 doi:10.1088/1742-6596/1112/1/012011

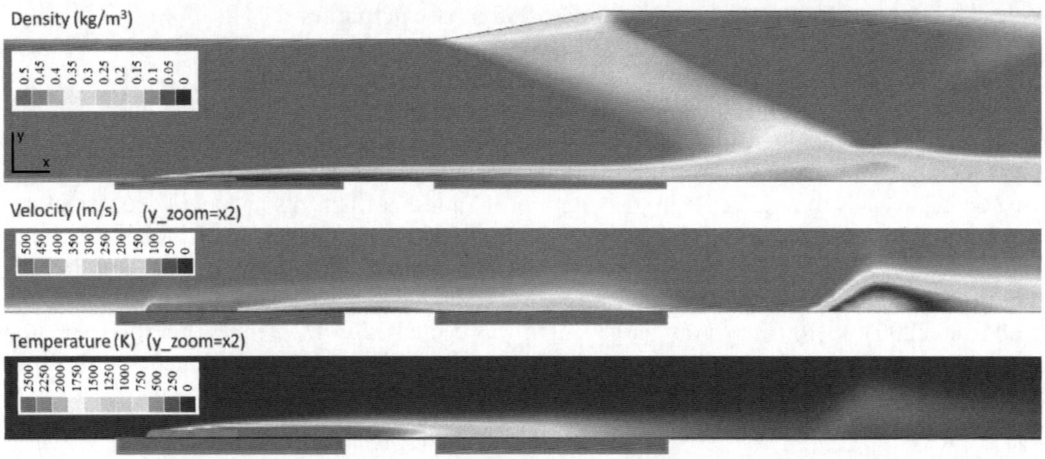

Figure 3. Simulation of influence of 1st actuator operation on flow parameters in the area of location of second actuator. Flow is directed from left to the right. Density field is presented for all cross-section of test section, velocity and temperature are presented for layer near wall and the images are enlarged in y-direction at 2 times.

3.2. Ignition and flame-holding by two actuators

Experiments with fuel ignition using simultaneous operation of two actuators were performed at following conditions. After start of fuel injection with mass flow rate 3,8 g/s to the supersonic flow the first actuator was activated with pulse duration of 120 ms. The second actuator was enabled 60 ms later with pulse duration of 60 ms, so last 60 ms both actuators operate together. Work of plasma actuators without fuel has no perceptible effect on static pressure distribution along the test section, so all detected changes of static pressure distribution are caused only by combustion. During this experiment it was observed, that operation of only first actuator lead to weak combustion with medium increase of static pressure which is gradually increased along the test section, as it is shown in Figure 4. Additional activation of second actuator leads to significant improvement of combustion process: flame front moves upstream that is clearly seen by time dependency of pressure at 150mm point. This distance is measured from the injection position. And for all other points the pressure increase was detected that may confirm the improvement of combustion completeness. The displacement of the flame front to the first actuator area leads to a change in the nature of its operation. The length of the plasma filaments decreases, and the work of the discharge becomes less stable because the discharge filaments follow the streamlines in the separation zone caused by combustion.

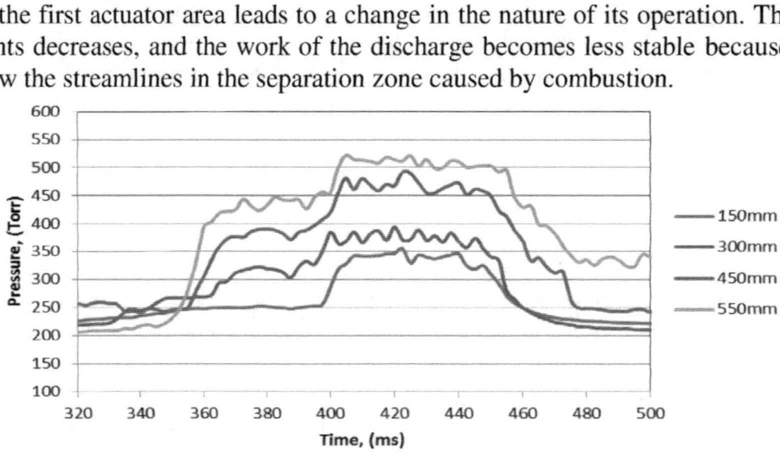

Figure 4. Pressure distribution during combustion for experiment with serial activation of plasma actuators

Obtaining active combustion in the case of operation of two actuators was expected based on the results of previous experiments. Also it was expected that turning of second actuator after obtaining active combustion with separation zone in the area of first actuator will keep the combustion quality at the same state because of important role of separation zone with Q-DC discharge as a plasma-chemical reactor. So turning off second actuator should reduce the power consumption of discharge system without decreasing the combustion efficiency. To test this hypothesis the following experiment was performed. Ignition of 3.8 g/s ethylene into the supersonic flow with Mach number M=2 was organized using two actuators and after obtaining the regime of active combustion with separation zone in the area of first actuator the second actuator was disabled. Time dependencies of electrical characteristics of both actuators are presented in Figure 5. After activation of first actuator (I=2A for one loop, U=0.9-1.2kV, total power W=4.2kW) medium pressure increase was detected. Next activation of second actuator 40 ms later (I = 3A for one loop, U=300-400V, total power W=1.8kW) results in pressure increase and change in character of 1st actuator operation. Length of discharge decreases together with voltage drop that can be seen in figure 5 (0.05-0.08 s). But after turning off the second actuator it was obtained that pressure caused by combustion restores to the values obtained at activation of only first actuator. And it is clearly seen in figure 5, voltage drop of first actuator also restored to the state before second actuator activation. So in the presented case, second actuator plays significant role for obtaining stable effective combustion even after the displacement of the flame front to the first actuator area. For a detailed explanation of this result, additional research is required.

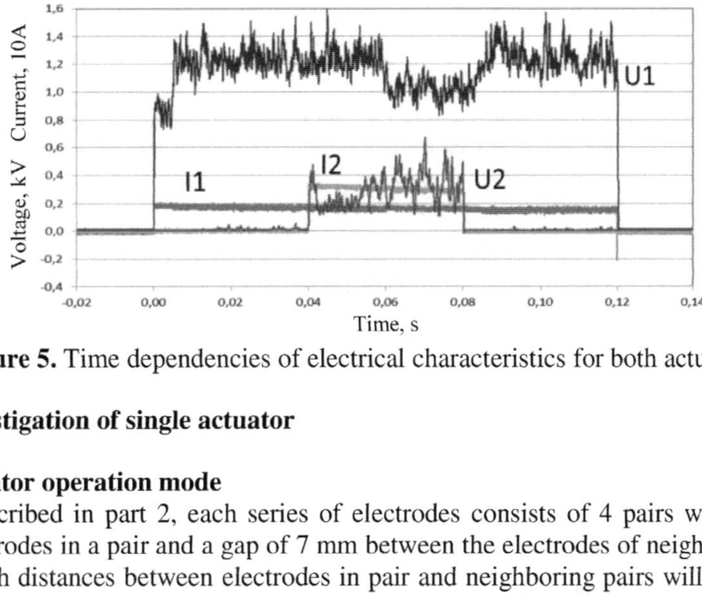

Figure 5. Time dependencies of electrical characteristics for both actuators.

4. Detailed investigation of single actuator

4.1. Single actuator operation mode

As it was described in part 2, each series of electrodes consists of 4 pairs with a gap of 5 mm between the electrodes in a pair and a gap of 7 mm between the electrodes of neighboring pairs. It was expected that such distances between electrodes in pair and neighboring pairs will provide barrier for the interaction of separated plasma loops. And as a result of such geometry of electrodes the configuration of plasma filaments presented in Figure 6(a) should be realized. But actually such configuration was realized only for short time after discharge activation. At the later time moments the plasma filament which is located close to neighboring earth electrode, could reconnect to this electrode. As a result the operation mode with shared ground electrode is realized. It was found that reconnection has no significant effect on the 1st loop which stays connected to the initial electrodes. But for the second loop reconnected to another ground the inter electrode gap increases, that lead to the increase of discharge length, voltage drop and power impact to the discharge, as it is shown in Figure 6(b-c). This mode also has a double current in one ground electrode, and this is a stable configuration, because the reconnection of second loop to initial ground electrode was not detected.

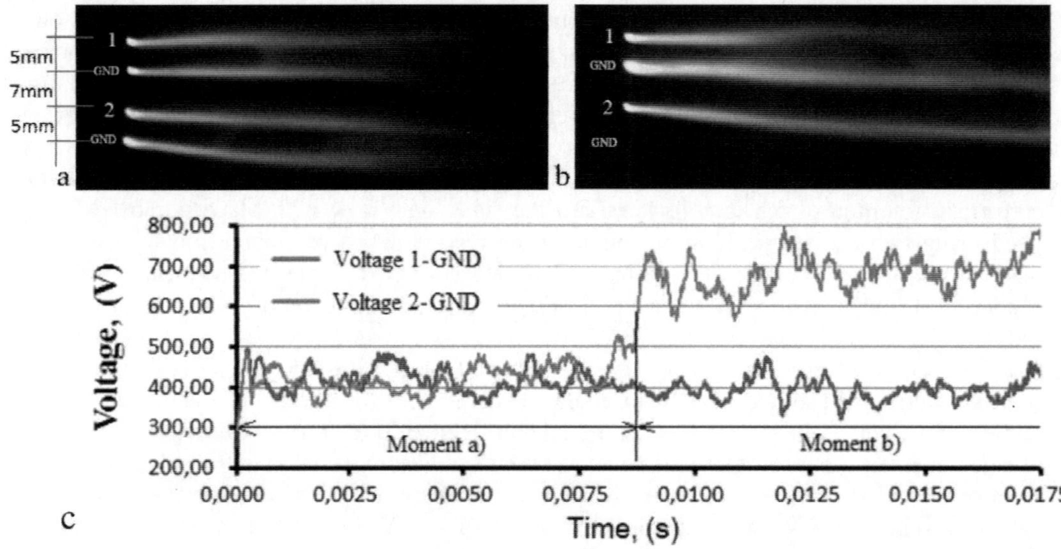

Figure 6 Change of plasma actuator operation mode. (a) Initial discharge configuration after activation of discharge (b) Stable configuration with shared ground electrode (c) Time dependence of voltage drop on discharge

4.2. Influence of discharge length on ignition

It was found that a configuration with a shared grounded electrode is capable to ignite the fuel mixture in several conditions then two non-interacting operated loops of one electrode row can't ignite the fuel, but both configurations have the same energy impact into the discharges. Next series of experiments were carried out to verify the influence of discharge length and local power increase due to the double current to the filament attached to the shared ground. For this purpose two following geometries of electrodes were used. To organize two independent loops with 5mm gap between electrodes (mode 1) two separate loops were spaced apart from each other by a distance of 19 mm, which made it possible to exclude the possibility of a breakdown into shared ground (see figure 7a). Second circuit (mode 2) was realized using 5 mm distance between adjacent loops, and the gap between the electrodes was 7 mm (see figure 7b). In order to prevent the breakdown of discharge to the (non ground in this case, the arrangement of electrodes was changed by the following way: pair c le/cathode + pair cathode/anode, so gap between two cathodes was 5 mm. Also ballast load for limitation of discharge current was installed on cathode side of discharge gap (normally it was installed on anode side). Also the possibility of fuel ignition in the supersonic flow by one plasma loop with 7mm gap between electrodes was tested (mode 2b).

Comparison of three presented cases was performed at fuel mass flow rate 3.8g/s. Stable ignition of fuel was obtained for all tested cases. Pressure distributions along the test section for main cases are presented in Figure 8. Two significantly different variations of combustion activity were obtained during series of experiments. First one is characterized by low pressure increase during significant distance of 300 mm after discharge, and only after point of 320mm the significant pressure gradient was detected. This point corresponds to the separation area induced by oblique shock from the top wall of test section, as it is shown in simulation results in Figure 3. In this case the discharge has a stable operation and long straight filaments. The second variation is characterized by immediate pressure increase directly after discharge area and increase of pressure gradient along the length of test section after point of 320mm. In this case the discharge has an unstable length, shape and position of discharge filaments because of operation inside the separation area caused by flame front. It was found that obtaining of second variation of combustion is possible at following conditions: 6.5kW of power necessary at mode1, 5.6kW of power necessary at mode2 and 3.5kW of power was used in mode2b. The typical discharge power in a series of previous experiments was about 12-18 kW for similar experimental conditions [7], so obtained reduction of power required for realization of stable active combustion is significant. Apparently, obtained reduction of power caused by an increase in the discharge length that provides the increase of the interaction time of the plasma with the elementary volume of the mixture.

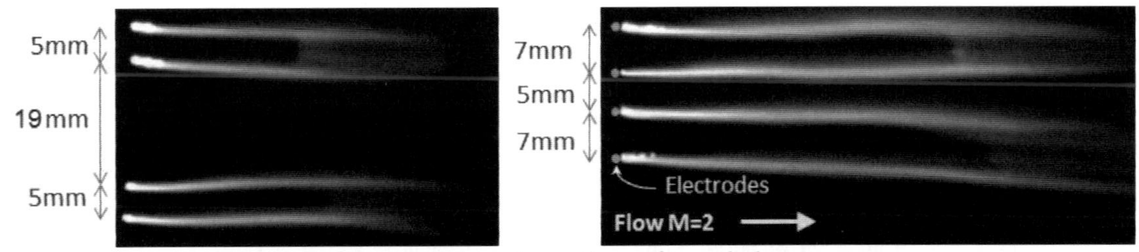

Figure 7. Geometrical configurations of tested plasma actuators.
Mode1 – two plasma loops with 5mm discharge gap,
Mode2 – two plasma loops with 7mm discharge gap.

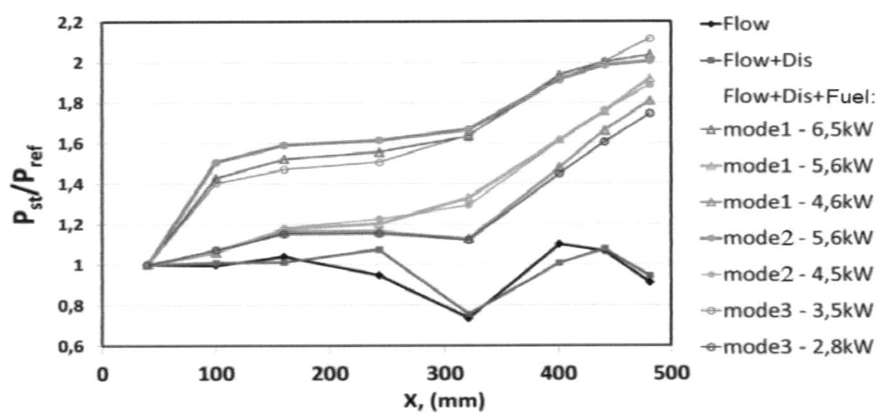

Figure 8. Pressure distribution along the test section for different discharge regimes at fuel mass flow rate 3.8g/s.

Conclusions

Experimental study of distributed plasma system for ignition and flameholding of ethylene in supersonic flow was performed. It was found that the first plasma actuator affects the operation of the second one: the activation of the first actuator leads to the voltage drop on second actuator on ~40%, and discharge length decreases in 1.8-2 times. For understanding of this effect, the numerical simulation of the supersonic flow with volumetric heat source in the location of the first actuator was performed. It was found that this heat source leads to decrease the density and velocity of gas in the area of second actuator that results in decrease the discharge length and the average discharge voltage. The tests of distributed plasma system demonstrate that activation of second actuator after ignition by the first actuator improves of combustion efficiency. However, a following turn-off of the second actuator leads to the restoration of the flame front position and the distribution of wall pressure over the test section.

The detailed study of operation of a single actuator shows that the reconnection of one discharge loop to the ground electrode of adjacent loop results in increase the discharge length and power deposition in the discharge. To verify the effect of discharge length on ignition and combustion pattern, the comparison of plasma actuators with different discharge gap (5mm and 7mm) was performed. It was found that, for intensive combustion, the actuator with 2 plasma filaments (4 electrodes) and 5mm gap should provide the power deposition of 6.5kW and 5.6kW for a weak combustion. However, operation of the actuator of same geometry but 7mm gap results in intensive combustion already at 5.6kW. To observe the intensive combustion, the minimal plasma power of 3.5kW was achieved using one plasma loop (2 electrodes activated) with 7mm discharge gap that is significantly lower than typical discharge power in the experiments with plasma-assisted combustion.

Acknowledgments
The work of E. Dolgov and A. Firsov was supported by Russian President Grant МК-1734.2017.8

References

[1] Ju Y Sun W 2015 Plasma assisted combustion: Dynamics and chemistry *Prog. Energy Combust. Sci.* **48** 21–83
[2] Leonov S 2018 Electrically Driven Supersonic Combustion *Energies* **11 (7)** 1733
[3] Sato Y, Sayama M, Ohwaki K, Masuya G, Komuro T, Kudou K, Murakami A, Tani K, Wakamatsu Y, Kanda T, Chinzei N and Kimura I 1992 Effectiveness of Plasma Torches for Ignition and Flameholding in Scramjet *J. Prop. Power* **8 4** 883–9
[4] Jacobsen L, Carter C, Baurle R A, Jackson T A, Williams S, Bivolaru D, Kuo S, Barnett J and Tam C-J 2009 Plasma-assisted ignition in scramjets *J. Prop. Power* **24 4** 641–54
[5] Takita K, Shishido K and Kurumada K 2011 Ignition in a supersonic flow by a plasma jet of mixed feedstock including CH4 *Proc. Combust. Inst.* **33 2** 2383–89
[6] Li F, Yua X-l, Tong Y-g, Shen Y, Chen J, Chen L-h and Chang X-y 2013 Plasma-assisted ignition for a kerosene fueled scramjet at Mach 1.8 *Aerosp. Sci. Technol.* **28 1** 72–8,
[7] Firsov A, Savelkin K, Yarantsev D and Leonov S 2015 Plasma-enhanced mixing and flameholding in supersonic flow *Philos. Trans. R. Soc.,* A **373** 20140337
[8] Firsov A, Dolgov E, Leonov S and Yarantsev D 2017 Advanced Ignition in Supersonic Airflow by Tunable Plasma System *IOP Conf. Ser.: Mater. Sci. Eng* **249 1**
[9] Leonov S B, Kochetov I V, Napartovich A P, Sabel'nikov V A and Yarantsev D A 2011 *IEEE Trans. Plasma Sci.* **39 2** 781–7

IOP Conf. Series: Journal of Physics: Conf. Series **1112** (2018) 012012 doi:10.1088/1742-6596/1112/1/012012

The Influence of the Geometry and Power of a Pulsed Capillary Discharge on the Properties of Erosive Plasma

A S Pashchina* and V F Chinnov

Joint Institute for High Temperatures, Russian Academy of Sciences, Moscow, 125412, Russia

*fgrach@mail.ru

Abstract. The results of the spectral study of a pulsed capillary discharge in the range of the relative capillary lengths of $l/d = 5$ and the discharge power densities of $q_S = 1 - 100$ kW/mm^2 are presented. The longitudinal profiles of the electron number density and temperature inside and on a short distance outside the capillary are measured. A good qualitative and quantitative agreement between the measured parameter profiles and the results of calculation based on the model of the expendable nozzle, which is valid for discharges in the "long" capillaries, was found. It is shown that a decrease in the relative length of the capillary, starting from $l/d < 2.5$, leads to a discrepancy between the experimental and calculated parameters profiles due to the increasing contribution of the electrode's evaporated substance to the total mass balance.

1. Introduction

A pulsed capillary discharge is known as a relatively simple and efficient method of obtaining plasma over a wide range of pressures, temperatures, and charged particles concentrations [1]. Depending on the method of the working body supply, one can distinguish the discharge in a gas-filled or in an initially empty capillary. In the latter case, the plasma composition is determined by the substance ablated from the capillary wall and electrode surface. The partial contribution of each source in the total mass balance depends on the power fluxes to the capillary wall and electrode, the ratio of their surfaces and the physico-chemical properties of the used materials. In two limiting cases, when the contribution from one source dominates, a capillary discharge with an evaporating wall (CDEW) and a capillary discharge with an evaporating electrode (CDEE) are distinguished. The conditions for these types of discharges are realized in "long" ($l/d > 3$) and "short" ($l/d < 1$) capillaries, respectively. In the intermediate case - $1 < l/d < 3$ - the partial contribution of both sources to the total mass balance is comparable.

Most of the theoretical and experimental works have been devoted to the study of discharges in "long" capillaries, in which the ablated substance from the capillary wall act as the main source of the working body. Gas dynamics of plasma in such capillary is described by the model of expendable nozzle [2]. One-dimensional CDEW models for the parameters averaged over the capillary cross section (the mass density and the temperature) were considered in [3,4]. The uniform distribution of the plasma temperature and mass density over the radius is taken as the initial prerequisite in these models. However, these prerequisites are not met in real conditions. So the spatial separation of the discharge onto a central high-temperature zone and a low-temperature layer adjacent to the capillary wall and containing a weakly ionized plasma is necessary to achieve a balance of mass and energy introduced into the capillary and carried away from it by plasma jets [5,6]. Subsequently, one-dimensional models of a two-zone capillary discharge were proposed in [7–10]. Recently, two-dimensional models have been proposed that take into account the features of heat, mass and radiant energy transfer for a spatially inhomogeneous plasma inside the capillary [11,12].

At the same time, the number of experimental works devoted to the study of the discharge structure inside the capillary is small. This situation, first of all, is connected with large optical thickness of the erosive plasma, whose blackbody radiation at high pressures (p>50 bar) restrict the use of spectral methods to study its internal structure. Another limiting factor is the capillary geometry, whose cylindrical surface (through which the observation is provided) acts as a scattering lens that leads to a distortion of the real spatial picture of plasma emission. Therefore, the use of spectral methods for studying the discharge structure inside the capillary is limited to diagnostic of the paraxial discharge zone if the optical thickness of plasma is not too large.

The aim of the work is the experimental study of the influence of the capillary geometry and the discharge power on the properties of erosive plasma, as well as an assessment of the suitability of the used spectral diagnostic methods for determining the main plasma parameters (the electron number density and temperature) inside the capillary.

2. The object and the methods of research

The object of research is a capillary arrester, whose detailed description is given elsewhere [13]. Perspex is used as a capillary wall material (chemical formula $C_5H_8O_2$). The depth and the initial diameter of a capillary are respectively $h = 5$ mm and $d = 1$ mm. A capacitive storage device with series-connected inductor is used for power supply of discharge. The algorithm of power supply approximately corresponds to the sine half-wave. The main parameters of discharge circuit are the following: the stored energy $Q = 80$ J, the amplitude of discharge current $I_m = 60 - 450$ A, duration of discharge pulse $\tau_d = 1 - 9$ ms. Researches were carried out in the range of the capillary diameter $d = 1 - 3.2$ mm that corresponds to the range of the discharge surface power density (the power divided by the cross-section of the capillary) $q_S = 1 - 100$ kW/mm^2.

During the experiments, the 2D-matrix spectra of plasma radiation emitted from the paraxial zone of discharge inside and outside the capillary were recorded on the ICCD matrix of the fast gate Andor iStar camera (minimal camera gate 100 ns) mounted in the output focal plane of the MS-257 spectrograph (width of the entrance slit $\delta = 20$ μm). The spectra observed through a previously polished external wall of the capillary. The spectral interval $\Delta\lambda = 650 - 685$ nm was chosen for 2D-matrix spectra recording. This interval contains the hydrogen line H_α (used for determination of the electron number density by the linear Stark effect), the doublet of the ionic carbon lines C II 657.8 nm and 658.3 nm, the multiplet of ion C II lines of 678-682 nm, and the rather intense continuum (inside the capillary), which can be used to determine the electron temperature. The ability to use the method of "Boltzmann exponent" to determine electron temperature in the paraxial zone of the plasma jet from the relative population of excited states of carbon ions, and also the rather high accuracy of this method was shown in [14]. The application of this method inside the capillary is difficult because the strong broadening and merging those lines with the continuum. Therefore, the method of the relative intensities of the H_α line and the continuum is mainly used in the present work to estimate the electron temperature. Evaluation of the radiation reabsorption from both the H_α line and the continuum under experimental conditions indicates the suitability of this method for estimating the electron temperature inside the capillary with an error of less than 20% increasing with power input.

3. The results of research

The longitudinal profiles of the main plasma parameters (the electron number density and temperature) along the discharge centerline are obtained in the range of surface power density of $q_S = 1 - 100$ kW/mm^2 (figure 1) using the recorded 2D spectra and the ratio $\frac{\varepsilon_{H_\alpha}}{\varepsilon_{cont}} = f(T_e)$ [15] (ε_{H_α}, ε_{cont}

IOP Conf. Series: Journal of Physics: Conf. Series **1112** (2018) 012012 doi:10.1088/1742-6596/1112/1/012012

– the intensities of H_α line and continuum, respectively). Change in discharge power density in the selected range permits to reveal the main features in the spatial distribution of plasma parameters in a supersonic, transonic and subsonic outflow mode from the capillary. It should be noted that the surface power density q_S, expressed as the power divided by the cross-section of the capillary is more convenient parameter for analyzing the gas dynamics of plasma flow, rather than volumetric power density (power divided by the capillary volume). In particular, the threshold value of the discharge power density q_S^{thr}, corresponding to the transition from subsonic to supersonic mode, is very weakly dependent parameter on the capillary length and the discharge circuit parameters, which makes it possible to compare the results of experiments obtained by different authors under different conditions. This was first noticed by the authors of [16]. The threshold value of surface power density can be easily determined from the diagrams in figure 1 by the position of the Mach disk, which closely approaches the capillary edge (the position indicated as $M = 1$). The threshold surface power density determined in this way, $q_S^{thr} = 15 - 20$ kW/mm², is in good quantitative agreement with the results obtained in [16].

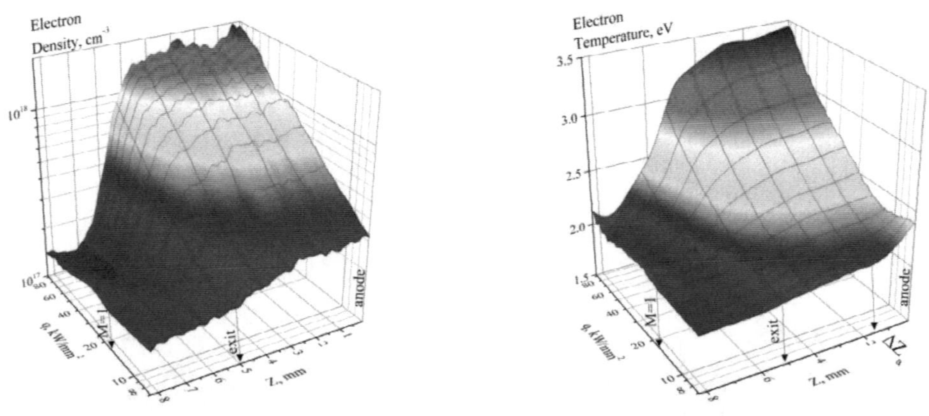

(a) (b)

Figure 1 The profiles of (a) electron number density and (b) temperature along the discharge centerline. The arrows indicate: the threshold power density q_{crit}, which corresponds to a transition from supersonic to subsonic flow regime of the jet ($M = 1$); the position of the capillary outlet (indicated as "exit"); the margin of spatial domain where the anode vapours prevail (ΔZ_a).

The results of temperature measurements are in good quantitative agreement with the results obtained by other methods. In particular, the temperature in the expansion section of the plasma jet is $T_e = 1.7 - 2.5$ eV, which agrees with enough accuracy (~20%) with the results obtained by the method of relative intensities of ionic carbon lines [14]. The electron temperature inside the capillary (at maximum values of the surface power density $q_S = 100$ kW/mm²) reaches $T_e = 3.4$ eV, which is in good agreement with the results obtained in [5,6]. Also, the measured temperature is in good quantitative agreement with the estimates based on Saha equation, which is an additional confirmation of the fulfillment of the local thermodynamic equilibrium, at least inside the capillary in the paraxial discharge zone.

Inside the capillary, whose relative length exceeds $l/d > 2.5$, the profiles of electron number density and temperature and also the pressure profile, restored from these data (taking into account the relation $p \approx 2n_e kT_e$ valid for a strongly ionized paraxial zone of discharge consisting mainly of electrons and hydrogen ions [13,17]), agree qualitatively with theoretical dependences obtained in [3] for the model of expendable nozzle. The gas dynamics of such nozzle, according to the assumptions made in [3], is determined solely by the flux of the evaporated substance from the capillary wall, and the contribution of the evaporated substance from the electrode surface is neglected. So, the

IOP Conf. Series: Journal of Physics: Conf. Series **1112** (2018) 012012 doi:10.1088/1742-6596/1112/1/012012

relationships obtained within the framework of this model are valid for "long" capillaries only. The model is based on the relationships for the balance of mass and energy introduced into the capillary and carried away from it by a plasma jet. According to the basic assumptions [3], the electric power introduced into the discharge is equal, on the one hand, to the power carried away by the plasma jet, and on the other hand - to the power emitted by the arc and expended for the plasma creation. The mutual cancelling of these terms in the energy equation causes its transformation into the adiabatic equation, so the plasma flow inside the capillary is adiabatic. An example of the pressure profiles inside the capillary, restored from the spectral data and calculated for the model of expendable nozzle, is shown in figure 2. Good qualitative and quantitative agreement between the experimental profile and the model one is a strong argument in favor of the correctness of the theoretical model basic assumptions [3] that allow us to use relations obtained in this model for evaluation the plasma parameters in the paraxial discharge zone if the relative length of capillary exceeds $l/d > 2.5$.

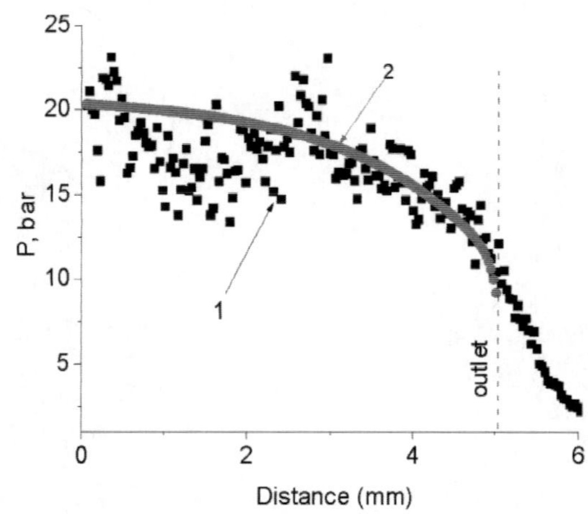

Figure 2 Pressure profiles inside the capillary along the discharge centerline: (1) restored from the spectral data, (2) calculated for the model of expendable nozzle [3]. Discharge parameters: $l/d = 3.6$, $q_S = 55$ kW/mm^2.

At the same time, a decrease in the relative length of the capillary leads to a discrepancy between the measured and calculated parameters' profiles. This discrepancy becomes particularly noticeable when the relative length of the capillary becomes less than $l/d < 2.5$. The discrepancy between the experimental and calculated profiles is observed in the vicinity of the internal electrode (anode) at a considerable distance ($\Delta Z_a = 1.5 - 2$ mm) from its surface. The electron temperature profile is subjected to the greatest distortion: the measured temperature near the anode surface exceeds its calculated value by approximately $\Delta T_e \approx 0.4$ eV (see figure 1b). At the same time, there is no visible distortion of the electron number density profile (see figure 1a). The analysis shows that the distortion of the electron temperature profile, increasing when the relative length of the capillary decreases, is due to increased fraction of the substance evaporated from the electrode surface in the total mass balance. In this case, the relationships obtained for "long" capillaries [3] cease to be satisfied. This conclusion is confirmed by the fact that the distortion of the temperature profile is reliably reproduced only in the case when the anodic spot of a large enough area (comparable with the capillary cross-section) is formed, which provides powerful jets of anode vapor. Such stationary anode spot is formed at high discharge currents (more than 300 A under experimental conditions). In the other cases - in the absence of anode spots or the formation of small spots randomly moving along the anode surface - the distortion of the temperature profile is not observed at all, or recorded irregularly, depending on how accurately the image of the anode spot coincides with the position of the spectrograph slit.

In the supersonic flow mode ($M_a = 1$ at the capillary outlet), the ratios of the pressure, electron number density and electron temperature at the bottom and at the outlet cross-section of the capillary retain a constant value. These data, taking into account the relations obtained in [3], were used for the effective adiabatic index in the paraxial zone of discharge evaluation. Its value estimates as $\gamma = 1.3 - 1.4$.

We found that the static pressure at the capillary outlet in the subsonic regime ($M_a \leq 1$), as well as in the plasma jet (irrespective of the flow regime) systematically exceeds the ambient gas pressure. This excess is greater the shorter the duration of the discharge pulse. For example, the pressure in the jet reaches up to 3 bar for a short-pulse discharge ($\tau_d = 1$ ms), while for a long-pulse discharge ($\tau_d = 9$ ms) it almost equal to atmospheric pressure. The obtained result indicates that the main reason for the increased pressure in the plasma jet is the transient process of pressure equalization inherent to nonstationary flows, that is observed regularly in experiments [18,19] and finds the justification in the theoretical calculations [20].

4. Conclusions

The results of the spectral investigation of a pulsed capillary discharge in the range of the relative capillary lengths of $l/d = 1.5 - 5$ and the discharge surface power densities of $q_S = 1 - 100$ kW/mm^2 are presented. The longitudinal profiles of the electron number density and temperature along the discharge centerline inside and on a short distance outside the capillary are measured. A good qualitative and quantitative agreement between the measured parameters profiles and their calculations based on the model of the expendable nozzle, which is valid for discharges in "long" capillaries, was found. It is shown that a decrease in the relative length of the capillary leads to a discrepancy between the experimental and calculated parameters profiles due to an increase in the fraction of the substance evaporated from the electrode surface in the total mass balance.

The obtained results confirm the principal possibility of using the combination of the half-width of the H_α line and the method of relative intensities of the H_α line and the continuum as an operative means of measuring the electron number density and temperature of plasma in the range of the studied parameters. The established relationship between the electrophysical and gas dynamic parameters of the plasma constitutes a good basis for the subsequent optimization of the plasma formation regimes in the capillary discharge, as well as the improvement of the spectral methods for microplasma objects diagnostics, in particular, for estimating important gas-dynamic parameter - the pressure - that is difficult to measure under these conditions using traditional methods.

References

[1] Min'ko L Ya 1970 *Obtaining and study the pulsed plasma flows* (Minsk: Nauka i tekhnika) (in Russian)
[2] Vulis L A 1950 *Thermodynamics of the gas flows* (Moscow: Gosenergoizdat) (in Russian)
[3] Belov S N 1978 *Journ. Appl. Spectroscopy* **28** 605-11 (in Russian)
[4] Bondarev B V, Muravenko V G and Shirokov M F 1977 *Teplofiz. Vys. Temp.* **15** 465–70 (in Russian)
[5] Ogurtsova N N, Podmoshensky I V and Shelemina V M 1974 *Teplofiz. Vys. Temp.* **12** 5–9 (in Russian)
[6] Ogurtsova N N, Podmoshensky I V and Smirnov V L 1976 *Teplofiz. Vys. Temp.* **14** 1–8 (in Russian)
[7] Kovitya P and Lowke J J 1984 *J. Phys. D. Appl. Phys.* **17** 1197–212
[8] Muller L 1993 *J. Phys. D. Appl. Phys.* **26** 1253–9
[9] Niemeyer L 1978 *IEEE Trans. Power Appar. Syst.* **PAS-97** 950–8
[10] Ruchti C B and Niemeyer L 1986 *IEEE Trans. Plasma Sci.* **PS-14** 423–34
[11] Keidar M and Boyd I D 2006 *J. Appl. Phys.* **99** 53301
[12] Wang W, Kong L, Geng J, Wei F and Xia G 2017 *J. Phys. D. Appl. Phys.* **50** 74005
[13] Pashchina A S, Efimov A V and Chinnov V F 2016 *High Temp.* **54** 488–502
[14] Efimov A V, Pashchina A S, Chinnov V F and Kazansky P N 2018 *Applied Physics* 24–9 (in Russian)

[15] Lochte-Holtgreven W 1968 *Plasma Diagnostics* Ed. North-Holland (Amsterdam; New York:Wiley–Interscience)

[16] Ender A Ya, Kuznetsov V I and Kolyshkin I N 2015 *Tech. Phys.* **60** 1720-24

[17] Pashchina A S, Efimov A V and Chinnov V F 2017 *High Temp.* **55** 650–64

[18] Sharikov I and Surzhikov S 2005 Experimental Study of Unsteady Supersonic Underexpanded Ablative Plasma Jet Dynamics (AIAA 2005-4930) *36th AIAA Plasmadynamics and Lasers Conference* (Toronto, Ontario Canada) pp 1–11

[19] Belov S N, Ogurtsova N N and Podmoshensky I V 1975 *Journ. Appl. Spectroscopy* **22** 396–402 (in Russian)

[20] Kuzenov V V, Ryzhkov S V, Gavrilova A Y and Skorokhod E P 2015 *Trudy MAI* **83** 51–74 http://trudymai.ru/upload/iblock/314/kuzenov_ryzhkov_gavrilova_skorokhod_rus.pdf (in Russian)

IOP Conf. Series: Journal of Physics: Conf. Series **1112** (2018) 012013 doi:10.1088/1742-6596/1112/1/012013

The influence of spatial inhomogeneity of pulsed capillary discharge on the gas dynamics of multicomponent plasma

A S Pashchina

Joint Institute for High Temperatures, Russian Academy of Sciences, 125412, Moscow, Russia

fgrach@mail.ru

Abstract. The results of experimental research of a pulsed capillary discharge with an evaporating wall in the power density range of $q_S = 10 - 100$ kW/cm^2 obtained by using independent methods of diagnostics, including optical spectroscopy, the ballistic pendulum method, and direct measurement of pressure inside the capillary are presented. The analysis of experimental results, carried out with the help of a two-zone model of a capillary discharge, revealed the decisive influence of the discharge spatial inhomogeneity on the gas dynamics of a multicomponent plasma, which turns out to be substantially different in the high-temperature paraxial and low-temperature peripheral discharge zones. It is shown that the gas dynamics of the paraxial discharge zone corresponds to the regime of the ideal expendable nozzle with good accuracy, and the paraxial zone itself provides the main contribution to the creation of the jet momentum, in spite of the fact that more than 80% of the evaporated substance is carried away from the capillary through the peripheral zone. It is shown that high losses in the plasma jet stagnation pressure, reaching up to 80% relative to the pressure measured inside the capillary, belong to the peripheral discharge zone, in which the density and velocity, respectively, are two orders of magnitude higher and an order of magnitude lower than the corresponding values in the paraxial zone.

1. Introduction

The pulse capillary discharge with an evaporating wall (CDEW) is known as a method for obtaining highly ionized dense plasma [1]. The composition of the CDEW plasma is determined by the substance ablated from the capillary wall and electrodes surface. The presence of several chemical elements with different properties (mass, excitation and ionization energies, etc.) makes the physics of such a plasma more diverse in comparison with a single-component one. A feature of this type of discharge is the injection of the evaporated substance towards the capillary axis, and as a consequence, the spatial separation of the discharge into a paraxial high-temperature zone and a relatively low-temperature layer adjacent to the capillary wall. The difference between the temperatures is the reason of various component and ionization composition of plasma in these zones, namely: localization of electrons and ions of high ionization degree in the high-temperature zone [2] and the dominance of atomic and molecular components in the low-temperature periphery [3]. The processes of thermal diffusion that provide localization of lightweight components in the high-temperature zone and heavy components in the peripheral one contribute to the additional separation of the discharge by the masses of the constituent components. The near-wall layer plays an important role in the balance of mass and energy introduced into the discharge and carried away from it by the plasma jets [1,4,5], and its area is

IOP Conf. Series: Journal of Physics: Conf. Series **1112** (2018) 012013 doi:10.1088/1742-6596/1112/1/012013

comparable with the cross-section of the paraxial zone [6,7]. According to estimates [4,6,7], the temperature in the near-wall layer is $T = 3000 - 5000$ K that provides optimum conditions for the formation of polyatomic molecules and condensed particles [8,9].

Such spatial inhomogeneity of the CDEW in radial direction inevitably influences the gas dynamics of the plasma inside and outside the capillary. For example, low frictional losses in the paraxial zone of the jet because the sharp decrease in viscosity at $T > 10000$ K provide its deep penetration into the dense gas [10], and high velocities in this zone contribute to its stability in perturbed atmosphere [11]. At the same time, the localization of condensed particles in the peripheral zone, and, as a consequence, decrease in viscosity, is able to destabilize the boundary layer and lead to the development of turbulence [12]. The low speed at the peripheral zone can be the reason of the jet momentum losses [1].

Despite the large number of experimental works devoted to the problems of the CDEW, the gas dynamics of the plasma in such discharge is studied insufficiently. In particular, there are no reliable data on such important gas-dynamic parameter as the plasma velocity. The use of direct methods of measuring the flow velocity by tracing the trajectories of plasma inhomogeneities cannot provide sufficient accuracy due to the different nature of these inhomogeneities and their spatial position. The use of the PIV method also does not lead to success due to the displacement of micron-sized particles onto the low-temperature peripheral zone. Moreover, the presence of these particles causes development of turbulence.

These difficulties cause the necessity of using non-intrusive diagnostic methods that do not disturb the flow pattern. Such methods, in particular, include the optical spectroscopy, the measurement of a jet momentum by the ballistic pendulum method, and direct measurements of the total pressure inside the capillary. These methods were used to perform complex diagnostics of a pulsed capillary discharge in the power density range of $q_S = 10 - 100$ kW/cm^2 and to obtain the data necessary for estimating the most important gas-dynamic and thermodynamic parameters of a multicomponent plasma (pressure, density, temperature, velocity, Mach number) that is one of the main goals of this work. The set of experimental results obtained with the use of independent mutually complementary diagnostics methods creates the opportunities for estimating the effects connected with spatial inhomogeneity of the capillary discharge and their role in gas dynamics of plasma that is of particular interest for scientific and practical applications.

2. The object and the methods of research

The object of research is a capillary arrester, whose detailed description is given elsewhere [3]. Perspex is used as a capillary wall material (chemical formula $C_5H_8O_2$). The depth and the initial diameter of a capillary are respectively $h = 5$ mm and $d = 1$ mm. A capacitive storage device with series-connected inductor is used as a power supply. The values of capacity, inductance, voltage and the storage energy of a power supply are the following: $C = 470$ μF, $L = 210$ μHn, $U = 600$ V, $Q = 80$ J. The algorithm of power supply approximately corresponds to the sine half-wave. Typical parameters of the discharge are the following: pulse duration - $\tau_d = 1$ ms, peak current - $I_m = 350 - 450$ A, peak power $N = 80 - 100$ kW. During the preliminary experiments, the mass of substance evaporated from the capillary wall, as well as the pressure inside the capillary, have been measured. The mass of the evaporated substance was determined by monitoring the diameter of the capillary after each discharge pulse. The pressure inside the capillary was measured by a Honeywell strain gauge 19C200PG3K. Researches were carried out in the range of the capillary diameter $d = 1 - 3.2$ mm that corresponds to the range of the discharge surface power density (the power divided by the cross-section of the capillary) $q_S = 1 - 100$ kW/mm^2.

The pressure inside the capillary varies during the discharge pulse in accordance with the discharge power algorithm. The parameters of the plasma flow, which can be subsonic or supersonic outside the capillary (depending on the total pressure ratio $N = P_0/P_\infty$, where P_0, P_∞ - the total pressure inside the capillary and the ambient gas pressure, respectively), are changed in an appropriate manner. The instantaneous image of the discharge obtained for $N \sim 20$, on which one can see the boundary of the high-temperature paraxial zone of discharge inside the capillary, as well as the main elements of the shock-wave structure of the supersonic plasma jet (intercepting shock, reflected shock, Mach disk, triple configuration point), is presented in figure 1.

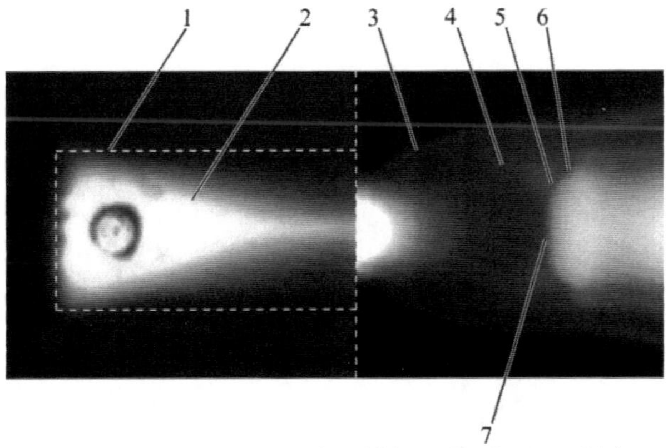

Figure 1 The instantaneous image of the pulsed capillary discharge: (1) boundary of capillary, (2) boundary of paraxial zone of discharge, (3) boundary of plasma jet, (4) intercepting shock, (5) triple configuration point, (6) reflected shock, (7) Mach disk. Total pressure ratio $N = P_0/P_\infty \approx 20$, camera exposure time $t = 1$ μs.

The ballistic pendulum method [13] is used to determine the momentum of the plasma jet from the measured values of deviation angle φ and oscillation period T of the pendulum: $p = \sqrt{2(1-\cos\varphi)}Mg\frac{s}{r}\frac{T}{2\pi}$ (M – mass of the pendulum, g – acceleration of gravity, s – distance from the rotation axis to the center of mass, r – distance from rotation axis to center of force application). The ballistic pendulum is a massive body suspended on needle supports. The capillary arrester and current-carrying buses are rigidly fixed on the ballistic pendulum. The parameters of the ballistic pendulum satisfy the condition $T \gg \tau_d$. The thrust force is determined by the ratio

$$F = (P_0 - P_\infty)S_a, \quad (1)$$

from which the stagnation pressure P_0 can be determined (S_a – cross-section of the capillary).

The longitudinal profiles of the electron number density and temperature in the paraxial discharge zone inside and outside the capillary are obtained using optical spectroscopy. The 2D-spectra recording technique is described in detail in [3]. The spectral interval $\Delta\lambda = 650 - 685$ nm was chosen for spectra recording. This interval contains the hydrogen line H_α (used for determination of the electron number density by the linear Stark effect), the doublet of the ionic carbon lines C II 657.8 nm and 658.3 nm, the multiplet of ion C II lines of 678-682 nm, and the rather intense continuum (inside the capillary), which can be used to determine the electron temperature. In the present work, the method of the relative intensities of the H_α line and the continuum is mainly used to estimate the electron temperature. Evaluation of the radiation reabsorption from both the H_α line and the continuum under experimental conditions indicates the suitability of this method for estimating the electron temperature inside the capillary with an error of less than 20% increasing with power input. The results of spectral diagnostics were used to estimate the plasma parameters in the paraxial discharge zone: the pressure $P_1 \approx 2n_e kT_e$, the mass density $\rho_1 = \frac{\mu_1 P_1}{RT_e}$ and the sound velocity $c_1 = \sqrt{\frac{\gamma RT_e}{\mu_1}}$.

3. The results of research

Dependences on the discharge power density of the pressures on the bottom (P_0) and on the outlet (P_a) of the capillary, as well as the capillary pressure ratio ($N_{cap} = P_0/P_a$), reconstructed from the results of spectral diagnostics, are presented in figure 2. The capillary pressure ratio remains approximately constant in the power density range corresponded to the supersonic regime of the plasma jet (the Mach number at the capillary outlet is $M_a = 1$). The threshold power density, at which the flow regime changes, is approximately equal to $q_S^{thr} \approx 18$ kW/mm^2 and quantitatively consistent with the results obtained in [14]. It is easy to verify that the capillary pressure ratio is described with good accuracy by the relation $P_0/P_a = 1 + \gamma M_a^2$, valid for the expendable nozzle [15]. This makes it possible to estimate the effective adiabatic index, whose value is approximately $\gamma \approx 1.3 - 1.4$ (see figure 2), as well as the Mach number at the capillary outlet in the subsonic flow regime ($q_S < q_S^{thr}$). Note that the pressure in the jet during its evolution exceeds the ambient gas pressure and only in the end of the discharge pulse approaches to the atmospheric pressure. Such feature is observed both in the supersonic and in subsonic flow regime. In particular, the exit pressure ratio in the subsonic flow regime ($M_a \leq 1$) reaches up to $n = P_a/P_\infty = 2 - 3$ (see figure 2). Moreover, the exit pressure ratio is the greater, the shorter the duration of the discharge pulse and vice versa. Such situation, obviously, is the consequence of nonstationarity of the pulsed flows, that is observed regularly in experiments [16,17] and finds the justification in the theoretical calculations [18].

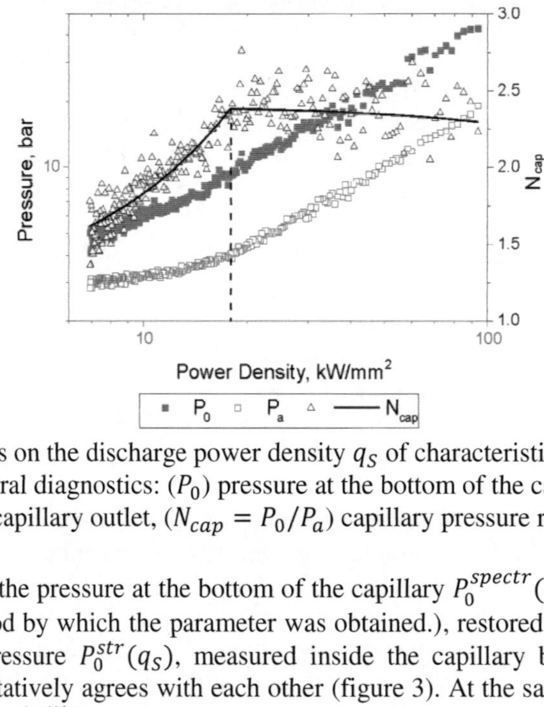

Figure 2 Dependences on the discharge power density q_S of characteristic pressures, reconstructed from the results of spectral diagnostics: (P_0) pressure at the bottom of the capillary, (P_a) pressure at the capillary outlet, ($N_{cap} = P_0/P_a$) capillary pressure ratio.

The dependences of the pressure at the bottom of the capillary $P_0^{spectr}(q_S)$ (The superscript is used for indication the method by which the parameter was obtained.), restored from the results of spectral diagnostics, and the pressure $P_0^{str}(q_S)$, measured inside the capillary by means of a strain gage qualitatively and quantitatively agrees with each other (figure 3). At the same time, the dependence of the stagnation pressure $P^{*ballist}(q_S)$ of the plasma jet, restored from the data obtained by the ballistic pendulum method, differs from the $P_0^{spectr}(q_S)$ and $P_0^{str}(q_S)$ dependences both quantitatively and qualitatively. The ratio of these pressures reaches $P_0^{spectr}(q_S)/P^{*ballist}(q_S) = 2 - 5$. Moreover, the dependence $P^{*ballist}(q_S)$ contains a singularity that is expressed in a sharp change in its slope in the vicinity of the discharge power density of $q_S \approx 35$ kW/mm^2. The noted feature directly indicates that the results of the jet's momentum measurement are determined by a combination of various factors, and the contribution of each individual factor is not predetermined and depends on the discharge power density. Both the processes inside and outside the capillary can act as such a factors. The latter ones, as a rule, are the sources of systematic errors of the plasma jet momentum measurement.

Nonstationarity of the pulsed jet and, as a consequence, the increased pressure inside the jet P_{Jet} in comparison with the ambient pressure P_∞, is one of the most significant external factors leading to systematic errors. Accounting for this factor, however, does not influence significantly on the pressure ratio $P_0^{spectr}(q_S)/P^{*ballist}(q_S)$, because replacing the last term in equation (1) from P_∞ to P_{Jet} allows us to reduce the gap between P_0^{spectr} and $P^{*ballist}$ less than 20%. At the same time, account of this and other external factors, allows to compensate only for the systematic error in measuring the jet momentum, but cannot significantly affect the very character of the dependence $P^{*ballist}(q_S)$. Therefore, it seems that the loss in the jet momentum (and also the specific form of the dependence $P^{*ballist}(q_S)$ related to this fact) is largely determined by the gas dynamics of the plasma inside the capillary, primarily by the spatial inhomogeneity of the flow.

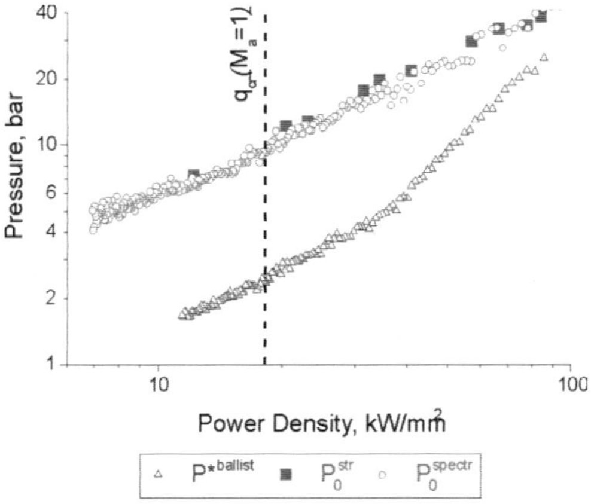

Figure 3 Dependences on the discharge power density q_S of the characteristic pressures obtained by various methods: (P_0^{spectr}) pressure at the bottom of the capillary, restored from the results of spectral diagnostics, (P_0^{str}) pressure, measured inside the capillary by a strain gage, ($P^{*ballist}$) stagnation pressure of the plasma jet, restored from the data obtained by the ballistic pendulum method.

To verify this assumption, estimates of the plasma parameters at the section inside the capillary were made, taking into account the spatial separation of the discharge into the high-temperature paraxial and low-temperature peripheral zones. The initial system of equations includes the thrust force equation and the continuity equation

$$\frac{F}{S_a} = G_{a1}v_{a1}\phi + G_{a2}v_{a2}(1-\phi) + P_a - P_{Jet} \quad (2)$$

$$G_a = \rho_{a1}v_{a1}\phi + \rho_{a2}v_{a2}(1-\phi) \quad (3)$$

where $\phi = S_1/S$ – reduced cross-section of the paraxial zone of discharge, S_1, S_a – cross-section of the paraxial discharge zone and the total capillary cross-section, respectively, $G_{ai}, \rho_{ai}, v_{ai}$ – mass flow rate, mass density and velocity of plasma at the capillary outlet for i-th zone of discharge (the values of the index $i = 1,2$ are used to denote the paraxial and peripheral discharge zones, respectively), P_a, P_{Jet} – pressures at the capillary outlet and in the plasma jet, respectively.

The initial data include:

- the results of measuring the thrust force $F(q_S)$ by the ballistic pendulum method;
- the results of measuring the mass flow rate $G_a(q_S) = \frac{\Delta m(q_S)}{\tau_d S}$ (Δm – the mass of evaporated substance during the discharge pulse, τ_d – discharge pulse duration);
- the results of spectroscopic measuring the electron number density and temperature, and also the estimates of pressure restored from these data. These data are used to calculate the density, velocity of sound, and plasma velocity in the paraxial discharge zone at the capillary outlet (in the local thermodynamic equilibrium approximation). The value of the effective molecular mass in the paraxial zone $\mu_1 = 1.43$ was used in calculations, which corresponds to a fully ionized plasma whose stoichiometric composition is determined by carbon and hydrogen in the proportion $[C]:[H] = 1:5$.

Taking into account the results of spectral diagnostics (see figure 2), it is assumed that the gas dynamics of the plasma in the paraxial discharge zone corresponds to the model of an ideal expendable nozzle assumed the absence of the pressure losses. Under these conditions, the dynamic pressure of the paraxial zone at the capillary outlet is determined by the relation

$$H_{a1} = P_0 - P_a = \gamma M_{a1}^2 P_a = \rho_{a1} v_{a1}^2 \quad (4)$$

This condition means that, in the absence of static pressure gradients over the capillary cross-section ($grad_r P = 0$), the dynamic pressure losses occurs exclusively in the peripheral discharge zone. Thus, the condition for the equality of the dynamic pressures and Mach numbers in the paraxial discharge zone and the peripheral one, which is the basic assumption of the capillary discharge two-zone models [6,7], is not satisfied in our case, so $\rho_{a1} v_{a1}^2 > \rho_{a2} v_{a2}^2$ and $M_{a1} > M_{a2}$.

The temperature and effective molecular mass in the peripheral discharge zone, which determine the density and speed of sound in this zone, are specified as additional conditions necessary for solving a system of two equations (2)-(3) with three unknowns. The calculation was carried out for the following values of these parameters: $T_{a2} = 3500$ K, $\mu_2 = 13.5$. The condition for the complete dissociation of the ablating substance of the capillary wall, as well as estimates of the temperature in the peripheral zone, obtained in [6,7] were used as criterion for choosing the temperature. The value of the effective molecular mass in the peripheral zone corresponds to a weakly ionized plasma whose stoichiometric composition is determined by carbon and oxygen in the proportion $[C]:[O] = 3.4:2$.

The solution of the system (2)-(3) makes it possible to determine the reduced area of the paraxial discharge zone ϕ, the velocity v_{a2} and the Mach number M_{a2} in the peripheral zone, the mass flow rate and dynamic pressure for each discharge zone. Results of calculation of some parameters are presented in figure 4. The qualitative course of the obtained dependences is determined primarily by the initial data $F(q_S), G_a(q_S)$, and $P_a(q_S)$, and, within the framework of accepted approximations, is not very sensitive to variations of the temperature and effective molecular mass in the paraxial and peripheral zones, which affect only the quantitative values of the calculated parameters. The analysis shows that the main contribution to the momentum of the plasma jet is provided by the paraxial discharge zone, the partial dynamic pressure of which ($h_{a1} = H_{a1}\phi$) is for 2-6 times higher than the corresponding value ($h_{a2} = H_{a2}(1-\phi)$) for the peripheral zone (see figure 4). At the same time, more than 80% of the ablated mass is carried away from the capillary through the peripheral discharge zone. Such situation, in general, is due to a significant difference in the flow parameters in the two mentioned zones: by more than order higher speed and more than two orders lower mass density of plasma in the paraxial discharge zone if to compare with the peripheral one (figure 5). A significant difference in these parameters is facilitated by the inequality of Mach numbers ($M_{a2} < M_{a1}$), as well as the comparable cross-sections of these zones, whose ratio decreases with discharge power density decrease (see figure 4). We note that the course of the calculated dependence $\phi(q)$ agrees with the experimental observations. Analysis of the images and spatial spectra of the discharge shows that the reduction in the power density (due to the increase in the capillary diameter, the other parameters (the capillary length and the energy deposited into the discharge) remain fixed) is accompanied not only by a decrease in the ratio of the cross-sections of the paraxial and peripheral zones predicted by the model, but also leads to the fact that this ratio varies along the capillary, and, the stronger, the lower the discharge power density (see figure 1).

The limitations of the two-zone model, as well as the assumptions about the constancy of the component composition in each zone and the temperature in the peripheral zone, do not allow us to indicate specific physical mechanisms responsible for the jet momentum losses. We suppose that the following processes provide the main contribution to the losses of the jet's momentum: viscous friction in the boundary layer (the plasma viscosity in the low-temperature peripheral and high-temperature paraxial zone differ by more than order of magnitude [10,19]), as well as plasma-chemical transformations in the peripheral zone, proceeding with the absorption of energy. Among the latter processes, in particular, is the growth of cluster particles, a significant concentration of which in the erosive plasma finds experimental confirmation [20,21]. The contribution of both mechanisms to the jet momentum losses is determined by the thickness of the low-temperature peripheral zone that, in a first approximation, makes it possible to explain the results obtained in ballistic pendulum experiment.

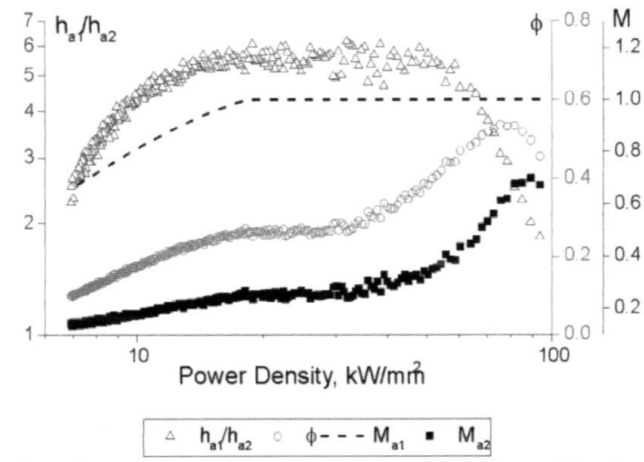

Figure 4 Calculated dependencies on power density q_S: M_{a1}, M_{a2} – Mach numbers, h_{a1}/h_{a2} – ratio of partial dynamic pressures, ϕ – reduced cross-section of the paraxial discharge zone. Indices 1, 2 related to the paraxial and peripheral zone, respectively.

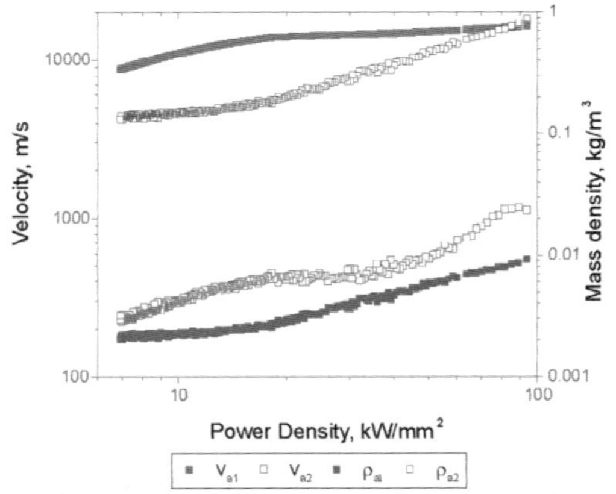

Figure 5 Calculated dependencies on power density q_S of plasma parameters at the capillary outlet in the paraxial and peripheral discharge zone: v_{a1}, v_{a2} – velocities, ρ_{a1}, ρ_{a2} – mass densities.

4. Conclusions

Thus, the analysis of the results of complex research of pulsed capillary discharge with an evaporating wall, obtained on the basis of independent diagnostic methods, allows us to conclude that the spatial inhomogeneity of the discharge exerts a decisive influence on the gas-dynamic parameters of multicomponent plasma. Spatial separation of the discharge by the masses of its components, localization of lightweight components in the high-temperature paraxial zone, and heavy ones in the low-temperature peripheral zone, leads to the fact that the gas dynamics in these zones is significantly different. Estimates show that the detected high losses of plasma jet stagnation pressure should be attributed, mainly, to the peripheral discharge zone, the density and velocity in which, respectively, are two orders of magnitude higher and an order of magnitude lower than the corresponding values in the paraxial zone. The paraxial discharge zone, whose gas dynamics with good accuracy corresponds to the mode of an ideal expendable nozzle, provides the main contribution to the creation of a jet momentum, in spite of the fact that more than 80% of the evaporated substance mass is carried away from the capillary through the peripheral zone. We believe that the main sources of the stagnation pressure losses are viscous friction in the boundary layer and the endothermic processes of plasma-chemical transformations in the peripheral zone. A study of the role of these factors is the subject of future research.

IOP Conf. Series: Journal of Physics: Conf. Series **1112** (2018) 012013 doi:10.1088/1742-6596/1112/1/012013

References

[1] Ogurtsova N N, Podmoshensky I V and Smirnov V L 1976 *Teplofiz. Vys. Temp.* **14** 1–8 (in Russian)
[2] Kocharyan A E, Bobrova N A and Sasorov P V. 2006 *Plasma Phys. Reports* **32** 887–95
[3] Pashchina A S, Efimov A V and Chinnov V F 2016 *High Temp.* **54** 488–502
[4] Ogurtsova N N, Podmoshensky I V and Shelemina V M 1974 *Teplofiz. Vys. Temp.* **12** 5–9 (in Russian)
[5] Ogurtsova N N, Podmoshensky I V and Rogovtsev P N 1971 *Teplofiz. Vys. Temp.* **9** 468–74 (in Russian)
[6] Ruchti C B and Niemeyer L 1986 *IEEE Trans. Plasma Sci.* **PS-14** 423–34
[7] Kovitya P 1987 *IEEE Trans. Plasma Sci.* **PS-15** 294–301
[8] Eremin A V. 2012 *Prog. Energy Combust. Sci.* **38** 1–40
[9] Reznikov B I, Alekseev N I, Bobashev S V, Zhukov B G, Ponyaev S A, Kurakin R O and Rozov S I 2011 *Tech. Phys.* **56** 1106
[10] Leonov S B and Luk'yanov G A 1994 *Prikl.Mekh.Tekh.Fiz.* 13–8 (in Russian)
[11] Pashchina A S and Klimov A I 2014 *Khim. Fiz.* **33** 78–86 (in Russian)
[12] Pashchina A S 2017 The Influence of Capillary Sizes and Discharge Pulse Parameters on the Plasma Properties of the Capillary Type Discharge *16th International Workshop on Magneto-Plasma Aerodynamics* ed V A Bityurin (Moscow, Russia: JIHT RAS) pp 76–9
[13] Sivukhin D V 1989 *General cours of physics. Vol. 1* (Moscow: Nauka) (in Russian)
[14] Ender A Ya, Kuznetsov V I and Kolyshkin I N 2015 *Tech. Phys.* **60** 1720-24
[15] Vulis L A 1950 *Thermodynamics of the gas flows* (Moscow: Gosenergoizdat) (in Russian)
[16] Sharikov I and Surzhikov S 2005 Experimental Study of Unsteady Supersonic Underexpanded Ablative Plasma Jet Dynamics (AIAA 2005-4930) *36th AIAA Plasmadynamics and Lasers Conference* (Toronto, Ontario Canada) pp 1–11
[17] Belov S N, Ogurtsova N N and Podmoshensky I V 1975 *Journ. Appl. Spectroscopy* **22** 396–402 (in Russian)
[18] Kuzenov V V, Ryzhkov S V, Gavrilova A Y and Skorokhod E P 2015 *Trudy MAI* **83** 51–74 http://trudymai.ru/upload/iblock/314/kuzenov_ryzhkov_gavrilova_skorokhod_rus.pdf (in Russian)
[19] Kovitya P 1984 *IEEE Trans. Plasma Sci.* **PS-12** 38–42
[20] Pashchina A S, Karmatsky R E and Klimov A I 2017 *Tech. Phys. Lett.* **43** 1033–6
[21] Pashchina A S and Valyano G E 2017 *J. Phys. Conf. Ser.* **891** 12299

IOP Conf. Series: Journal of Physics: Conf. Series **1112** (2018) 012014 doi:10.1088/1742-6596/1112/1/012014

Visualization of interaction between gas-dynamic perturbation caused by spark discharge and detached supersonic airflow

A S Saveliev[1,2]

[1] Joint Institute for High Temperatures of the Russian Academy of Sciences, Izhorskaya 13 Bd 2, Moscow, 125412 Russia
[2] Moscow Institute of Physics and Technology (State University), Institutskii lane 9, Dolgoprudny, Moscow Region, 141700 Russia

fisteh@mail.ru

Abstract. The method of long spark discharge initiating near the surface of the aerodynamic model, streamlined by supersonic airflow, was investigated. It has been shown experimentally that when a special construction discharger is used, a spark discharge can be initiated with a frequency of the order of 1 kHz at energy in one pulse of the order of 0.1 J and duration of the order of 0.1 μs. The gas-dynamic perturbation caused by the discharge was visualized, instantaneous patterns of the flow in the presence of a discharge were obtained, and a technique for reconstructing the dynamics of the flow change after the discharge was proposed.

1. Introduction

Initiation of an electrical discharge in the air flow can lead to a significant change in the flow pattern and the aerodynamic characteristics of the streamlined model [1,2]. Among the problems that researchers are trying to solve are decreasing of aerodynamic resistance [3], decreasing the effects of stall and separation of the flow from the surface of the body and, thus, improving the aerodynamic characteristics of the body [4], increasing the efficiency of mixing fuel with the incoming airflow and igniting the combustible mixture by discharge [5], controlling of laminar-turbulent transition [6], controlling of jet noise and tangential discontinuities in flow [7], creating addition forces for control [8] and other problems. It should be noted that mostly at studies process of discharge initiation are considered when the supplied electric power is much less than the energy of the incoming airflow. It is assumed that the hydrodynamic perturbations caused by discharge in the flow thus lead to a redistribution of the flow parameters (for example, a local increase in temperature and pressure) or when constant action is not required and the discharge can be initiated in a pulse-periodic mode. In this case the instantaneous power released in the incoming gas flow turns out to be much greater than the incident flow energy per second, however, in average at time the input power is less than the incident flow power. This paper presents the results of visualization of the process of long spark discharge initiating of submicrosecond duration in an incoming supersonic airflow with Mach number $M = 2$, at discharge initiation frequencies up to 2000 Hz and with an average energy input of less than 300 W. Earlier it was shown [9] that the initiation of a spark discharge on the surface of a body (airfoil model) can significantly change the flow pattern near the surface, including the displacement of the separation point downstream.

At high values of the angle of attack, near the trailing edge of the airfoil, a separated flow can be formed, and it characterizes by the presence of reversal or chaotic flow with respect to the main airflow. At the same time, in the separation region near the surface, the pressure is greater than it would be without flow separation. In order to act on the separation region it is proposed to initiate a long spark discharge near the surface of the airfoil upstream. This creates a gas-dynamic perturbation with increased pressure relative to the static pressure of the main flow, which moves downstream and interacts with the separation zone.

Upon initiation of a long spark discharge of submicrosecond duration on the surface a cylindrical shock wave is formed in the oncoming stream. Depending on the energy released in the discharge the propagation speed of the front of this wave can be different. At present work the energy in the discharge is less than 0.1 J for a spark discharge length of 10 cm. In this case the Mach number of the front of the cylindrical shock wave is slightly higher than 1, and the main airflow blows downstream the shock wave and the gas-dynamic perturbation (hot gas and high pressure region). When this perturbation interacts with the separation zone the separation point shifts downstream.

2. Experimental setup

A quasi-two-dimensional airflow around an aerodynamic body (NACA airfoil) with a pulsed-periodic instantaneous energy release near the surface in front of the separation zone near the trailing edge is experimentally simulated. The energy is released when a spark discharge of submicrosecond duration is initiated near the airfoil surface.

To initiate a spark discharge in a supersonic airflow, an electric discharger of a special design was used (see figure 1), so called segmented discharger [9]. To create long spark discharge, a set of several conducting segments 9, that are installed in the streamlined surface of the aerodynamic body (NACA airfoil), of the characteristic size of 3 mm are used, separated from each other by a thin dielectric film 10 of 0.1 mm thickness. In such a design, it becomes possible to create a discharge on a significant gap between the outer electrodes, i.e. the discharge is localized near the surface of the conducting segments, and the oncoming airflow does not lead to the blowing of the spark downstream. This is due to the redistribution of the electric field and the increase in its local strength near the surface of the conducting segments. It should be noted that the dielectric film 10 can be removed between segments located at the ends of the discharger 11. Since the number of segments 11 can be changed, in the experiment it is possible to set the length of the discharge gap, i.e. the length of the spark and the length of the region near the surface of the body, in which a rapid release of energy occurs.

The aerodynamic body with a flush-mounted discharger on its upper surface is an asymmetric NACA64a212 airfoil with a chord length of 10 cm and a span of 15 cm. The conducting segments of the discharger are 38 mm from the trailing edge of the airfoil. Fused silica is the material of airfoil, and it is characterized by good resistance to discharge plasma action (heat, chemical reactions). The airfoil was placed in a supersonic jet with Mach number $M = 2$, formed in an atmospheric-vacuum wind tunnel (see figure 1). The supersonic flow is formed when air 1 passes from the atmosphere to the vacuum system of the installation through de Laval nozzle 2 with a diameter of the output section of 11 cm. The supersonic jet of air forms and it has a static pressure approximately equal to $p_{st} = 0.15$ atm, and the flow velocity is close to 500 m/s. When the wind tunnel is operating, three modes of supersonic flow can be outlined: underexpansion, overexpansion, the matched regime when the pressure surrounding the jet p_a is less than p_{st}, $p_a > p_{st}$, $p_a = p_{st}$ respectively. The pressure p_a constantly increases as air from the atmosphere fills the vacuum system, and the filling rate is $Q = 1$ kg/s. At the same time as long as supersonic regime exists, the static pressure p_{st} remains constant. The underexpanded mode is characterized by the presence of rarefaction waves in a supersonic jet, which leads to a nonuniform distribution of the components of the flow velocity vector. The case of overexpansion is characterized by the presence of shock waves, which also disrupt the uniformity of the flow parameters, and when an aerodynamic body is placed in such a jet, flow separation may occur due to the interaction of the shock wave and the boundary layer. Because of the relatively large volume of the vacuum system ($V_0 = 240$ m^3), the time for supersonic flow maintaining is about 60 seconds, however, to measure the flow parameters of the aerodynamic models it is necessary to work on a much narrower time interval containing the time moment when $p_a = p_{st}$. This moment of time is characterized by uniform distribution of the velocity vector components in the volume of the jet (at a distance from nozzle out section of the order of the nozzle diameter). Thus in presented experiments on the airflow around the airfoil were carried out near the matched regime and no longer than 10 seconds.

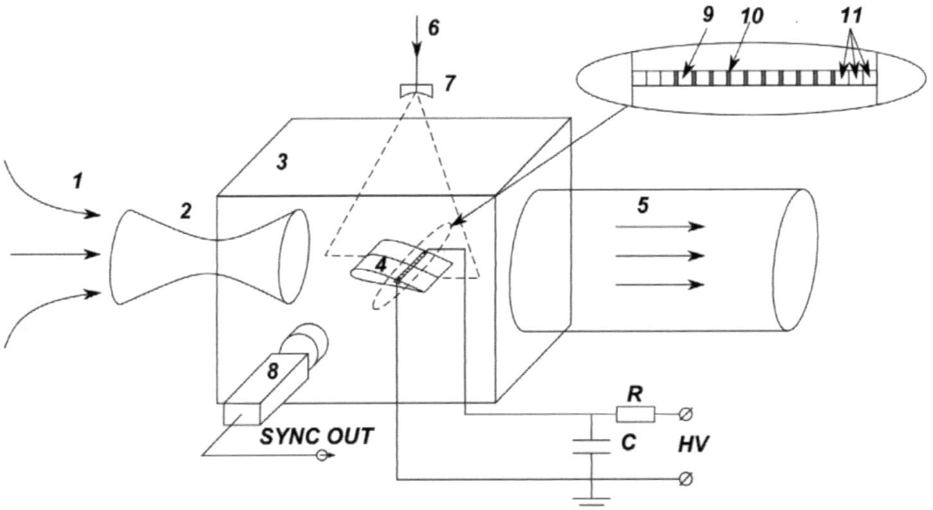

Figure 1 Schematic representation of the experimental setup.

The high voltage from the capacitor C (capacitance C = 4.1 nF), which is charged from a high voltage power supply HV (Spellman SL15P2000) through the ballast resistor R (resistance R = 90 kΩ) is applied to the electrodes of the discharger. When the charge voltage of the capacitor C exceeds the breakdown voltage, a discharge occurs on the surface of the NACA airfoil. With the help of a 12-bit high-resolution oscilloscope LeCroy 66Zi, the voltage across the capacitor C is recorded for 10 seconds, which allows tracing the stability of the spark discharge parameters: breakdown voltage, discharge energy, discharge duration.

To visualize the separated flow near the trailing edge of the NACA airfoil the Particle Image Velocimetry (PIV) technique is used, see figure 1. Air 1 together with oil droplets of submicron diameter from the atmosphere enters the de Laval nozzle 2 where it accelerates to supersonic speed and a supersonic jet in the working chamber 3 flows around the NACA airfoil with the spark discharger 4. Then the air jet flows to the vacuum system 5. The oil aerosol is formed using a generator based on the Laskin nozzle [10], and the used liquid is vegetable oil. The working chamber 3 has several transparent windows of optical glass, so it is possible to perform visualization. The laser beam 6 passes through the laser sheet optics 7, then through the transparent window gets inside the working chamber 3. The laser sheet (in figure 1 is shown in dashed lines) illuminates droplets of oil aerosol in the flow area above the upper surface near the trailing edge. The plane of the laser sheet contains the axis of the nozzle 2. The laser radiation is scattered on the oil droplets and scattered light is recorded by the Imager Pro X 2M double exposure video camera. Laser system consists of two Nd:YAG lasers, so the PIV system allows getting two photographs with a resolution of 1600x1200 pixels and with a frequency of up to 14 Hz. The delay between the pulses of lasers in the experiment was 0.8 μs. Operation of the laser and video camera is controlled by the internal generator of the PIV system, and the process of initiating the discharge occurs at a random time. To compare each time moment of PIV-photograph of the laser radiation scattered on the drops and the time moment of discharge the SYNC OUT output of the camera is connected to the second channel of the oscilloscope. Also in the experiment, the pressure values p_a and p_{st} are recorded in order to compare the frames to the flow regime from the nozzle. The typical size of the visible video camera area is 10 cm.

IOP Conf. Series: Journal of Physics: Conf. Series **1112** (2018) 012014 doi:10.1088/1742-6596/1112/1/012014

3. Experimental results

The main problem in visualizing of interaction between separated flow near the trailing edge and the gas-dynamic perturbation that occurs when a spark is discharged on the surface of a NACA airfoil is the unsteadiness of the detached flow. One of the methods of obtaining data that allows evaluating the effect of this disturbance on the position of the separation point position is to keep constant all parameters and make multiple photographs (PIV measurements). If the energy released in the discharge, the place where the discharge was initiated, the time t between the discharge time moment and the time moment of photograph taken for PIV measurements are fixed, then flow patterns can be averaged and a picture of the interaction between the gas-dynamic perturbation and the separation zone at time t is obtained.

The place of initiation of the discharge, as already mentioned above, is fixed due to the design of the discharger. It should also be noted that the surface is not destroyed by the discharge plasma due to the selected airfoil material. Figure 2 shows a photograph of the spark discharge in a supersonic air flow on the surface of the airfoil made by a video camera with an exposure of 100 microseconds (which is much longer than the flying time). It can be seen that a plasma spark 10 cm long has sharp contours and is directly above the spark discharger.

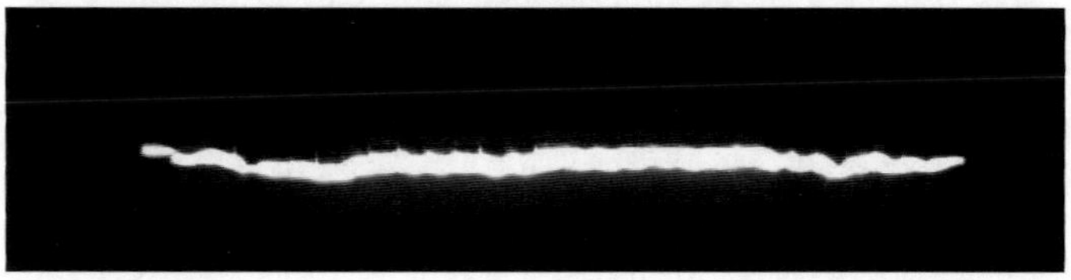

Figure 2 Typical photograph (top view) of a spark discharge in a supersonic airflow with Mach number $M = 2$. On presented photograph flow is from top to down.

By the measurements taken using a high-resolution oscilloscope, it is shown that the breakdown voltage, energy, discharge initiation period, discharge duration have sufficient stability during the experiment time (10 seconds). Figure 3 shows the histogram of the breakdown voltage obtained by processing the voltage waveform on the high-voltage capacitor C for the case of the airfoil angle to the oncoming flow of 14 degrees (RF is the relative frequency of occurrence of the breakdown voltage value).

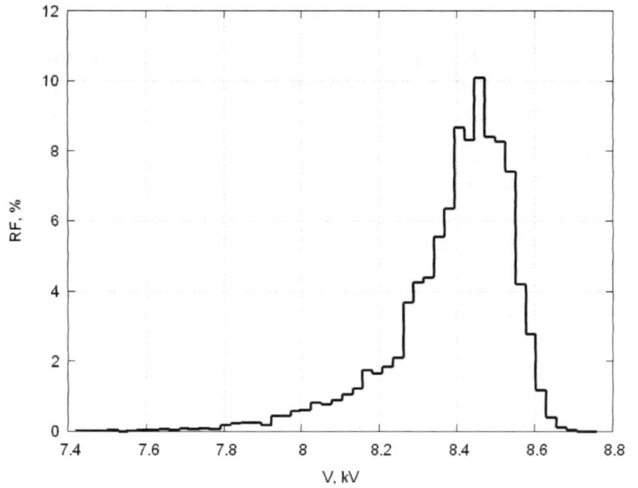

Figure 3 Typical histogram of voltage V on capacitor C at the moment of breakdown of the discharger.

Similar histograms can be obtained for the energy E stored in the capacitor at the time moment of discharge, both for the discharge duration τ, and for the discharge frequency f (or period T). The discharge duration τ is defined as the time during which the voltage drops from 0.9 to 0.1 of the voltage across the capacitor at the time moment of breakdown. The energy E is defined as the difference between the energy in the capacitor at the moment of breakdown minus the energy in the capacitor after the breakdown of the discharge gap. In reality, the second value (the residual energy in capacitor C) is much less than E (by two orders of magnitude). It should be noted that, under the experimental conditions, the discharge initiation frequency was about 2 kHz. This means that the histograms are built with a significant number of measurements (of the order of 20,000).

Such histograms are characterized by the value of the parameter that is most often encountered in this experiment, as well as the width of the histogram measured at its half-height. For example, it follows from figure 3 that a breakdown voltage of about 8.5 kV is most often encountered (in 10% of the total number of discharges), and the width of the histogram at a height of 50% is about 0.2 kV. Figure 4 shows the results, the most frequent values of the energy E in various experimental launches. The vertical error corresponds to the half-width of the histogram, and the dashed line represents the average value of E for the histogram maxima in experiments No. 1–10.

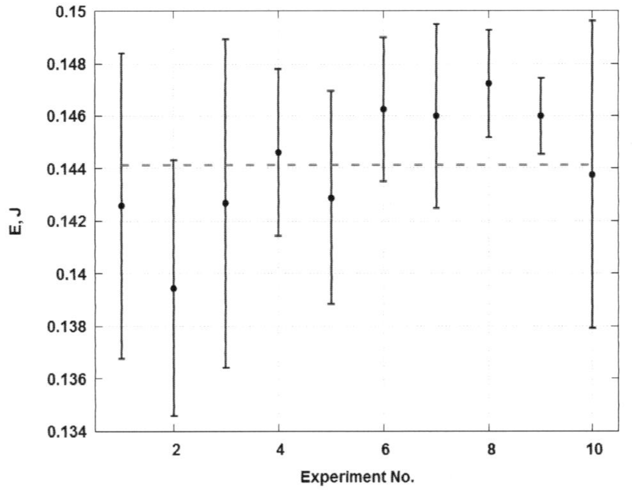

Figure 4 The maxima of the histograms E obtained in different experimental launches

For the PIV measurements of the process of interaction between the gas-dynamic perturbation caused by the discharge and the separation zone, the following parameters were chosen: the length of the discharge gap is 10 cm, the capacitor $C = 4.1$ nF, the angle of attack is 14 degrees, the resistance of the ballast resistor is $R = 90$ kΩ. A number of experiments were performed (10), at each voltage oscillograms on the capacitor were recorded, and then histograms of the discharge parameters were plotted for: energy stored in capacitor E, breakdown voltage V, discharge duration τ, period between serial discharges T, repetition frequency f. For each of the 10 histograms a conclusion was made about the stability (repeatability) of the initiation parameters of the discharge V, E, τ, T, f. Also from the obtained data the average value of these parameters was calculated and the ranges of variation were determined. The data obtained are as follows: $V = 8.4$ kV (+/- 5%), $E = 0.14$ J (+/- 10%), $\tau = 0.3$ μs (+/- 10%), $T = 450$ μs (+/- 20%), $f = 2.2$ kHz (+/- 20%). Thus, the energy E and the duration of the release of energy in the flow τ vary within ± 10%. This gives reason for averaging the resulting spatial flow patterns corresponding to a certain delay range t from the moment of discharge to the moment of PIV measurement. Due to the low speed of the PIV system, and also to increase the number of spatial distributions of the velocity vector for averaging, the value of t is selected from 0 μs to 150 μs with step of $\Delta t = 10$ μs. These spatial flow patterns are grouped if they correspond to the delay in the range from t to $t + \Delta t$. The spatial flow patterns corresponding to the time delay between the discharge and shooting moments lying in the range from t to $t + \Delta t$ are averaged, and the delay of $t + \Delta t/2$ is assigned to obtained average picture. Figure 5 shows the spatial distributions of the flow velocity vector over the upper surface of the airfoil near its trailing edge and at different time moments. For clarity the color indicates the region in which the flow velocity is subsonic (i.e. the separation zone), the white region is

IOP Conf. Series: Journal of Physics: Conf. Series **1112** (2018) 012014 doi:10.1088/1742-6596/1112/1/012014

the region of supersonic main flow, the green area corresponds to the position of the airfoil in space. The place of initiation of the discharge is not shown, because it is located upstream (at a distance of 38 mm from the trailing edge). The flow patterns presented are obtained by averaging 15 PIV measurements.

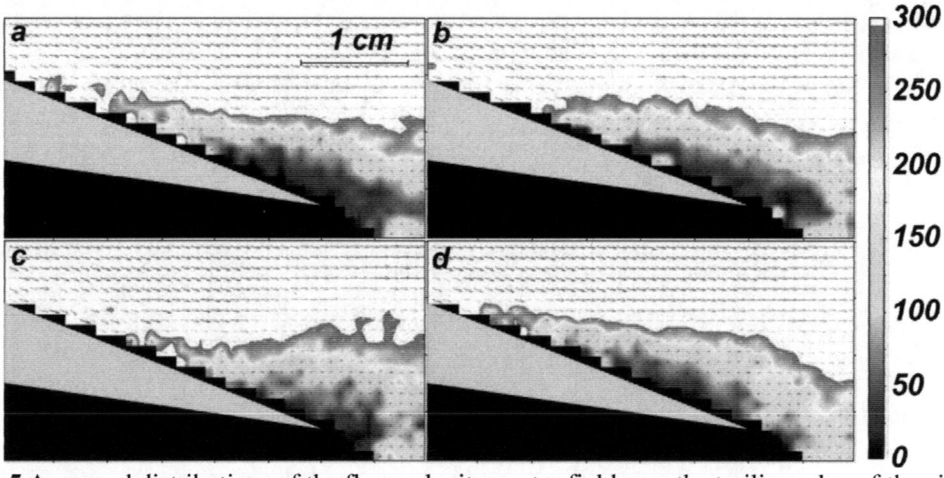

Figure 5 Averaged distributions of the flow velocity vector field near the trailing edge of the airfoil: a — time moments *t* in the range from 0 to 10 μs after the discharge, b — 60–70 μs, c — 90–100 μs, d — 140–150 μs. Values on the color scale are in m/s.

For a different time value *t* after the moment of the discharge similar pictures of the distribution of the velocity vector in the space above the airfoil are obtained. According to the average flow patterns the separation point position relative to the trailing edge (average distance from the separation point to trailing edge) is obtained and the root mean square deviation of the separation point position from the mean value is obtained from individual frames from the time range from t to $t + \Delta t$. Figure 6 shows the results of processing spatial patterns, where the width of the point corresponds to the width of the time range $\Delta t = 10$ μs, and the height is the root mean square of deviation in the position of the separation point for the selected time interval. It can be seen that for some time after the moment of discharge, the position of the separation point is displaced downstream, but then the position of the separation point returns to a position that is slightly upstream from the initial position.

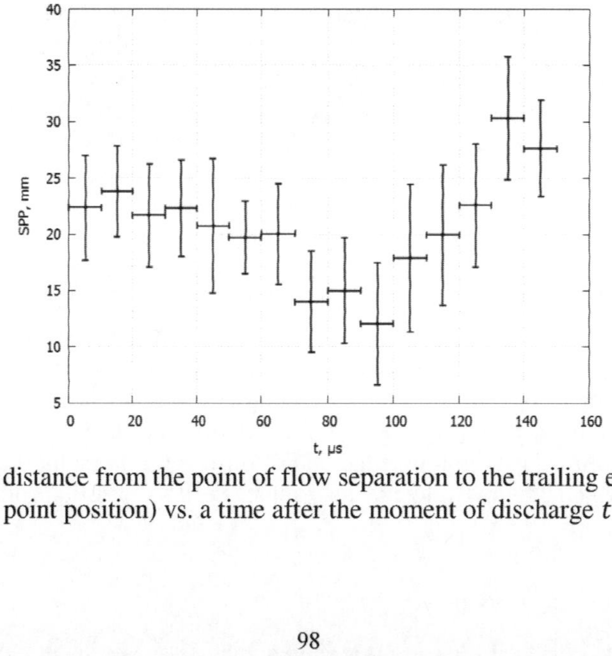

Figure 6 Change of distance from the point of flow separation to the trailing edge (SPP, separation point position) vs. a time after the moment of discharge t.

4. Conclusions

The visualization and measurement of the spatial distribution of the flow velocity over the upper surface of the airfoil, which is streamlined by the supersonic airflow, was carried out with initiating a rapid heat release and in the presence of a flow separation near the trailing edge. With the help of a special design of discharger a heat release occurs by initiating long spark discharge, which occurs when a voltage is applied to a discharger of a certain value. This generates a cylindrical shock wave and a long spanwise gas-dynamic perturbation that interacts with the separation zone. These circumstances allow concluding that a quasi-two-dimensional case is modeled in presented experiment.

The discharge initiation parameters were measured experimentally and it is shown that the self-initiation of the spark discharge occurs with relatively constant parameters, in particular the energy and duration of the discharge. These parameters vary from discharge to discharge, from experiment to experiment no more than in a range of +/- 10% of the average value.

The relative constancy of the discharge parameters made it possible to perform a detailed flow visualization using the PIV method, which unambiguously shows the position of the subsonic separated flow region. The position of the point on the upper surface of the airfoil is shifted to the trailing edge of the airfoil as the gas-dynamic perturbation from the discharge moves downstream. Then the separation point returns to its initial position and even upstream. This circumstance may be due to the fact that the leading front of a cylindrical shock wave propagating along the boundary layer can cause its detachment. The front of the cylindrical shock wave propagates along the flow, which moves with supersonic speed (Mach number is $M > 2$, flow speed is about 600 m/s), both upstream and downstream. The front of the shock wave leads to the appearance of a positive pressure gradient in the boundary layer upstream, which can lead to an earlier formation of separation than it was before the discharge was initiated. This circumstance should impose certain limitations on the energy to be released with discharge into the stream.

The obtained instantaneous PIV flow patterns with the delay t in range from 0 μs to 150 μs show that the interaction of the gas-dynamic perturbation caused by the spark discharge, characterized by the presence of an area of increased pressure relative to the static pressure of the main supersonic flow, leads to deformation of the separation zone and to the displacement of the separation point downstream. Apparently, the initiation of a spark discharge within the separation zone is not advisable, because will lead to even greater increase of pressure in the separation region and to the appearance of a positive pressure gradient in the boundary layer upstream.

Acknowledgments

The reported study was funded by RFBR according to the research project № 18-38-00803.

References

References

[1] Moreau E 2007 *J. Phys.* D*: Appl. Phys* **40** 605–36
[2] Bletzinger P, Ganguly B N, Van Wie D and Garscadden A 2005 *J. Phys. D: Appl. Phys.* **38** R33
[3] Bityurin V A, Efimov A V, Kazanskiy P N, Klimov A I and Moralev I A 2014 *High Temp.* **52 4** 483–89
[4] Correale G, Popov I B, Ratikin A E, Starikovskii A Y, Hulshoff S J and Veldhuis L L M 2011 *Proc. 49th AIAA Aerospace Sciences Meeting including the New Horizons Forum and Aerospace Exposition (Orlando FL, January 2007)* AIAA 2011-1079
[5] Leonov S B, Houpt A W and Hedlund B E 2017 *Proc. 48th AIAA Plasmadynamics and Lasers Conf. AIAA AVIATION Forum (Denver CO)* AIAA 2017-3673
[6] Moralev I, Sherbakova V, Selivonin I, Bityurin V and Ustinov M 2018 *Int. J. Heat Mass Transfer* **116** 1326–40
[7] Kopiev V F, Kazansky P N, Kopiev V A, Moralev I A and Zaytsev M Yu 2017 *J. Phys.* D*: Appl. Phys.* **50** 475204

[8] Elias P Q and Castera P 2013 *J. Phys.* D*: Appl. Phys.* **46** 365204
[9] Golub V V, Saveliev A S, Sechenov V A, Son E E and Tereshonok D V 2010 *High Temp.* **48 6** 903–9
[10] Raffel M, Willert C E, Wereley S and Kompenhans J 2007 *Particle Image Velocimetry. A Practical Guide* (Berlin: Springer)

IOP Conf. Series: Journal of Physics: Conf. Series **1112** (2018) 012015 doi:10.1088/1742-6596/1112/1/012015

Stimulation of instability of flow of endothermic reacting and ionized gas past curved surfaces by sonic waves

A S Baryshnikov

Ioffe Institute, Politechnicheskaya 26, Saint-Petersburg, 194021, Russia

al.bar@mail.ioffe.ru

Abstract. Stability of plane-parallel and weak curved chemically reacting gas flow is studied. It is shown that for endothermic processes instability is possible both for high-frequency disturbances and low frequencies. Instability of flow is not associated with viscous properties of matter and has a baroclinic character. Comparison with experiment and with previous similar theoretical investigations is made. Instability of flow with ionization in air can be stimulated by an external ultrasonic radiation. Research is important for control the transfer coefficients in gas during the transition to turbulent flow and for control the flow and aerodynamics of wrapping.

1. Introduction

It is well known that turbulent flow in many ways much more preferable than laminar. So, for example, it is improving the profile shape due to later detachment of the turbulent flow wrapping around the bodies. Flows inside the channels have more filled profile, and heat transfer on a wall is much higher.

This is connected with the fact that in turbulent flow transport coefficients (viscosity, thermal conductivity) is higher than in laminar [1]. However, it is important that turbulent flow would be developed, i.e. a characteristic scale, wavelength of perturbation would be low and the perturbation frequency would be high.

Study of influence of chemical processes in gas on flow stability and especially endothermic processes such as dissociation and ionization emerged from experimental and theoretical study of instability of flow behind bow shock wave in some polyatomic gases [2,3].

In the process of studying this effect it was understood, that in addition to influence of chemical processes on the effect, influence of several other factors should be taken into account. The first of these is the curvature of the current, because for fairly large curvature flow instability will happen even in the ideal gas.

The second factor is reduction of the ratio of specific heats of gas, as for specific heats equal to 1 the flow and shock wave become unstable [4]. Of course, perturbations wavelength should be taken into account also because for some wavelengths instability can manifest itself, while for others - can't.

Boundary conditions on the surface of solids, such as surface roughness certainly affect the stability of gas movement, as in other tasks of mathematical physics. However, in the problem stated the internal cause of flow instability explores, for example, state of the matter, chemical reactions in it. In the investigated problem it is important to find out exactly how the internal processes may affect the appearance of instability. Setting recalls Orr - Summerfield task of flow instability owing to viscous properties of gas [5]. However accounting boundary conditions only will complicate the understanding of the causes of instability.

2. Setting of the problem

The proposed formulation of the problem considers plane layer of gas flow. The transverse distribution of gas dynamic parameters and their transverse derivative in it, as well as the power of release or absorb energy through chemical-physical reactions are characteristic parameters, changing from task to task. Boundary conditions are lacking.

Viscosity of the gas is ignored so, because its accounting could complicate understanding, as accounting boundary conditions. Thus, the internal causes of instability are explored, which are not related to the influence of boundary conditions and viscosity. In this formulation, it is naturally to explore dispersion equation. This problem, of course, is easier than the general setting, with accounting the boundary conditions [6,7,8].

Incorporation of boundary conditions corresponds to the incorporation of the transverse velocity gradient in the layer, which occurs due to the zero velocity of the flow at the border. Without taking into account the transverse gradient, thus, emergence of instability cannot be realized. Curvature flow in persistent Cartesian coordinates also leads to a deviation from the vertical velocity profile. Therefore parameter of crookedness should be proportional to the deviation of transverse derivative speed profile from zero.

Of course, flow curvature should be weak so flat-parallel flow approximation could be applied for. This condition is feasible if the member that contains the parameter of crookedness will be a lot less than the rest members of the equations. It is clear that wavelength of perturbations must be among the task parameters.

Finally, the influence of the course of physical-chemical processes in gas on the emergence of the instability is most important for researched problem factor. The greatest influence on gas flow these processes have through heat exchange between internal and kinetics degrees of freedom, i.e. through gas heating or cooling as a result of physical-chemical processes. Therefore, the study should take into account the speed of release (release power) or absorption of heat as a result of processes in gas, more precisely, the partial derivatives of the power on temperature and density of gas.

In the below setting task is linearized by the curvature (by the transverse gradient of velocity profile), because investigation of a large curvature involves the necessity of direct numerical simulation of occurrence of instability, which today is nearly impossible in real flows [9].

3. Used equations

The problem statement is based on equations derived in [10]. It is known that the instability of plane-parallel flows occurs when the Reynolds number Re is big enough and the source system of linearized hydrodynamics equations can be written in a shorter form [10]

$$\rho[i(U-c)f+U'\phi]=-\left(\frac{i}{\gamma M^2}\right)\pi$$

$$\rho i\alpha^2(U-c)\phi=-\left(\frac{i}{\gamma M^2}\right)\pi'$$

$$i(U-c)r+\rho'\varphi+\rho(if+\varphi')=0$$

$$\rho[i(U-c)\theta+T'\phi]=-(\gamma-1)(if+\varphi')P+\theta\frac{dQ}{dT}+r\frac{dQ}{d\rho}$$

$$\frac{\pi}{P}=\frac{r}{\rho}+\frac{\theta}{T}$$

Derivatives denoted with strokes are taken on the variable y; ρ is the density of the gas, U is the velocity of the gas, p - pressure, T is temperature, c is the speed of sound; M is the Mach number of the flow, γ is the ratio of specific heats, μ is dynamic shear viscosity; f,φ,r,θ,π-amplitude of enthalpy, gas flow rate, density, temperature and pressure respectively.

One takes into account derivatives of the volumetric capacity of heat source (or heat absorption) as on the temperature and gas density - $dTQ, d\rho Q$. Crookedness of current lines is modeled by specifying the speed gradient across layer for flat-parallel currents - U'. Transverse direction y is selected inside the layer, i.e., the direction of propagation of pressure perturbations. With this choice of coordinate system convex flow is modeled by positive gradient U', and concave - negative. The original system of equations for small perturbations of the flow is reduced, as it is known [6,7], to differential equation of the second order:

$$\pi'' = \pi' \frac{2U'}{U-c} + \pi\alpha^2 \left\{1 - \frac{(U-c)^2\gamma M^2}{T} \cdot \frac{\alpha(U-c)\rho + id_T Q}{\gamma\alpha(U-c)\rho + id_T Q(1-d)}\right\} \quad (1)$$

Here pressure perturbations $\delta p = \pi(y)\exp[i\alpha(x-ct)]$ and $T(y), U(y), \rho(y), P(y)$ - distributions of temperature, velocity, density, and pressure across the layer; c is the speed of sound in the gas, α and β - wave numbers of disturbances along the flow (x-coordinate) and transverse the flow (y coordinate). M - Mach number, calculated through characteristic values of U and $T, d = (\rho d_\rho Q)(T d_T Q)^{-1}$; i - the imaginary unit. As in the theory of boundary layer stability [10,11], in this work it is accepted that instability occurs when $Im\delta p > 0$. Solution of (1) is looked up as usual as follows [10,11]: $\pi_0 \exp(i\beta y)$. So$\pi' = i\beta\pi_0 \exp(i\beta y)$, $\pi'' = -\beta^2\pi_0 \exp(i\beta y)$.

Substituting these expressions as well as expressions for π in equation (1) and reducing the exponent, we get:

$$-\beta^2 = \left(\frac{2U'\alpha\rho}{iZ}\right) i\beta + \alpha^2 - \left(\frac{\gamma M^2}{T\rho^2}\right)(iZ)^2 \left[\frac{iZ + d_T Q}{\gamma iZ + id_T(1-d)}\right],$$

$$\delta p = \pi(y)\exp[i\alpha(x-ct)] = \pi(y)\exp[i\alpha(x-c)t] = \pi(y)\exp[(ZKs\rho^{-1})t].$$

So, the study on stability is reduced to dispersion equation of 4-th degree relative $Z = i\alpha\rho\,(U - c)(Ks)^{-1}$:

$$Z^4 + \kappa Z^3 + Z^2 + \left[\frac{\kappa(1-d)}{\gamma} + \mathrm{g}\right] Z + \mathrm{g}\frac{\kappa(1-d)}{\gamma} = 0 \quad (2)$$

that depends on the 4 parameters: $\gamma, d = (\rho d_\rho Q)(T d_T Q)^{-1}$ $(K^2 = \alpha^2 + \beta^2, s^2 = P\rho M^{-2} = \gamma P^2 U^{-2})$. Weak crookedness of flow means that $|\mathrm{g}| \leq 0.1$. I.e. wavelength of perturbations must be more than 20 times less than the typical distance of influence different areas of flow each to other, that justifies the using of the local flat-parallel setting in the problem statement.

4. Results of computations.

Complex roots of the equation of the fourth degree (2) were calculated by the method of Ferrari. Four complex roots were obtained in the calculations. However, three of them are trivial-obvious.

One solution is proved to be more informative. For endothermic physical-chemical processes with normal dependence of reactions speed on the pressure and temperature $\kappa > 0, 1 > d > 0$ and for convex currents area of stability ($ReZ < 0$) are realized for average values of κ (figure 1). Reduction of γ and d or increasing of curvature parameter g as expected leads to a decrease in the area of stability. Calculations showed that the equation (2) for the specified parameters has a "wave" solution (2 complex conjugate solution) and 2 purely real solutions, one of which $\approx -\mathrm{g}$ for any κ, γ and d (it could be called the "gradient" solution). The width of the field of stability depends on values of d, γ and g. Increase in γ, and d (for constant g) leads to expand the range of stability. A particularly sharp increase can occur if d is little. For instance, stability range increases almost in 2 times with the growth of γ from 1.2 (polyatomic gases) to 1.4 (air) when $d \leq 0.1$. For zero g, as can be seen from figure 1, instability disappears.

It should be noted that the instability of the flow has inviscid nature and does not require that the second derivative of the velocity profile equals zero. This means that instability is not barotropic and has baroclinic nature [5].

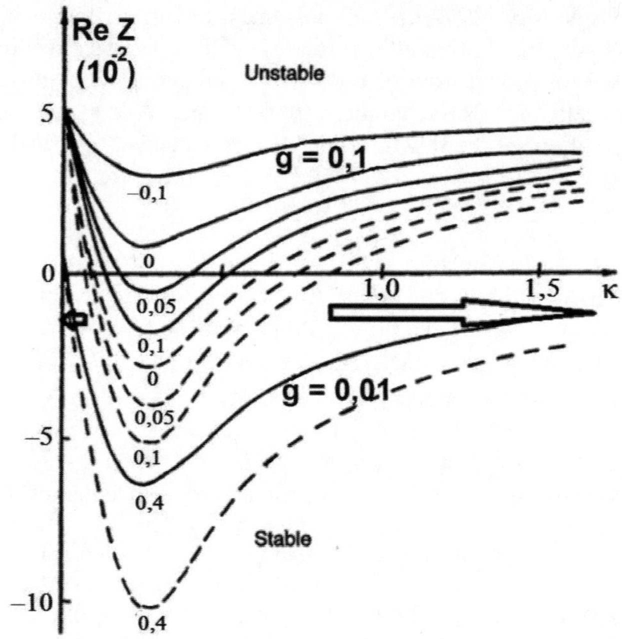

Figure1 Real part of complex roots of equation (2) as function of dimensionless perturbations wavelength κ in depending parameters d, γ and g. Number under curves are values of d. Solid curves $-\gamma = 1.2$. Dashed lines $-\gamma = 1.4$. Instability corresponds $Re(Z) > 0$. Shaped white arrows show areas of wavelength κ, for which unstable gas motion is realized in the plane-parallel weak curved flows for g $= 0.1$ and $d = 0.1$ as well as for g $= 0.01$ and $d = 0.4$.

The results of the calculations are consistent with calculations of L M Mack [6] and G V Petrov [7] that have nothing to do with crookedness flows. In those calculations, the gradient of the transverse profile speed was greater than zero due to the boundary conditions. In these works, the instability of plate-parallel flow of chemically dissociating gas around a plate is discovered when posting boundary conditions. They are second and third modes according to the classification of L M Mack [6] and not the primary viscous mode of perturbations. Perturbation increments are consistent with increments in the above presented calculations.

5. Evaluation of instability regimes

The equation parameters should be expressed through the basic values of medium. The parameter $d = (\rho d_p Q)(T d_p Q)^{-1} = n/u$ is proportional to the reaction order, n, and is inversely proportional to the relative energy of activation, $u = \Delta E(R_g T)^{-1}$ because absorption of energy Q is proportional to the rate of reaction $Q = \left(\Delta E\left(R_g\right)^{-1}\right) A\rho'' \exp(-u)$, where ΔE is the warmth of one act of reaction, R_g - gas constant. The parameter κ is also easy to find: $\kappa = d_T Q(Ks)^{-1} = \left(M(P\rho)^{-1/2}\right) A u^2 \rho^n e^{-u} \lambda$ $(K\,\lambda^{-1}, \lambda$ is wavelength of perturbations, $s = (P\rho)^{1/2}(M)^{-1}$, M is the Mach number of the flow and A – preexponent of constant of reaction.

Consider plasma flow in the neck of the air intake. The radius of curvature of the surface R_k equals 0.17 m and g $= 0.01$. It appears from calculations for g $= 0.01$ that instability occurs when $\kappa < 0.02$ or when $\kappa > 10^4$. For weak shock waves in the neck (flow Mach number $M \approx 0.5$) pressure $p \approx 6 \cdot 10^5$ Pa, density $\rho \approx 0.1$ kg·m^{-3}, the temperature of the atoms and ions $T_a \approx 10^3$°K, and the temperature of electrons $T_e \approx 1.2 \cdot 10^4$°K. At this regime, the ionization process goes through associative ionization, involving electronically-excited levels [12]:

$$N + O + 2{,}8\text{ ev} \rightarrow NO^+ + e$$

$$k_r = 0.43 \cdot 10^{11} \exp\left(-\frac{32535}{T}\right)\ \mathrm{m^3(kg\ s)^{-1}}$$

It could be considered that ΔE equal to difference between associative ionization energy of NO and electronically-excited levels energy of $O_2(a_1\Delta)$, so $u = \Delta E(R_g T)^{-1} \approx 22.3$, and for binary reaction parameter $d \approx 0.1$. Then: $\kappa \approx$ "B" $\cdot M\lambda$ [λ – in meters], g $\approx$ "C" $\cdot M^2\lambda/R_k$. Coefficient "*C*" depends on the direction of propagation of the perturbation in the layer of the flow. It is maximal at the direction of 45° to the direction of flow. Herewith "C" = 1.41. At sound speed 600 м с$^{-1}$ behind the shock wave, for reaction constant of associative ionization k_r and density $\rho = 0.1$ кг м$^{-3}$ coefficient "B" is equal to 0.75 m^{-1}.

Thus, the instability of the air plasma flow in the neck of the air intake should occur for short waves (wavelength $\lambda < 0.05$ m, frequency $f > 8$ kHz), which is close to the boundary of the range of ultrasound in air (20 kHz). Another boundary corresponds to infrasound vibrations and rise increment (ReZ) for it grows very slowly with increasing wavelength, as can be seen from figure 1 (right branch).

Ultrasonic disturbances in gas and plasma rapidly damped, so to stimulate shortwave instability one should use external exposure by ultrasonic waves.

Consider the flow instability behind bow shock wave when gaseous CF_4 flows around segmental body [2]. Constant of decomposition of CF_4 can be found in [13]:

$$CF_4 + M \rightarrow CF_3 + F + M;$$

$$K_{f1} = 6.15 \cdot 10^{31} T^{-4.64} \exp\left(-\frac{122421}{RT}\right)\ \mathrm{cm^3(mol \cdot s)^{-1}}$$

The radius of curvature of blunting segmental body $R_k = 0.04$ m, so $g = 0.1, \Delta E = 122$kcal·mol^{-1}. Then constant "B" equals $1.64 \cdot 10^5$ m^{-1}. Gas temperature $T = 1800$ K, gas density behind the shock wave is equal to 8 kg·m^{-3}. The parameter $u = 30.6$. The process of dissociation is of the second order in density, so the *d* parameter is 0.066. For $g = 0.1$ of high-frequency instability corresponds to $\lambda < 1.2 \cdot 10^{-5}$ m (frequency $3 \cdot 10^8$ Hz or more) that is unrealistic.

More realistic case is low-frequency instability. For $g = 0.1, d = 0.1, \gamma = 1.2$ according to figure 1 border instability: $\kappa = 5$ (wavelength $\lambda = 5 \cdot 10^{-5}$ m). But for this λ we have $g = 4.4 \cdot 10^{-4}$ and border instability $\kappa = 1000$, not value 5 estimated for $g = 0.1$. It is easy to verify that the maximum value of wavelength λ where g will remain constant, equal to 10 mm, which corresponds to the observed perturbations in the experiment [2]. The parameter κ depends both on the wavelength and on the parameter g, therefore, reducing of the wavelength border will increase border of *g*. Unstable wave length of perturbations must be defined for each task by selection of appropriate parameters, if this selection will converge.

6. Conclusions.

One of the solutions to the dispersion equation of disturbances in flat-parallel weak curved reacting gas flow shows the instability for endothermic reactions, as for shortwave disturbances and for longwise disturbances, which disappear with the disappearance of crookedness.

The detected effect depends on the ratio of specific heats of gas and on the curvature of the streamlined surfaces, as well as on the speed of release or absorption of energy due to physical and chemical processes. Instability has baroclinic nature, and for example, flows of air plasma unstable at high-frequency of disturbances, order 8 kHz or more, regardless of whether there is a bend of speed profile or not ($U'' = 0$), as it is required for the barotropic instability in the theory of Orr-Summerfield.

Because the attenuation of ultrasonic frequency perturbations in plasma is great, stimulation of plasma flow instability needs to be realized through the outside influence of ultrasonic waves.

As possible application it should be noted three positions. All of them are only hypothetical opportunities. The first is rising of compression in hypersonic intake, because shock wave in neck should be unstable due to instability of plasma flow. The second is possibility of reduction of

IOP Conf. Series: Journal of Physics: Conf. Series **1112** (2018) 012015 doi:10.1088/1742-6596/1112/1/012015

aerodynamic drug due to the later separation of boundary layer of turbulent plasma flow. And third is an improvement of efficiency of plasma-chemical technologies due to more uniform conditions of turbulent plasma flow across the section of tube and greater thermo flux to reactor walls.

References

[1] Loitsyanskii L G 1987 *Mechanics of Liquids and Gases, Int. Series of Monographs in Aeronautics and Astronautics: Division II: Aerodynamics* ed. R T Jones, W P Jones (Oxford: Pergamon)

[2] Baryshnikov A S, Basargin I V, Bobashev S V, Monakhov N A, Popov P A, Sakharov V A and Chistyakova M V 2016 *J. Eng. Phys. Thermophys.* **89** 1232

[3] Baryshnikov A S 1996 *Tech. Phys.Lett.* **22** 667

[4] Gurevich L E and Rumyantcev A A 1970 *J. Exp. Theor. Phys.* **58** 1395

[5] Monin A S 1986 *Sov. Phys. Usp.* **29** 843

[6] Mack L M 1975 *AIAA J.* **13** 278

[7] Petropv G V 1979 *Development of perturbations in boundary layer* (Novosibirsk: ITPM, Siberia Dep. of RAS) p104 [in Russian]

[8] Molevich N E 1999 *Izvestia RAS, Mechanics of fluids and gases* №5 82 [in Russian]

[9] Schumann U, Grötzbach G, and Kleiser L 1980 *Prediction methods for turbulent flows* ed. V Kolman (Washington, Hemisphere hublishing corporation) p103

[10] Gaponov C A and Maslov A A 1980 *Development of perturbations in compresstd fluxes* (Novosibirsk: Nauka) [in Russian]

[11] Schlichting H 1974 *Grenzschicht-Theorie.* (Karlsruhe: Verlag G.Braun)

[12] Gladkov A A, Polyanskiy O Yu, Agafonov V P, Vertushkin V K 1972 *Nonequilibrium physicochemical processes in aerodynamics* ed. G I Maykapara (Moscow: Mashinostroenie) [in Russian]

[13] Modica A P and Sillers S J 1968 *J. Chem. Phys.* **48** 3283

IOP Conf. Series: Journal of Physics: Conf. Series **1112** (2018) 012016 doi:10.1088/1742-6596/1112/1/012016

Effect of Electric Discharges in Magnetic Field on Hypersonic Flow around Bodies

V P Fomichev [1,2] and M A Yadrenkin [1*]

[1] ITAM SB RAS, Novosibirsk, Russia
[2] NSU, Novosibirsk, Russia

*yadrenkin@itam.nsc.ru

Abstract. To investigate the possibilities of magnetohydrodynamic (MHD) effects on a hypersonic flow structure, it is necessary to ionize the gas flow. It is possible to create a local region of conductivity of the flow using electrical discharges under experimental conditions. The paper considers the MHD interaction of a high-speed air flow and electrical discharges (high-voltage pulse and high-frequency) in a homogeneous magnetic field when flowing around test models. It is shown that the use of discharges in a magnetic field makes it possible to significantly change the shock-wave structure of the flow near a streamlined surface: to change the angle of the oblique shock, to generate new shock, to transform attached oblique shock to the bow shock. Systematization of the obtained data on the value of the hydromagnetic interaction parameter S characterizing the strength character of the observed effects is performed. Thus, $S < 0.05$ corresponds to the weak MHD-interaction, $S = 0{,}05 - 0{,}15$ – moderate MHD-interaction, $S >$ 0.15 – strong MHD-interaction.

Introduction

The control of the aerodynamics of advanced hypersonic vehicles is a crucial task requiring new approaches and principles. Under conditions of hypersonic flight for the realization of effective MHD interaction become advantageous. This way of controlling the structure of the flow avoids the use of massive and complex mechanical systems for changing the shape of the aerodynamic surfaces of the hypersonic jets, and also reduces the response time of the system to control commands. Various experimental and numerical studies have shown the effectiveness of this type of interaction: from a weak deviation of the oblique shock waves to the generation of new and even a total reorganization of the shock-wave structure of the flow [1-15].However, so far there are practically no attempts to classify the observed effects by some characteristic value. To estimate the force component of the MHD effect, this parameter can be the Stewart number, which shows the ratio of the work of volumetric electromagnetic forces to the work of the inertial forces of the high-speed flow. In the paper, an example of the realization of MHD interaction near models of various geometries and corresponding values of the Stewart parameter are presented.

Problem statement

The flow near a plate, a wedge and a blunt body with a hypersonic airflow M=6-7 is considered. Figure 1 shows the experimental schemes for the test models.

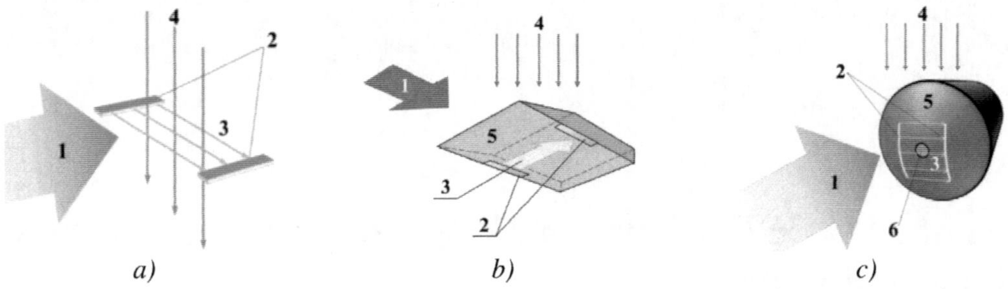

Figure 1 Experimental schemes for the investigation of the local MHD-interaction. 1 – hypersonic flow, 2 – electrodes, 3 – electrical discharge, 4 – magnetic field, 5 – model, 6 – pressure sensor.

Flow parameters: speed about 2 km / s, density 0,02 kg / m3, temperature 250 K. The discharge current between the electrodes is directed across the air flow and across the magnetic field so that the force determined by the product of the electric current density value by the magnitude of the magnetic induction, was directed upstream for the local gas braking.

MHD test rig and experimental techniques

To simulate the MHD interaction in a hypersonic flow, MHD test rig, based on a shock tube, is used (Figure 2). After the diaphragm break, the gas from the high-pressure chamber *1* compresses the working gas in the low-pressure channel *3*, and behind the reflected shock quasistationary parameters of the working gas at the supersonic nozzle inlet are formed. The test rig allows simulating high-speed flows of various gases with Mach number $M = 6 - 10$. When using air, the flow parameters are simulated correspond to conditions at an altitude of 30-50 km. The time of existence of the quasi-stationary mode is about 1 ms. The working chamber 5 is located inside a powerful electromagnet 6, generating a constant magnetic field of 0-2.5 T. It is also possible to install various ionization devices for the flow. In the experiment, the ionization of the flow was carried out using a high-voltage pulse discharge and a radio-frequency (RF) discharge. To generate a pulsed discharge, a long line of capacitors was used, and for a RF discharge a special generator was used. To observe the structure of the flow near the model, an optical schlieren-system with an adaptive visualizing transparency is used instead of Foucault knife [16], That approach allows recording density gradients at low density of the working gas. Recording of images is carried out on a high-speed camera.

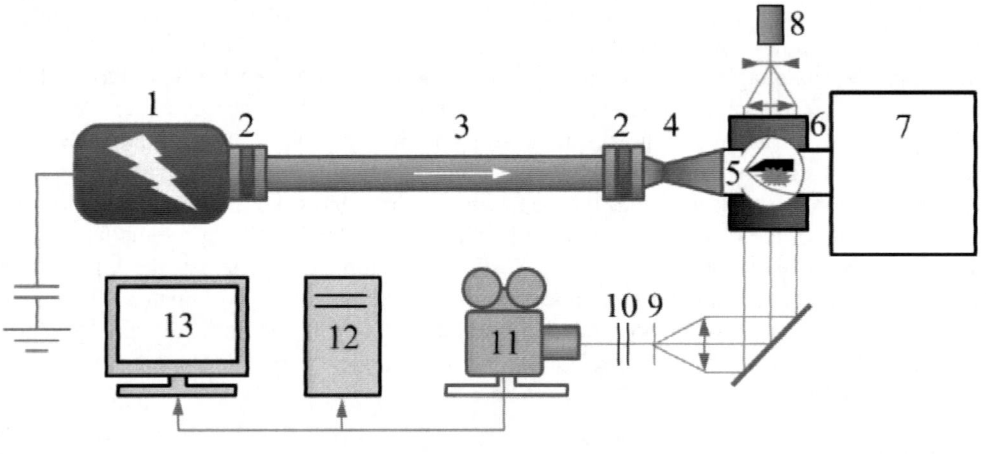

Figure 2 Test rig with optical system: 1 – high pressure chamber, 2 – diaphragm, 3 – low pressure channel, 4 – nozzle, 5 – working chamber, 6 – electromagnet, 7 – receiver, 8 – laser, 9 – AVT, 10 – filters, 11 – camera. 12 – diagnostic section, 13 – PC.

IOP Conf. Series: Journal of Physics: Conf. Series **1112** (2018) 012016 doi:10.1088/1742-6596/1112/1/012016

Experimental study of the pulse discharge in a hypersonic flow

When a discharge was initiated between two narrow electrodes (Fig.1a), a volumetric electromagnetic force appeared in the hypersonic air flow with $M = 6$ and in a magnetic field, directed toward the incoming flow. As a result of the MHD interaction at $B \geq 1$ T, the arc of the discharge moved toward the flow and the appearance of a head shock wave in front of the interaction region (Figure 3). This can be explained by local gas deceleration as a result of strong MHD interaction. The duration of the electric discharge is 120 μs. During this time, a quasistable region of the inhibited flow has time to form, but in general the observed process can not be considered stationary.

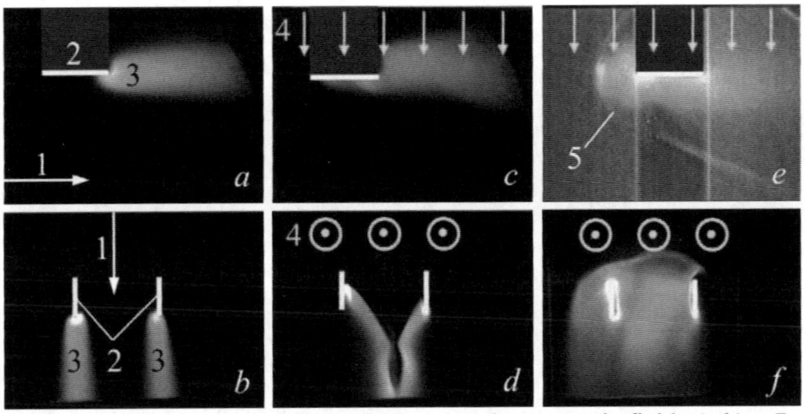

Figure 3 Pulsed electrical discharge in a flow of $M = 6$ and a magnetic field. a), b) – $B = 0$ T, c), d) – $B = 0.34$ T, e), f) – $B = 2$ T; 1 – flow direction, 2 – electrodes, 3 – discharge plasma, 4 – magnetic field direction, 5 – bow shock.

Experimental study of the MHD interaction near the plate model

When the MHD interaction is realized on the plate surface (figure 1*b*) at similar test conditions, the initiation of the discharge does not occur in the free flow, but between the electrodes flush with the model surface in the region behind the attached shock wave generated by the leading edge of the model (Figure 4*a*).

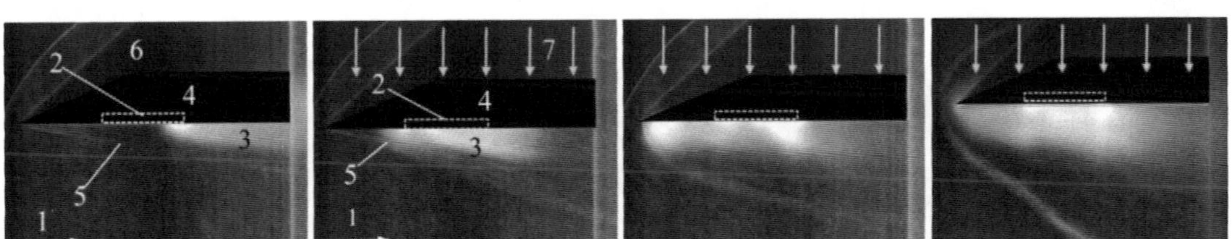

a) *b)* *c)* *d)*

Figure 4 Pulse discharge on the plate surface in the $M = 6$ flow and in a magnetic field. *a)* – $B = 0$ T, *b)* – $B = 0.1$ T, *c)* – $B = 0.34$ T, *d)* – $B = 0.8$ T; 1 – flow direction, 2 – electrodes, 3 – discharge plasma, 4 – model, 5 – shock wave, 6 – shield plates, 7 – direction of magnetic field.

In this case, even when the magnetic induction reaches 0.1 T, it is possible to stabilize the position of the arc of the discharge in the region of the electrodes, while the attached shock deflects to a larger angle (figure 4*b*). And if the value of the magnetic induction is 0.3 T and higher, the discharge protrudes toward the leading edge and the shape of the attached shock significantly changes (figure 4*c*). At a magnitude of the magnetic induction above 0.6 T, the attached shock transforms to the detached bow shock, since the discharge arc is pulled out towards the flow in front of the model under the action of electromagnetic forces. It should be noted that at $B > 0.6$ T, pulsations in the position of the shock are observed and the current and voltage pulsations are associated with this. This process is similar to the shunting process in plasmatrons with a self-setting arc length. The shunting frequency was about 35 kHz under the test conditions, and the speed of the arc element's movement in upstream direction was up to 700 m/s at a flow rate of about 2000 km/s.

When the RF discharge with a frequency of 900 kHz was using for the local ionization on the plate model, the discharge burning time was 300 μs. In this case, the process of flowing occurs differently. Unlike a pulsed discharge, the RF discharge has not a force effect on the flow, but only support a sufficient level of ionization. Force action occurs as a result of the appearance of an induced current in the discharge circuit as a result of the motion and deceleration of a conducting gas closed to electrodes in a magnetic field.

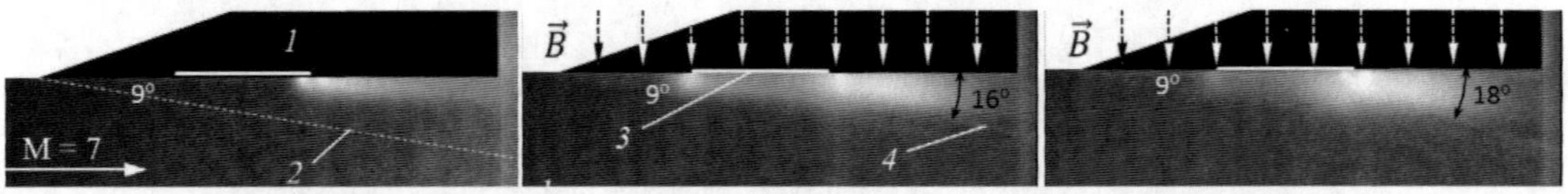

Figure 5 Generation of the oblique shock on the MHD-region with RF-discharge: 1 – plate model, 2 – oblique shock generated by the leading edge, 3 – electrodes, 4 – oblique shock formed on the MHD interaction region.

At $B \geq 0.9$ T, the appearance of a new shock wave in the region of the MHD interaction, is observed (figure 5*b*). As B-field induction was increasing, the shock angle was also increasing (figure 5*c*). In Figure 6 shows the dependence of the shock wave angle on the magnitude of the magnetic field at flow Mach number $M = 7$. A similar effect is observed when using a pulsed discharge, but at very low value of $B \sim 0.05$ T. This is explained by the significant difference in the current value of electric discharges obtained from various sources.

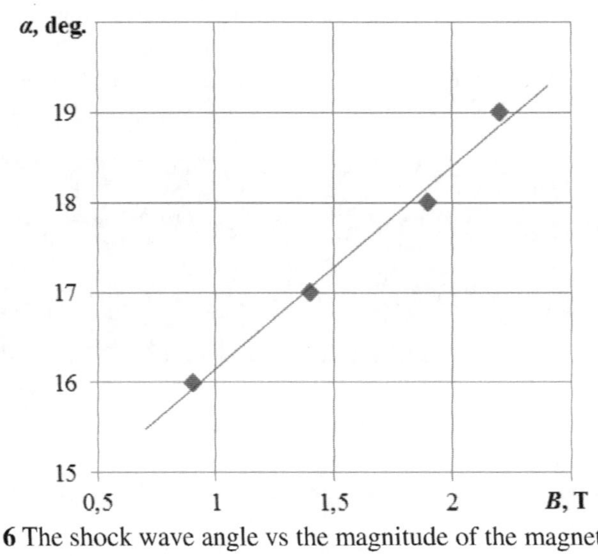

Figure 6 The shock wave angle vs the magnitude of the magnetic field.

Experimental study of the MHD interaction near the blunt body

A similar configuration of the electrodes was mounted in the surface of the model of the blunt body (the model of the reentry capsule). Then when the discharge is initiated in a magnetic field $B > 1$ T in a hypersonic airflow, the discharge plasma concentrates in front of the model will be observed and the head shock wave will move upstream (Figure 7*c*).

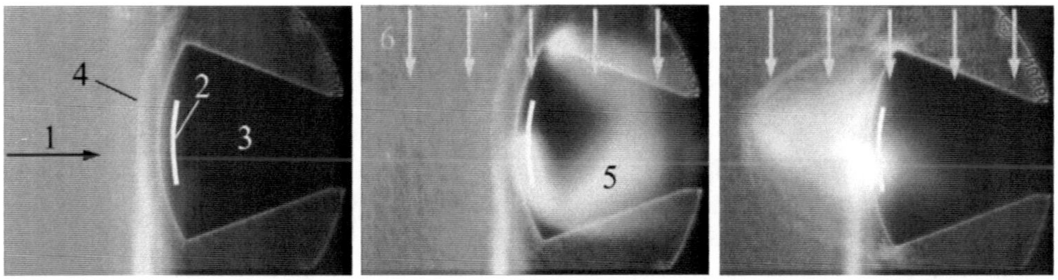

a) *b)* *c)*

Figure 7 Pulse discharge in front of the blunt body in the flow $M = 6$ and in a magnetic field: a) $B = 0$ T, b) $B = 0.34$ T, c) $B = 2.2$ T; 1 – flow direction, 2 – electrodes, 3 –model, 4 – bow shock, 5 – discharge, 6 –B-field direction.

Methods for determining the parameter of hydromagnetic interaction

The effectiveness of the force effect of plasma in a magnetic field on the structure of the incoming flow is most conveniently expressed in terms of the ratio of the work of electromagnetic forces to the work of inertia forces (the Stuart number):

$$S = \frac{\vec{j} \times \vec{B} l}{\rho_\infty v_\infty^2}$$

where $\vec{j} \times \vec{B}$ is the volume electromagnetic (the product of the current density by the magnitude of the magnetic induction), l is the characteristic size along the march coordinate, $\rho_\infty v_\infty^2$ is the work of inertia forces. To determine the parameter S under test conditions, it is necessary to adapt it to the measured characteristics of the process.

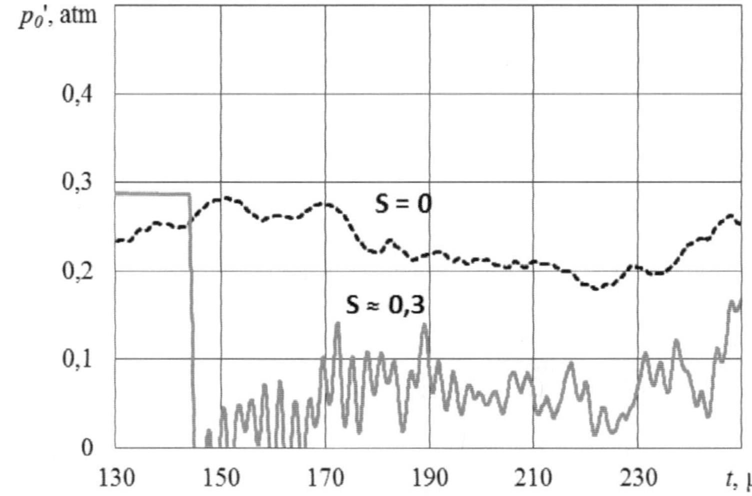

Figure 8 Data from the pressure sensor installed at the critical point on the model of a blunt body without MHD interaction (S = 0) and for strong MHD interaction (S ≈ 0.3)

When analyzing the data of experimental studies of a pulsed discharge in a hypersonic air flow on free electrodes and on a plate model, the parameter S was estimated from the total discharge current I and the size of the interaction region across the flow and along the magnetic field b:

$$S = \frac{I B}{\rho_\infty v_\infty^2 b}$$

IOP Conf. Series: Journal of Physics: Conf. Series **1112** (2018) 012016 doi:10.1088/1742-6596/1112/1/012016

In the case of using a high-frequency discharge, it is difficult to determine the magnitude of the induced current. Therefore, it was assumed that all work of electromagnetic forces is expressed in the work of pressure forces Δp, which leads to the formation of an oblique shock with an appropriate angle of inclination:

$$S = \frac{\Delta p}{\rho_\infty v_\infty^2}$$

A similar approach is applicable when analyzing the data of experimental studies of MHD interaction on a blunt body. At the critical point of the model, a high-frequency piezoelectric pressure sensor was installed, which fixed the pressure drop Δp behind the bow shock (Figure 8).

Discussions and conclusions

Experiments have shown that by varying the discharge current and the magnitude of the magnetic induction at a constant flow rate, a weak, moderate, and strong MHD effect on the flow can be realized. Thus, $S < 0.05$ corresponds to the weak MHD-interaction, $S = 0{,}05 - 0{,}15$ – moderate MHD-interaction, $S > 0.15$ – strong MHD-interaction.

With a relatively weak effect, $S \approx 0.05$, the angle of inclination of the attached oblique shock, generated by the leading edge of the model, changes.

With a moderate MHD interaction, increasing S to 0.15 results in generation of a new shock wave and an increase in the angle of inclination of the attached oblique shock generated by the leading edge and a change in its shape.

A further increase in S to 0.2-0.25 leads to the formation of a detached shock wave in front of the leading edge of the model or to the removal of the bow shock wave from the surface in the case of a model of a blunted body - a strong MHD interaction. At $S > 0.2$, shunting of the discharge in the electrode region was observed, which leads to oscillations in the shock position.

It have shown that different types of electric discharges can be used to efficiently change the shock-wave structure of flow around various bodies in a hypersonic flow and in a magnetic field.

A strong MHD interaction at $S > 0.2$ leads to the appearance of a bow shock both in the case of free electrodes flowing and in the case of flowing various bodies with electrodes built into their surface. For comparable values of the parameter S, the observed shock-wave structure of the flow practically no depends on the shape of the model (Figure 9).

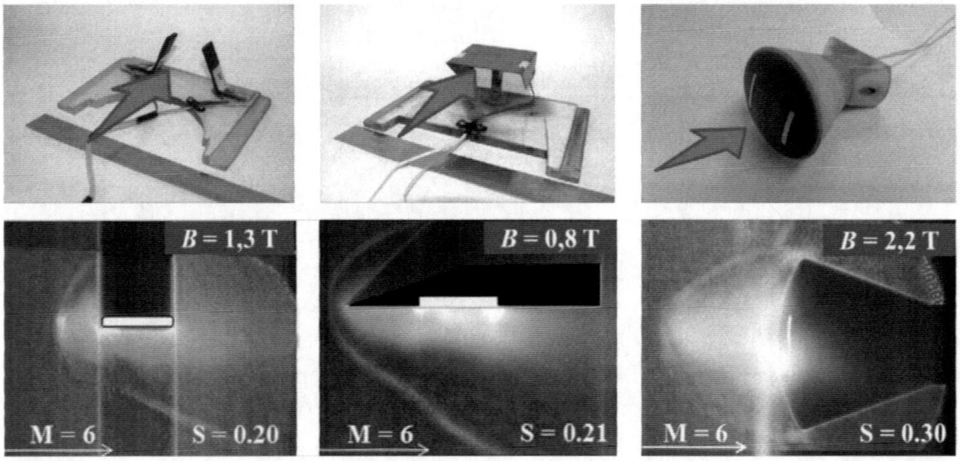

Figure 9 Photos of the shock waves at strong MHD-interaction near different systems of electrodes

This paper shows the principal possibility of using electric and magnetic fields to control the aerodynamics of aircraft moving in the upper layers of the atmosphere at a hypersonic speed. Obviously, the maximum efficiency of the method can be achieved under conditions of natural thermal ionization at the surface of the apparatus, since the energy will be spent only on the creation of a magnetic field. However, in some cases it is worth considering the use of electrical discharges for solving local problems.

Acknowledgments
The study was carried out at the expense of a grant from the Russian Science Foundation (project No. 17-19-01375).

References

[1] Bityurin V A, Botcharov A N, Lineberry J MHD Flow Control in Hypersonic Flight *AIAA Pap.* 2005-3225
[2] Bityurin V A, Baranov D S, Bocharov A N, Lineberry J 2003 MHD interaction at a cylinder in hypersonic flow *Abstracts: 5th Workshop on Magnetoplasma Aerodynamics* ed. V.A. Bityurin (Moscow: JIHT RAS)
[3] Bisek N J, Boyd I D, Poggie J 2010 Numerical Study of Magnetoaerodynamic Flow Around a Hemisphere *J. Spacecr. Rockets* **47** **5** 816-27
[4] Cristofolini A, Borghi C A, Neretti G., Schettino A, Trifoni E 2012 Experimental investigations on the magneto-hydro-dynamic interaction around a blunt body in a hypersonic unseeded air flow *J. Appl. Phys.* **112** 093304
[5] Shneider M N, Macheret S O, Miles R.B 2002 Nonequilibrium Magnetohydrodynamic Control of Scramjet Inlet *AIAA Pap.* 2002-2251.
[6] Brichkin D L, Kuranov A L, Sheflda E G 1999 The Potentialities of MHD Control for improving Scramjet Pereformance *AIAA Pap.* 1999-4969.
[7] MacCormack R W 2008 Numerical Simulation of Aerodynamic Flow Including Induced Magnetic and Electric Fields *AIAA Pap.* 2008–4010.
[8] Bobashev S V, Erofeev A V, Lapushkina T A, Poniaev S A, Vasil'eva R V, Van Wie D M 2003 Experiments on MHD Control of Attached Shocks in Diffuser *AIAA pap.* 2003-0169
[9] Korotaeva T A, Fomichev V P, Shashkin A P, Yadrenkin M A 2001 Investigation of magnetohydrodymamic interaction in a supersonic air flow at M=8 *Tech. Phys.* **56** **3** 327-34 DOI: 10.1134/S1063784211030108
[10] Fomichev V P, Yadrenkin M A 2013 Experimental investigation of the magnetohydrodynamic parachute effect in a hypersonic air flow *Tech. Phys.* **58** **1** 144-7 DOI: 10.1134/S1063784213010052
[11] Fomichev V P, Yadrenkin M A 2013 Pulsations of location of the bow shock as a result of strong MHD interaction during the hypersonic flow around the plate *Tech. Phys. Lett.* **39** **1** 68-70 DOI 10.1134/S1063785013010082
[12] Fomichev V P, Yadrenkin M A 2013 Investigation of a high-frequency discharge initiated across a hypersonic air flow and a magnetic field *Tech. Phys. Lett.* **39** **1** 71-3. DOI 10.1134/S1063785013010094
[13] Fomichev V P, Yadrenkin M A 2013 MHD-Control of a Shock-Wave Structure at a Hypersonic Flow *5th European Conference for Aeronautics and Space Sciences. Munich, 1-5 July 2013, Abstracts*
[14] Fomichev V P, Yadrenkin M A, Shipko E K 2016 Effect of volumetric electromagnetic forces on shock wave structure of hypersonic air flow near plate *AIP Conf. Proc.* **1770** **1** 030068. DOI: 10.1063/1.4964010
[15] Fomichev V P; Yadrenkin M A 2017 The Structure of a Hypersonic Air Flow near a Plane Surface at Various Intensities of Magnetogasdynamic Interaction *Tech. Phys. Lett.* **43** **12** 1063-6 DOI: 10.1134/S1063785017120057
[16] Pavlov A A, Pavlov Al A, Golubev M P 2008 Use of AVT for gas flow visualization *14th International Conference on the Methods of Aerophysical Research (ICMAR-2008): Proc.* Pt.1, (Novosibirsk: Parallel) p. 92-3

IOP Conf. Series: Journal of Physics: Conf. Series **1112** (2018) 012017 doi:10.1088/1742-6596/1112/1/012017

Flow characteristics in the water-cooled channel of the MHD accelerator

D S Baranov[1], V A Bityurin[1], A N Bocharov[1], S S Bychkov[1*], V A Grushin[1], N V Tretyakova[1], N I Batura[2], E B Vasilevsky[2] N G Zhurkin[2] and N M Kolushov[2]

[1]JIHT RAS, Moscow, Russia
[2]TsAGI, Zhukovsky, Russia

*serg-bychkov@yandex.ru

Abstract. Many problems related with the development of a hypersonic vehicle and development of new thermal protection materials are directly related to the possibility of conducting a wide range of laboratory researches under conditions that are close to the real flight conditions as possible. In this connection, the work on the formation of high-enthalpy and high-speed gas flows in ground-based installations is of special importance. It is shown in this paper that the flow characteristics required can be achieved using the experimental setup with a magnetohydrodynamic accelerator after its updating. The feature of this research facility is that more than half of the electric energy supplied from electric circuit is converted directly into the kinetic energy of the flow. Preliminary analytical and experimental studies of the MHD accelerator with a water-cooled channel allow one to conclude that a hypersonic flow with duration up to 10 s is formed at the SMGDU path. The total enthalpy in the core of the flow is close to 20 MJ/kg, while a required high level of total pressure is being maintained. The facility with such flow parameters becomes unique and allows one to significantly expand the range of ground-based testing, for example, testing of thermal protection systems, and testing of novel contactless flow control methods for hypersonic flight conditions.

Introduction

The hypersonic flight control and thermal protection of a hypersonic vehicle surface is one of the topics of special interest. Apart from traditional tools for vehicle trajectory control, the plasma technologies become more and more attractive. Attractiveness of plasma technologies relies on their feasibility to change flow characteristics by means of electric and magnetic field in so called non-intrusive manner. The latter implies that no control elements are subject to direct flow action. In addition to feasibility for flow control, there are two other topics related with applying of on-board magnet: namely, surface heat load mitigation and direct conversion of kinetic energy of the hypersonic flow to electric power. To study all these possibilities in ground-based facilities close-to-real flight conditions should be created, which is a problem of great challenge. The free-stream velocities should be as high as 6 – 8 km/s, the specific enthalpy of the flow should exceed 10 MJ/kg, the time duration of test run should exceed 1s (1-10 s).

The TsAGI Magneto Hydro Dynamic (MHD) Wind Tunnel Facility has been created to meet the needs for hypersonic flow conditions as close as possible [1,2]. For example, the free-stream flow velocities in the range of 8 – 15 km/s have been reached in earlier studies. Later, in 1999-2006, principal studies on magnetic field flow control have been carried out on this facility. The results of these studies are summarized in [3]. While high-velocity flow has been realized, the drawback of the facility is small duration time of the run (0.5-0.7s). The current efforts are intended to increase this time and make the facility suitable for both MHD flow control research and experimental studies on thermal protection systems. The upgraded facility is thought to exceed some key parameters achieved on currently the most powerful Facility PWT-Scirocco [4-6], in which the hypersonic flow (Mach number – 6, total pressure is up to 6 Bar, specific enthalpy is up to 20 MJ/kg) is generated by arc heater. The objectives of the current efforts on TsAGI MHD Facility is to obtain flow with total pressure up to 10 Bar, specific enthalpy more than 20 MJ/kg, and duration time up to 10s.

The high velocity, high enthalpy flow is achieved by means of using a) preliminary heating of the flow by the arc rotating in magnetic field, and b) consequent magneto hydrodynamic acceleration of the flow followed by the nozzle.

In the current paper, the first experimental results obtained on the upgraded facility are presented. The physical and computational model developed to analyze the flow in the facility duct and in the test section is formulated. The preliminary numerical results on MHD flow in the facility duct are given.

Experimental data

The segmented Faraday-type MHD channel is the unit that experiences the highest electrical and thermal loads. For this reason, the duration of the supersonic flow was limited by 0.7 seconds in the standard mode. The research potential of the setup can be significantly improved by means of the using a new channel with water cooling. For this purpose, a test section of the channel was developed and manufactured in the JIHT RAS. In new channel, the copper electrodes and side walls are cooled by water. Some characteristics of both channels are given in Table 1.

Table 1. Characteristics of MHD channel of TsAGI MHD Wind Tunnel Facility.

Type of channel	Number of electrode pairs	Electrode width, mm	Insulator gap, mm	Cross-section of the channel at the inlet/outlet, mm
standard	35	5	3	15x15/25x15
water-cooled	21	10	3	15x15/25x15

Figure 1 shows the results of electrical parameters measurements for two runs of TsAGI MHD Wind Tunnel Facility with channels of different design. The number of electrode sections is indicated on the abscissa axis. The red lines correspond to the run of standard channel for 0.7 s. The blue lines correspond to the run of the water-cooled channel for 3 s. Both runs were conducted under the same conditions. The working gas was air. The gas flow rate was 12.4 g/s. The weight fraction of K-Na seed was 1%. The electrodes were loaded with the same voltage sources. It can be seen from the figure that the distributions of the quantities measured for two channels are close in form but differ in absolute values. The difference is explained by the fact that in the water-cooled case the magnetic field inside the MHD accelerator was weaker due to the design features. Thus, total inputted electrical power reduced to 21 electrode pairs was 270 kW for standard channel and 180 kW for water-cooled one, respectively. Series of runs with this power operation mode was carried out for water-cooled channel, and the time of trouble-free operation of the MHD accelerator was extended from 0.7 to 10 seconds.

IOP Conf. Series: Journal of Physics: Conf. Series **1112** (2018) 012017 doi:10.1088/1742-6596/1112/1/012017

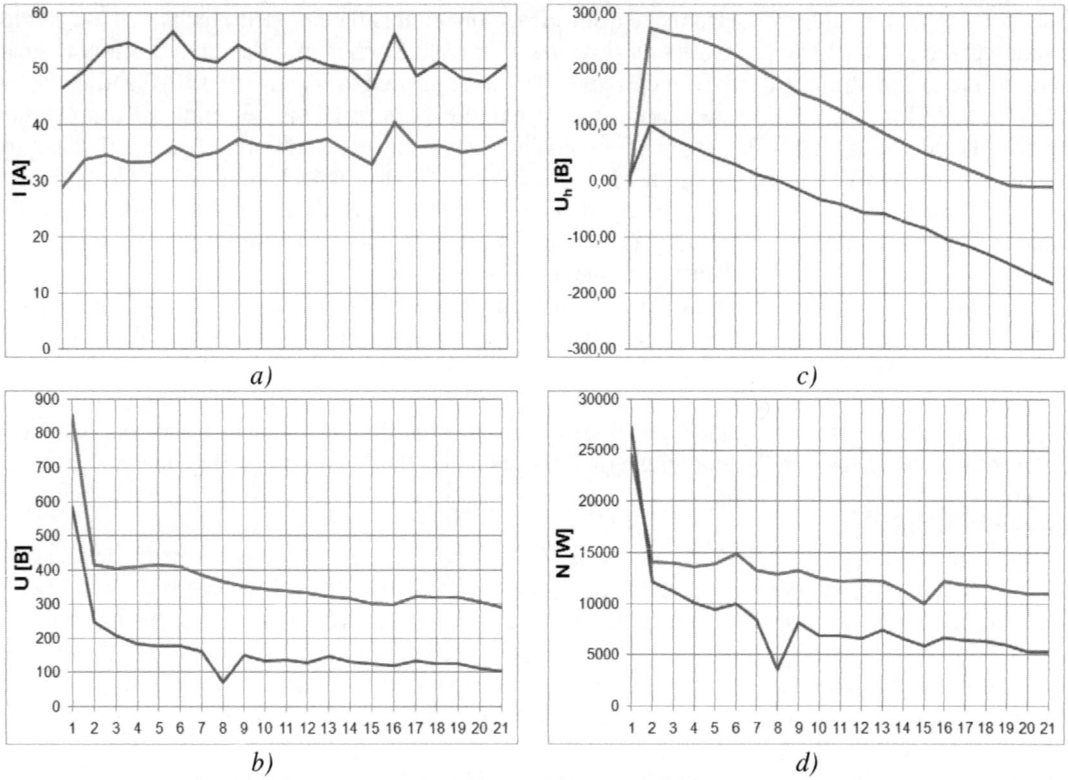

Figure 1 Comparison of electrical characteristics of two channels.

a) – electrode current, A

b) – electrode Faraday voltage, V

c) – electrode Hall voltage, V

d) – electrical power input, W

Computational model

The layout of the TsAGI MHD Wind Tunnel Facility is presented in Fig.2. 250 kWt arc heater provides a supersonic flow at the inlet of MHD acceleration section. To provide functioning of MHD accelerator NaK eutectic is injected.

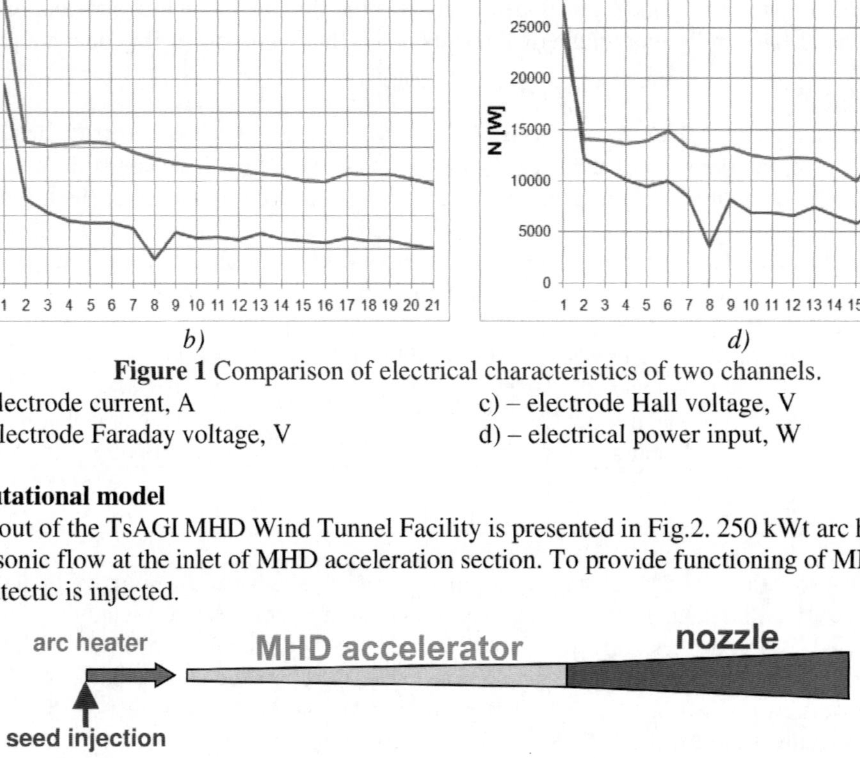

Figure 2 Layout of the flow in TsAGI MHD Wind Tunnel Facility duct.

Some parameters of the Facility and flow are given below. MHD accelerator inlet cross-section is 15 ×15 mm, outlet cross-section is 25 ×15 mm (extension in Y direction), length of MHD accelerator (in flow direction, X) is 550 mm, nozzle outlet cross-section is 70 ×100 mm, nozzle length is 350 mm. The flow characteristics at the MHD accelerator inlet are taken to be as follows:

static pressure is 25 kPa,
static temperature is 2500 K,
velocity is 2000 m/s,
Mach number is 2. (1)

Chemical composition in mole fractions:

$$x_{N2} = 0.78, x_{O2} = 0.20,\ x_e = x_{K+} = 0.0133\,, \quad (2)$$

IOP Conf. Series: Journal of Physics: Conf. Series **1112** (2018) 012017 doi:10.1088/1742-6596/1112/1/012017

We shall consider the flow of chemically reacting air with the seed. So, seven chemical substances are assumed to take part in the chemical conversions: N_2, O_2, NO, N, O, K, K^+, e (electron). Reduced chemical kinetics of Park [9] is used to simulate the flow in the duct.

Governing equations

To numerically simulate the flow in the facility duct, consider the following set of governing equations.

$$\frac{\partial \rho}{\partial t} + \nabla(\rho \mathbf{U}) = 0 \quad (3)$$

$$\frac{\partial \rho \mathbf{U}}{\partial t} + \nabla(\rho \mathbf{U}\mathbf{U}) + \nabla \boldsymbol{\tau} = -\frac{\partial P}{\partial \mathbf{r}} + \mathbf{J} \times \mathbf{B} \quad (4)$$

$$\frac{\partial \rho e^0}{\partial t} + \nabla(\rho \mathbf{U} h^0) + \nabla(\mathbf{U}\boldsymbol{\tau}) + \nabla \mathbf{q} = \mathbf{JE} \quad (5)$$

$$\frac{\partial \rho_i}{\partial t} + \nabla(\rho_i \mathbf{U}) + \nabla \boldsymbol{\Gamma}_i = \dot{\omega}_i, \quad (6)$$

$$\frac{\boldsymbol{\partial}}{\boldsymbol{\partial}\mathbf{r}} = \boldsymbol{\eta}_x \frac{\boldsymbol{\partial}}{\boldsymbol{\partial} x} + \boldsymbol{\eta}_y \frac{\boldsymbol{\partial}}{\boldsymbol{\partial} y}, \; \boldsymbol{\nabla} = \frac{\boldsymbol{\partial}}{\boldsymbol{\partial} x} + \frac{1}{y^\xi}\frac{y^\xi}{\boldsymbol{\partial} y} \quad (7)$$

$$h^0 = e^0 + \frac{P}{\rho}, e^0 = e + \frac{U^2}{2}, \quad (8)$$

$$h = e + \frac{P}{\rho} = \sum Y_i h_i \,, h_i(T) = h_{i,f} + \int_{T_{ref}}^{T} C_{pi}\, dT \quad (9)$$

$$P = \sum P_i, P_i = \rho_i R_i T, R_i = \frac{R^0}{W_i} \quad (10)$$

Equations (3) – (5) govern the transport of total mass, momentum and total specific energy. Equation (6) specifies the transport of the mass of each chemical species. The transport equations are closed with the equations of state, (8) – (10). In equations (3) – (10) ρ is density, $\mathbf{U}$ is velocity, P is pressure, T is translational temperature, e^0 is total specific energy, e is internal specific energy, h^0 is total specific enthalpy, h is internal specific enthalpy, ρ_i is the density of i-th chemical species ($\rho = \sum \rho_i$), $\dot{\omega}_i$ is i-th species rate of production in chemical conversions, Y_i is mass fraction of i-th species, $Y_i = \frac{\rho_i}{\rho}$, P_i is i-th species partial pressure, h_i is i-th species internal enthalpy, C_{pi} is i-th species heat capacity at constant pressure. Operators in (7) are gradient and divergence operator, respectively. $\boldsymbol{\eta}_x$, $\boldsymbol{\eta}_y$ denote basic unit vectors, index $\xi = 0$ corresponds to cartesian coordinate system, and $\xi=1$ corresponds to cylindrical one.

Electrodynamic part of the whole system is assumed to describe by the following equations including Maxwell's equations for quasi-steady electrodynamics and generalized Ohm's law.

$$\boldsymbol{\nabla}\mathbf{J} = \mathbf{0}, \mathbf{E} = -\frac{\boldsymbol{\partial\varphi}}{\boldsymbol{\partial}\mathbf{r}} \quad (11)$$

$$\mathbf{J} + \frac{\boldsymbol{\beta}}{\boldsymbol{B}}(\mathbf{J} \times \mathbf{B}) - \mathbf{2}(\mathbf{1} - \boldsymbol{s})^{\mathbf{2}}\frac{\boldsymbol{\beta\beta_i}}{\boldsymbol{B^2}}(\mathbf{J} \times \mathbf{B} \times \mathbf{B}) = \boldsymbol{\sigma}(\mathbf{E} + \mathbf{U} \times \mathbf{B}) \quad (12)$$

In equations (11) – (12) $\mathbf{J}$ is electric current denisty, $\mathbf{E}$ is electric field strength, φ is scalar electric potential, σ is electric conductivity, $\mathbf{B}$ is magnetic flux density (magnetic induction), β is Hall parameter for electrons, β_i is ionic Hall parameter, s is ionization degree . Electromagnetic field effects on the flow through electromagnetic force $\mathbf{J}\times\mathbf{B}$ (Equation 4) and through the energy yield $\mathbf{JE}$ (equation 5). Equations (11), (12) can be reduced to partial differential equation for electric potential [6].

The fluxes in equations (3) – (6) are determined as follows. Mass diffusion flux and heat flux read as:

$$\boldsymbol{\Gamma}_{\mathrm{i}} = -\rho D_i \frac{\partial Y_{\mathrm{i}}}{\partial r}, \mathbf{q} = -\lambda \frac{\partial T}{\partial \mathbf{r}} + \sum h_i \boldsymbol{\Gamma}_i \quad (13)$$

Here, D_i is effective diffusion coefficient of i-th species, λ is heat conductivity of the mixture. The viscous stress tensor is specified as follows:

$$\tau_{ij} = \frac{2}{3}\eta \delta_{ij} \nabla \mathbf{U} - \eta\left(\frac{\partial U_{\mathrm{i}}}{\partial \mathrm{x_j}} + \frac{\partial U_{\mathrm{j}}}{\partial \mathrm{x_i}}\right) \quad (14)$$

In (14) δ_{ij} is Kronecker symbol. The transport coefficients D_i, λ and η are calculated in accordance with the model presented, for example, in papers [9, 10]. Electric conductivity is determined by the following relationships:

$$\boldsymbol{\sigma = e\mu_e n_e,\ \mu_e = e\frac{\tau_e}{m_e} \equiv \frac{\beta}{B},\ \tau_e = \nu_e^{-1},\ \nu_e = \sum \nu_{ej},\ \nu_{ej} = n_j u_e Q_{ej}} \quad (15.1)$$

$$\boldsymbol{\mu_i = e\frac{\tau_i}{m_i} \equiv \frac{\beta_i}{B},\ \tau_i = \nu_i^{-1},\ \nu_i = \sum \nu_{ij},\ \nu_{ij} = n_j u_i Q_{ij}} \quad (15.2)$$

Here, e is electron charge, m_e and m_i are electron and ion mass, respectively, b_e and b_i are electron and ion mobilities, ν_e and ν_i are mean electron and ion collision frequencies, n_j is the number density of particles of j-th kind, u_e and u_i are mean thermal electron and ion velocities, Q_{ej} and Q_{ij} are collision cross-section for collisions of electrons and ions with particles of j-th kind.

Chemical kinetics model

The rate of production of i-th chemical component due to chemical reactions in gaseous phase is given as:

$$\dot{\omega}_i = W_i \sum_{r=1}^{N_r} \left[\chi_{i,r}^{"} - \chi_{i,r}^{'}\right]\left(k_{fr} \prod_{l=1}^{N} c_l^{\chi_{l,r}^{'}} - k_{br} \prod_{l=1}^{N} c_l^{\chi_{l,r}^{"}}\right) \quad (16)$$

$$c_i = \frac{\rho_i}{W_i}, k_{fr} = a \cdot T^b \cdot \exp(-E_a/T) \quad (17)$$

Here, c_i is molar concentration of i-th component, W_i is its molecular mass. $\chi_{l,r}^{'}, \chi_{l,r}^{''}$ are stoichiometric coefficients of reagents and products in r-th reaction. k_{fr} и k_{br} are forward and backward reaction rate constants, respectively; a, b, E_a are reaction parameters, N_r is total number of reactions, N is total number of components. For thermal kinetics, E_a is activation energy. $T_w = 1800K$

Initial and boundary conditions

The boundary conditions for transport equations (3) – (6) are set as follows. The supersonic conditions (1), (2) are specified at the inlet boundary. At the outlet boundary (nozzle exit), all variables are extrapolated from the interior. At the wall the temperature is assumed to given, $T_w = 1800K$, corresponding to water-cooled wall. Initial conditions are not needed when steady-state solutions are sought. However, they are required in unsteady calculations. Typically, solution obtained for some conditions of external electric loading is used as initial one when these external electric loading is changed.

In general, the boundary conditions for electrodynamics equations (11), (12) may be non-local (functional) and time-dependent. The inlet and outlet boundaries are considered as insulating boundaries. Here zero normal current is set to zero, $J_n = \mathbf{Jn} = 0$. The same conditions are set at the insulators separating electrodes. Electric potential is specified at the electrode boundaries. The value of potential at each electrode is calculated from the relationships determining external electric loading of the MHD channel. Currently, the electrode currents are assumed to be known as functions of time. Taking this into account, the boundary conditions for electrodynamic problem read as follows (17), (18):

$$\varphi = P_k \quad (17)$$

Electrode potential, P_k, are sought from the relations below:

$$\int \mathbf{Jn} dS = I_k(t), k = 1, \ldots, N_E \quad (18)$$

In (18) $I_k(t)$ is the total current into k-th electrode, t is time, N_E is the number of electrodes. We consider the case of water-cooled channel, for which $N_E = 42$, i.e. 21 electrodes on both bottom and top wall. Relationships (18) is one of the simplest manner to specify boundary conditions for MHD channel electrodynamics. As a role, more sophisticated relations are required to link electrodynamic characteristics in the channel and those in external circuits. The solution technique applied to solve for electrodynamic problem allows one to specify any physically reasonable relations between the currents and potentials.

Solution technique

The complete set of governing equations is solved using different techniques for different equations. Explicit time-integration technique is used to advance flow solution in time. The source-terms are treated as averaged over time-step value determined by gasdynamic characteristic times rather than chemical

IOP Conf. Series: Journal of Physics: Conf. Series **1112** (2018) 012017 doi:10.1088/1742-6596/1112/1/012017

ones. This implies that chemical kinetics equations (16, 17) are integrated over gasdynamic time-step using implicit stiff ODE solver.

At every time-step electrodynamics problem (11), (12), (17), (18) is solved with multi-grid solver [11]. LU-factorization is used as a driving method on each iteration of three-level flexible-cycle multi-grid solver, and direct method is applied on the coarsest grid. Acceleration technique can be applied due to the manner, in which the external circuit conditions are treated. This technique allows a stable solution of the whole equation set even when electrodynamic equations are solved only at every n-th time-step. Typically, n varied from 10 to 100, which depends on the external circuit conditions.

Discretization of governing equations is carried out with control-volume method on the structured computational grid. Inviscid fluxes are calculated with several AUSM-techniques [12, 13], which provide low-diffusion carbuncle-free solutions. Viscous fluxes in equations (3) through (6) are calculated with the scheme analogous to central-difference one (see, for example, paper [14]). A bit different discretization scheme is used in solving electrodynamic problem to accurately take into account Hall effect.

Preliminary numerical results on the flow in MHD Wind Tunnel Facility

First estimations of capabilities of upgraded facility were carried out in papers [15 – 17]. Here, we shall consider preliminary numerical results obtained with the computational model discussed above. Consider the MHD channel which characteristics are listed in Table 1. The electric loading is presented by 21 electrode pairs, which realizes the old scheme, i.e. Faraday-type MHD channel is considered. Every section of each pair consists of 7 mm electrode and 3 mm insulator. For the first time, 50 A current per electrode section was specified. Total current is 1050 A. Magnetic field was specified in a simplest manner: $\mathbf{B} = (0,0,B)$. Distribution of B is shown in Fig.3 or Fig.4. The distributions of wall potentials and current stream functions are shown in Fig.3 and Fig.4, respectively. It was assumed that the nozzle is maintained at grad potential. Horizontal parts of potential curves correspond to equipotential boundaries of electrodes. The stream function curve represents the current distribution in the channel. Notice that stream function is defined as: $\frac{\partial\psi}{\partial x} = -J_y, \frac{\partial\psi}{\partial y} = J_x$.

Integral of stream function over any line segment is the total current passing through this segment.

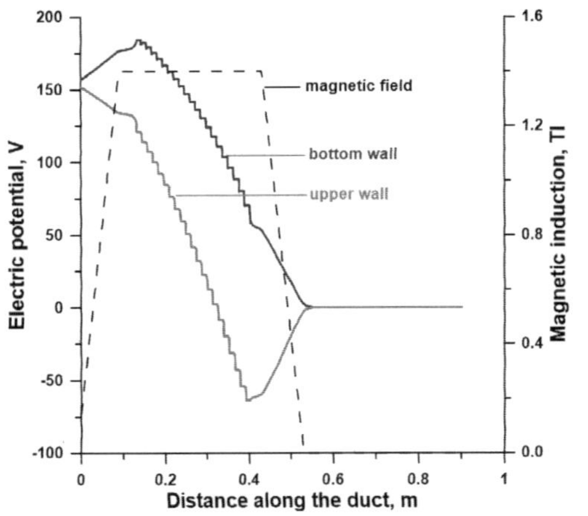

Figure 3 Distribution of electric potential (solid lines) over the bottom and top wall, distribution of magnetic induction along the duct axis (dashed).

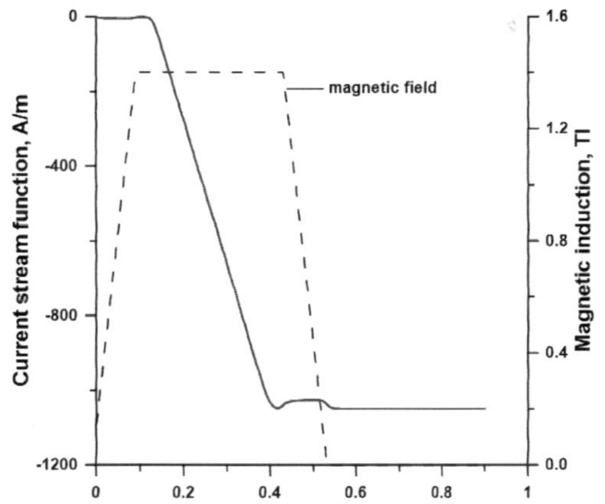

Figure 4 Distribution of electric current stream function (solid lines) along the duct axis (dashed).

The total power supplied to the MHD channel is determined by the expression $W = \int \mathrm{J}\mathrm{E}dV$. We consider an operation mode with moderate power supply $W = 106$ kW. The power of ponderomotive force $A = \int (\mathbf{J} \times \mathbf{B})\mathbf{U}dV$=80.6 kW. So, efficiency of MHD acceleration can estimated as $\eta_E = A/W =$

IOP Conf. Series: Journal of Physics: Conf. Series **1112** (2018) 012017 doi:10.1088/1742-6596/1112/1/012017

0.76. The operation mode just discussed is not an optimal one. The power supply is almost three times less than those allowed by the external circuit and desirable from the view point of hypersonic thermodynamics. Also, Hall effect seems to be significant, which diminishes the efficiency of MHD interaction. Further efforts could be done to optimize the external electric circuit.

Fig.5 presents the distribution of axial velocity and Mach number along the axis of the duct ($y = 0$). The effect of MHD interaction is exposed as drastic increase of these quantities in the region of MHD channel.

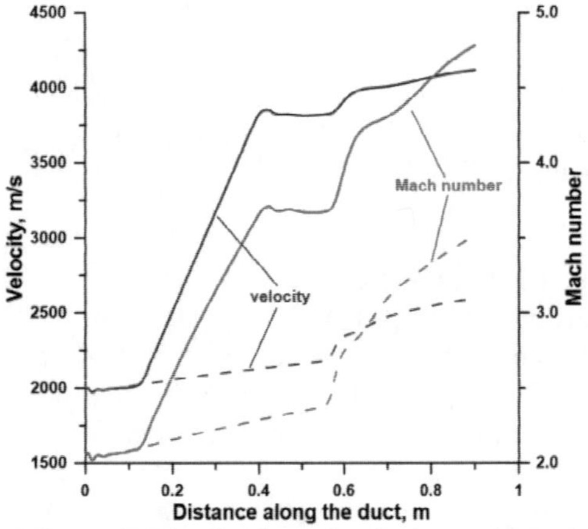

Figure 5 Distribution of velocity and Mach number along the duct axis. Solid lines – MHD on, dashed lines – MHD off.

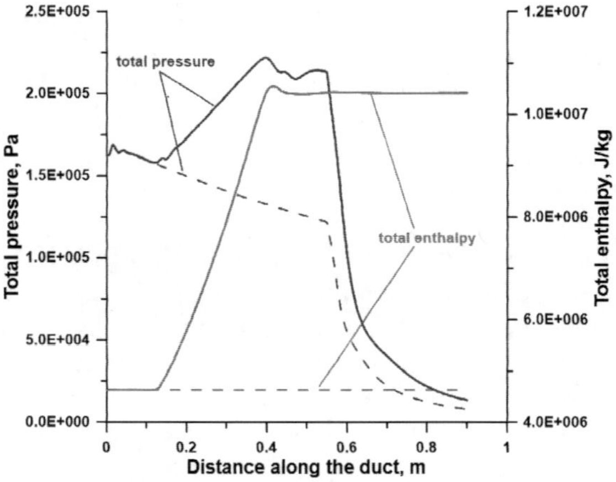

Figure 6 Distribution of total pressure and total enthalpy along the duct axis. Solid lines – MHD on, dashed lines – MHD off.

Distributions of total pressure and total specific enthalpy are shown in Fig.6. Even for non-optimal operation mode the total enthalpy reaches the value exceeding 10 MJ/kg, which is considered as the lowest one for hypersonic operation mode. Main characteristics of the flow at the nozzle exit station are presented in Fig.7 and Fig.8.

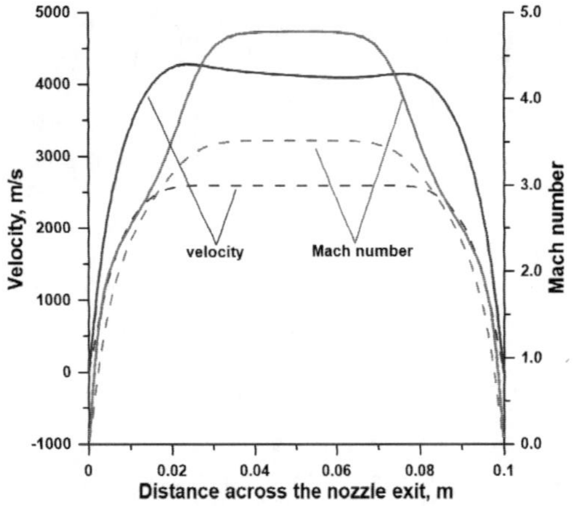

Figure 7 Distribution of axial velocity and Mach number across the nozzle exit. Solid lines – MHD on, dashed lines – MHD off.

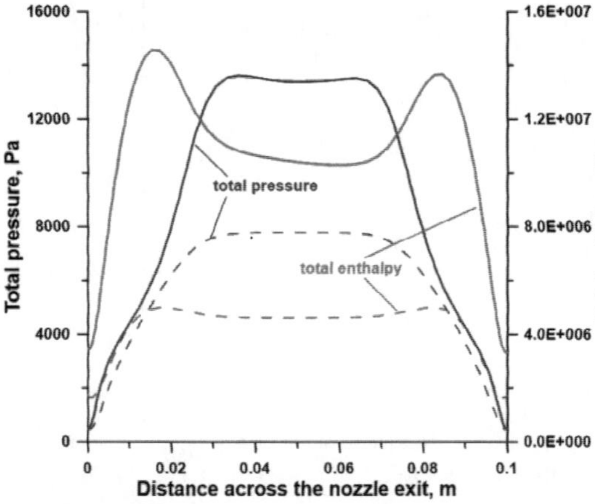

Figure 8 Distribution of total pressure and total enthalpy across the nozzle exit. Solid lines – MHD on, dashed lines – MHD off.

Finally, two-dimensional fields of basic flow quantities are presented in Figures 9 – 11

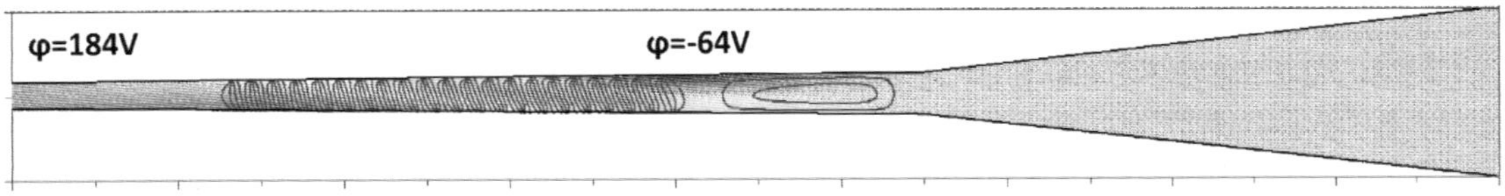

Figure 9 Current stream-lines (dark) and electric potential (filled). $B^* = 1.4$ Tl, $I_{tot} = 1050$A.

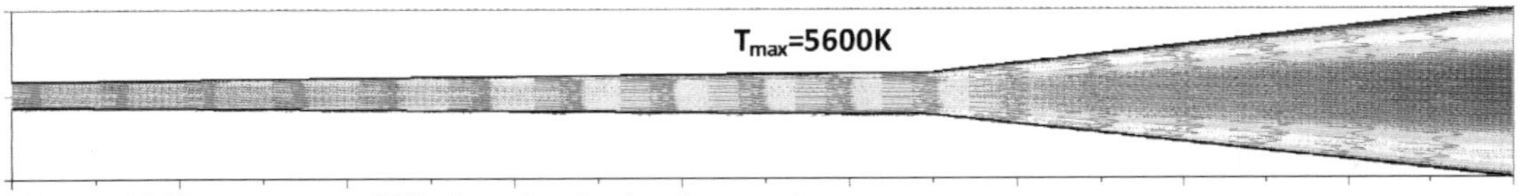

Figure 10 Temperature (filled) and velocity (vectors).

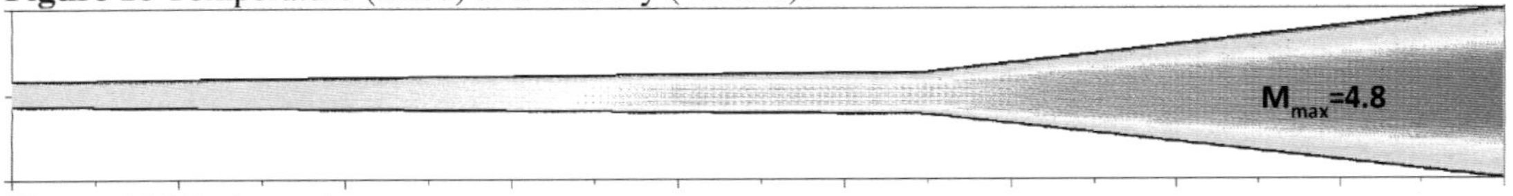

Figure 11 Mach number.

Concluding remarks

The following three circumstances could be noticed.

First. The flow in the MHD channel is not symmetric one. This is due to the presence of Hall current (or leakage current) directed against the flow. This leakage current generates the force directed from bottom to top, i.e. in the direction of main (Faraday) current. One of the means to prevent or to diminish the influence of the leakage currents (and to diminish the flow asymmetry) is to reconfigure the external circuit in favor of Hall-type or Montardi-type MHD unit. Another possible solution is to diminish Hall parameter, specifically in near-electrode regions, where Joule heating is especially high. This could be done by increasing the ionization degree. Increase in ionization degree leads to increase of mean electron collision frequency because of growing of the role of column collisions.

Second. The level of specific total enthalpy (see Fig.6 and 8, for example) is sufficient for required hypersonic flow. However, distribution of this quantity is quite non-uniform across the nozzle exit. This undesirable property of the flow is thought to be mitigated by optimization of the geometry and initial flow conditions.

Third. It can be seen from Figures 7, 8 that the hypersonic flow of interest takes place in the core of the flow which is of order of 3 – 4 cm in height. The near-wall layers, also rather thick, are both high-enthalpy and high- temperature ones (~6000K). We hope the new water-cooled channel will capable of managing with these hard thermal conditions.

These preliminary estimates show that the main result – generation of long-term high-enthalpy hypersonic flow by means of MHD acceleration – can certainly be achieved.

Acknowledgements

The work was carried out with the partial support of the Russian Foundation for Basic Research (project 18-08-00890 A).

References

[1] Alferov V I. 2000 Current Status and Potentialities of Wind Tunnels with MHD Acceleration *High Temp.* **38 2** 300-13.

[2] Alferov V I, Dmitriev L M, Egorov B V. et al. 2001 On the possibility of reproduction in the wind tunnels of the parameters of the hyperspeed flow and conditions in the combustion chamber of the SHVRD *High Temp.* **39 5** 743-47.

[3] Bityurin V, Bocharov A, Baranov D and Bychkov S 2007 Power Extraction Experiment with a Surface MHD Generator in Hypersonic Airflow *38th AIAA Plasmadynamics and Lasers Conference In conjunction with the 16th International Conference on MHD Energy Conversion, Miami, FL, June 25-28, 2007* AIAA-2007-3882.

[4] Scortecci F, Paganucci F, d'Agostino L, and Andrenucci M 1997 A new hypersonic high enthalpy wind tunnel *AIAA Pap.* 1997-3017

[5] Cristofolini A, Borghi C A, Carraro M R, Neretti G, Passaro A, G Fantoni G and Biagioni L 2008 Hypersonic MHD interaction on a conical test body with a Hall electrical configuration *IEEE Trans. Plasma Sci.* **36** **2** 530–41

[6] Cristofolini A, Borghi C A, Carraro M R, Neretti G, Schettino A, Trifoni E, Battista F, Passaro A and Baccarella D 2012 Experimental investigations on the magneto-hydro-dynamic interaction around a blunt body in a hypersonic unseeded air flow *J. Appl. Phys.* **112** 093304 DOI: 10.1063/1.4764105.

[7] Park C, Jaffe R I, Partridge H 2001 Chemical-Kinetics Parameters of Hyperbolic Earth Entry *J. Thermophys. Heat Transfer* **15** **1** 76-90

[8] Bityurin V A, and Bocharov A N 2006 Magnetohydrodynamic interaction in hypersonic flow around a blunt body *Fluid Dyn.* **41** **5** 188-203.

[9] Zubkov A I, Tirskiy G A, Levin V A, Sakharov V I 1998 Motion of bodies in Earth's and Planet's Atmospheres with Supersonic and Hypersonic Velocities under Conditions of Chemical Conversions, Heat Transfer and Radiation *Rep. No.4507, Institute of Mechanics of Moscow State University* (in Russian).

[10] Andriatis A V, Zhluktov S A, Sokolova I A 1996 Transport Coefficients for Chemical Nonequilibrium Components of Air Mixtures *Mat. Model.* **4** 1 44-64.

[11] Hatchinson B R, Raithby G D 1986 A multigrid method on the additive correction strategy Numerical Heat Transfer **9** 511-537.

[12] Wada Y and Liou M-S 1997 An Accurate and Robust Flux Splitting Scheme for Shock and Contact Discontinuities *SIAM J. Sci. Comput.* **18** **3** 633-57.

[13] Liou M-S 1996 A Sequel to AUSM: AUSM+ *J. Comp. Phys* **129** 364-82.

[14] Barth T J, and Jesperson D C 1989 The Design and Application of Upwind Schemes on Unstructured Meshes *AIAA Pap.* 1989-0366

[15] Bityurin V A, Bocharov A N, Bychkov S S, Baranov D S, Grushin V A, Kusmartseva E V, Zhurkin N G, Boushmin A S, Khandyrin A V 2016 Experimental and Theoretical Study of MHD Interaction at Flat Plate in Supersonic Flow *Abstracts of the 15th International Workshop on Magneto-Plasma Aerodynamics, Moscow, April 19-21, 2016*, p.70-1.

[16] Bityurin V A, Bocharov A N, Ivanov P P 2016 Optimization of TsAGI MHD accelerator operation mode for TPS experimental tests *15th International Workshop on Magneto-Plasma Aerodynamics, Moscow, April 19-21, 2016,* p.171-3.

[17] Batura N I, Vasilevsky E V, Kolushov N M, Zhurkin N G, Bityurin V.A, Bocharov A.N, Ivanov P P, Baranov D S, Bychkov S S, Grushin V A, Mozdykov V A, Zalkind V I, Tretyakova N V Design and the First Test Results of the Water-cooled MHD Accelerator of SMHDU TsAGI *Abstracts of the 16th International Workshop on Magneto-Plasma Aerodynamcs, April, 2015, Moscow*

IOP Conf. Series: Journal of Physics: Conf. Series **1112** (2018) 012018 doi:10.1088/1742-6596/1112/1/012018

Features of radiation heat transfer in a medium with variable refractive index

A V Galaktionov

Joint Institute for High Temperatures of Russian Academy of Sciences, 13 Bd.2 Izhorskaya str., Moscow 125412, Russia

E-mail: andrei.v.galaktionov@gmail.com

Abstract. In the literature one can find at least three different radiative transfer equations and various radiation diffusion equations for media with spatially varying refractive indices. We argue that the classical version of the transport equation seems to be valid. A generalized radiation diffusion equation, which coincides with the results of several authors, is derived. The equation allows one to take into account the reflection from the internal interfaces by means of the delta function in the radiation diffusion coefficient. This equation makes it possible to calculate heat exchange by radiation in a medium with a variable refractive index without additional computational costs. The physical meaning of the coefficients of the generalized radiation diffusion equation is discussed. The diffusion coefficient determines the momentum exchange between radiation and matter, and the absorption coefficient determines the exchange of energy.

1. Introduction

The radiative transfer equation (RTE) in absorbing-emitting and scattering media has applications in a wide variety of subjects, including optics, astrophysics, atmospheric science, neutron transport, biomedical optics, semitransparent crystal growth, fuel combustion, rocket propulsion, hypersonic shock layers, and heat transfer in porous media. The RTE for light refractive media was introduced and studied in astrophysics and physics of planetary atmospheres [1-3], physics of plasmas [4], and only recently attracted attention in biomedical optics and engineering applications [5-16].

Attempts to solve transport equation in a medium with a coordinate-dependent refractive index lead to considerable difficulties. Complexity arises as a result of the curvature of the rays due to contiguous change of the refractive index, and due to the occurrence of reflection from the surfaces of discontinuity of the refractive index. This occurs due to melting, pore collapse in insulations, gas ionization, and so on. Additional difficulties arise when the optical properties of the medium or the location of the boundaries are not known beforehand and are defined in the solution process, for example, when solving the Stefan problem. Refractive index alterations are especially significant in living biological tissues.

Because of the wide range of applications of the transport equation terminology varies between different fields. The RTE can be written out for intensity, specific intensity, brightness, radiance, and so on, with different definitions of those quantities. In a medium with a constant refractive index, this does not cause serious problems. However, discrepancies in notation cause problems in interpreting results when refractive index varies within the medium.

Interactions of light with the matter due to absorption, emission, and scattering are well understood, but the interaction of refraction is not. There are currently three [5] or even four [13] different versions of radiative transfer equation involving refraction for nonhomogeneous media. These equations are reduced to the same equation for homogeneous media when refractive index is assumed constant.

In classical monographs [1-4] and in most articles, the radiation intensity is defined either phenomenologically or is considered proportional to the photon's distribution function f. The (specific) intensity of radiation is defined as $I = h\nu(c/n)f$ $I = h\nu(c/n)f$, where $h\nu$ is the energy of a single photon, c is the speed of light in vacuum and n is the refractive index [3]. This leads to well-known radiative transfer equation [1-4]

$$\frac{n}{c}\frac{\partial I}{\partial t} + n^2 \frac{d}{ds}\frac{I}{n^2} = -(k+\beta)I + kn^2 I_p\big(T(r)\big) + \beta \int_{4\pi} \sigma(\Omega,\Omega')I(\Omega')d\Omega' , \quad (1)$$

where $\frac{d}{ds}$ is the derivative along the ray, k is the absorption coefficient, β is the scattering coefficient, $\sigma(\Omega,\Omega')$ is the scattering phase function, describing the probability of scattering from direction (Ω') to Ω, T is the temperature, I_p is the equilibrium (Planck) radiation intensity.

The equation (1) shows that the fundamental quantity $\frac{I}{n^2}$, the radiation intensity divided by the square of the refractive index, remains unchanged between collisions along the ray. The same quantity $\frac{I}{n^2}$ enters into the boundary conditions for the transport equation [1-4].

Serious difficulties arise when trying to determine the intensity in the geometric optics limit through the time averaged Poynting vector [17, 18]. From general considerations it is clear that the path from the time-reversible (at least in a vacuum) Maxwell equations to the irreversible kinetic equation of radiation transfer (1) cannot be simple and short. Having no place to discuss this issue here, we note that different definitions of radiation intensity lead to different conservation laws [12, 17], and apparently these problems are closely related to the Abraham-Minkowski controversy [22,23].

Below we show that the reflection from the internal boundary can be accounted for by introducing delta-functions in the radiation diffusion coefficient. In this case, the boundary thermal radiation flux is continuous, and the radiation energy density has a discontinuity. It is shown that the generalized diffusion coefficient of radiation determines the momentum exchange between radiation and matter, and the generalized absorption coefficient determines the exchange of energy.

2. New versions of radiative transfer equations

Recently, several new versions of the radiative transfer equation were proposed and compared in reviews [5, 13]. The radiation transfer equation can be rewritten for the specific intensity $L(r,\Omega,t)$ in the form [5]

$$\frac{n}{c}\frac{\partial L}{\partial t} + \Omega\cdot\nabla L + \Sigma + (k+\beta)L = \kappa n^2 L_p + \beta\int_{4\pi}\sigma(\Omega,\Omega')L(r,\Omega',t)d\Omega' , \quad (2)$$

where ∇ is the gradient with respect to position r. The specific intensity $L(r,\Omega',t)$ is the radiant power flux density at position r, direction Ω and time $L(r,\Omega',t)$. The intensity $I(s)$ and the specific intensity t, satisfy relation $I(s) = L(s)\delta\Omega(s)$ [5]. The term Σ on the left-hand side is the streaming term, different among the three formulations of RTE from [5-16], which describes the changes of the specific intensity because of refraction. The term Σ has three different forms:

RTE1 ref. [3, 4, 6-8, 12, 14]:

$$\Sigma^1 = \frac{1}{n}[\nabla n - (\nabla n\cdot\Omega)\Omega]\cdot\nabla_\Omega L - \frac{2}{n}(\nabla n\cdot\Omega)L , \quad (3)$$

where ∇_Ω is the gradient with respect to direction Ω

RTE2 ref. [9, 10, 12, 16]:

$$\Sigma^{21} = \frac{1}{n}[\nabla n - (\nabla n\cdot\Omega)\Omega]\cdot\nabla_\Omega L + \frac{1}{n}(\nabla^2 S\cdot\Omega\cdot\nabla n)L , \quad (4)$$

RTE3 ref. [5, 11]:

$$\Sigma^3 = \frac{1}{n}[\nabla n - (\nabla n\cdot\Omega)\Omega]\cdot\nabla_\Omega L - \frac{2}{n}(\nabla n\cdot\Omega)L + \frac{1}{n}(\nabla^2 S\cdot\Omega\cdot\nabla n)L , \quad (5)$$

where S is the eikonal which is to be found by solving the eikonal equation [17]

$$|\nabla S|^2 = n^2 \,. \tag{6}$$

Three terms are used to describe the change of the specific intensity because of refraction. The first term $\frac{1}{n}[\nabla n - (\nabla n \cdot \Omega)\Omega] \cdot \nabla_\Omega L$, is due to the differential change of the specific intensity along a ray path, which is the same between the three formulations. The second term $-\frac{2}{n}(\nabla n \cdot \Omega)L$, is due to the change of the directional differential along the ray path, which is in RTE1 and RTE3 but not in RTE2. The third term $\frac{1}{n}(\nabla^2 S \cdot \Omega \cdot \nabla n)L$, is the ray divergence term due to the refractive variation of the transverse section of a ray tube along the ray path [5].

Ref. [5] argues that only the radiative transfer equation RTE3 is consistent with the intensity law of geometric optics. Ref. [13] prefers RTE2 and Tualle [14] showed that RTE1 do not contradict to RTE2. Both ref. [5, 12] believe that the difference of those results is because the difference of the (specific) intensity definition.

In our opinion, the situation remains unclear and requires further research. Moreover, compliance with the laws of geometric optics is useful for diagnostic tasks. For radiative energy transfer problems, the integrated over solid angle characteristics of the radiation are more important. It is essential to fulfill the laws of conservation of energy and momentum in the whole heat and mass transfer problem [19, 20]. Therefore, we will proceed from the classical transport equation (1), RTE1 and investigate the radiation diffusion approximation.

3. Radiation diffusion equation

Radiation diffusion equation is one of the most popular methods of approximate solutions of the transport equation. This equation is always, at least qualitatively, correctly describes the characteristics of energy transfer by thermal radiation. To simplify the calculations, we consider the stationary one-dimensional problem and obtain the radiation diffusion equation in a medium with variable refractive index.

Strictly speaking, in the equation (1) $n(x)$ is a ray index of refraction [4]. However, in the case of an isotropic medium, it coincides with the usual refractive index. In an isotropic medium the Snell law $n(x)\sin\theta(x) = const$ is valid along the ray, where $\theta(x)$ is the angle between the ray and the positive direction of the coordinate axis. The Snell law is actually a consequence of the eikonal equation (6) and replaces it in the one-dimensional case.

From the Snell law follows

$$\frac{d}{ds} = \mu\frac{\partial}{\partial x} + \frac{1}{n}\frac{dn}{dx}(1-\mu^2)\frac{d}{d\mu}\,, \tag{7}$$

where $\mu = \cos\theta$.

Substituting (7) into (1) and integrating over the angle, both directly and after multiplication on μ we obtain

$$\frac{d}{dx}K = -U + kn^2U_p(T(x))\,, \tag{8}$$

$$\frac{d}{dx}K = -\frac{n'}{n}(K-U) + \big(k + \beta(1-\bar{\mu})\big)q\,, \tag{9}$$

where $U = \int_{4\pi} I d\Omega$ is the energy density of the radiation (without a factor equal to the inverse group velocity of light in the medium), $q = \int_{4\pi} I\mu d\Omega$ is the energy flux density, $K = \int_{4\pi} I\mu^2 d\Omega$, $U_p = 4\pi I_p$ is the energy density of the equilibrium radiation in a vacuum, and $\bar{\mu}$ is the average cosine of the scattering.

Using asymptotic equality $K \approx U/3$ that is valid in the diffusion limit, we exclude K from (9):

$$\frac{d}{dx}\frac{U}{n^2} = -\frac{3\big(k+\beta(1-\bar{\mu})\big)}{n^2}q\,, \tag{10}$$

which in a more conventional form can be rewritten as Fick's law for a medium with variable refractive index:

IOP Conf. Series: Journal of Physics: Conf. Series **1112** (2018) 012018 doi:10.1088/1742-6596/1112/1/012018

$$q = -Dn^2 \frac{d}{dx}\frac{U}{n^2} = -\frac{n^2}{3(k+\beta(1-\bar{\mu}))}\frac{d}{dx}\frac{U}{n^2}, \tag{11}$$

where $D = 1/3(k + \beta(1-\bar{\mu}))$ is the usual radiation diffusion coefficient.

With ordinary law of energy conservation (8), this expression gives the equation of radiation diffusion in a medium with variable refractive index.

$$-\frac{d}{dx}Dn^2\frac{d}{dx}\frac{U}{n^2} + kU = kn^2U_p(T(x)) . \tag{12}$$

The diffusion equation (12) should be supplemented by the usual boundary conditions at the outer boundaries, which can be written out in the form

$$-Dn^2\frac{d}{dx}\frac{U}{n^2} \pm \frac{1}{2}\frac{1-R_{le(ri)}}{1+R_{le(ri)}}\left(U - n^2U_p\left(T_{le(ri)}\right)\right) = 0 \text{ , at } x = x_{le(ri)} \text{ ,} \tag{13}$$

where $R_{le(ri)}$ is the hemispherical reflection coefficient of the left (right) boundary and $T_{le(ri)}$ is the temperature of the external medium.

This equation is obviously generalized to the multidimensional case:

$$-\nabla Dn^2\nabla\frac{U}{n^2} + kU = kn^2U_p(T(r)) . \tag{14}$$

Equation (14) coincides with the diffusion equations obtained in Refs [6-8]. All these works use the classical radiative transfer equation (1) i.e. RTE1. However, in Refs [12, 15, 16] three different equations for diffusion of radiation in a medium with a variable refractive index was obtained. But none of them can withstand criticism [7].

4. Accounting for reflection from internal boundaries

So far, we have assumed that refractive index depends smoothly on the coordinate. Now suppose that the refractive index changes abruptly, for example, during a phase transition, or assume that there is an internal interface between media with known reflection and transmission coefficients. For the radiation intensity at the interface, in the general case, one can write

$$I_{l(r)}^{-(+)}(\Omega') = \int_{2\pi} r^{l(r)}(\Omega',\Omega) I_{l(r)}^{-(+)}(\Omega)\mu d\Omega + \int_{2\pi} p^{r(l)}(\Omega',\Omega) I_{l(r)}^{-(+)}(\Omega)\mu d\Omega , \tag{15}$$

where $r(\Omega',\Omega), p(\Omega',\Omega)$ is the bi-directional coefficients of reflection and transmission of the boundary.

Integrating these relations with respect to the angle, and using the reciprocity relations for the transmittances, we obtain

$$\frac{U_r}{n_r^2} - \frac{U_l}{n_l^2} = -\frac{2(2-p_l-p_r)}{pn^2}q , \tag{16}$$

where $pn^2 = p_r n^2(\xi+0) = p_l n^2(\xi-0)$ and $p_{l(r)}$ is the transmission coefficient of the boundary when radiation propagates from the right (left) side of the boundary see Fig. 1.

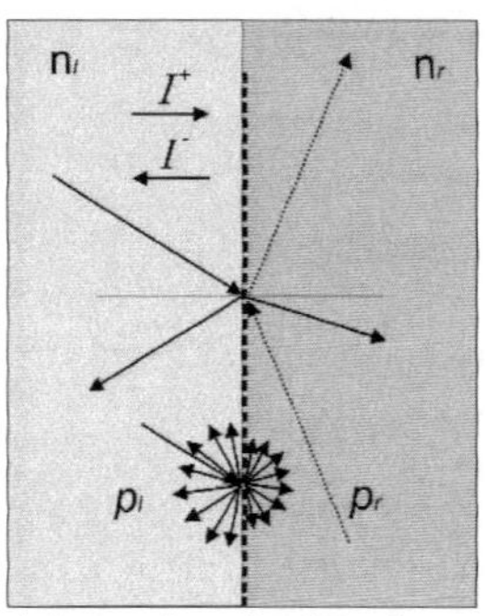

Fig. 1 A boundary inside the medium

Comparing expressions (10) and (16), it is easy to see that with the help of the delta function they can be written out in a uniform way

$$\frac{d}{dx}\frac{U}{n^2} = -\left[\frac{3(k+\beta(1-\bar{\mu}))}{n^2} + \delta(x-\xi)\frac{2(2-p_l-p_r)}{pn^2}\right]q , \tag{17}$$

where ξ is the interface coordinate. Obviously, if there are several boundaries in the layer, delta functions should be added at each interface coordinate.

The equation (17) with the energy equation (10) and ordinary boundary conditions (13) form a closed system of Caratheodory ordinary differential equations with generalized functions in the coefficients [24]. The existence and uniqueness theorems and the continuous dependence of solutions on the coefficients are proved for the Caratheodory equations. This means that if we approximate the delta function $\delta(x)$ by a "cap" function $\delta_\varepsilon(x)$ and solve the problem numerically with the help of ordinary homogeneous conservative difference schemes, we can expect good convergence to the discontinuous solution.

The radiation diffusion equations (10), (17) may be rewritten in the form (12) with new radiation diffusion coefficient

$$D_\varepsilon = \frac{1}{\left(3\left(k+\beta(1-\bar{\mu})\right)\right)} + Pn^2\delta_\varepsilon(x-\xi)\,, \tag{18}$$

where $P = 2(2 - p_l - p_r)/pn^2$. Such a diffusion equation uniformly describes not only a smooth change in the refractive index, but also possible reflection and scattering of radiation at jumps of the refractive index.

The relation at the boundary with the jump of the refractive index inside a medium can be generalized to the multidimensional case. We will focus on the simplest case and assume that there is only a single boundary present in the medium which is locally flat. That is, the radii of curvatures of the surface are much greater than the wavelength of the radiation.

Then, by analogy with equation (16), we can write the conjugation conditions on the boundary in the form

$$\mathrm{O}^T(r)\Theta(\mathrm{r})\mathrm{O}(r)\mathbf{q}(x) = -\nabla\frac{U(r)}{n^2(r)}\,, \tag{19}$$

Where $\mathrm{O}(r)$ is the rotation that combines the axis x^3 with the normal to the boundary at the point ξ; $\Theta(r)$is the covariant tensor with components

$$\begin{vmatrix} 1/Dn^2 & 0 & 0 \\ 0 & 1/Dn^2 & 0 \\ 0 & 0 & \frac{1}{Dn^2} + P\delta(x^3-\xi) \end{vmatrix}. \tag{20}$$

Approximating the δ function by a "cap" function and introducing the symmetric contravariant tensor by the relation $M_\varepsilon(r)\mathrm{O}^T(r)\Theta(r)\mathrm{O}(r) = 1$, we can write Fick's law (19) in a more familiar form

$$q(r) = -M_\varepsilon(r)\nabla\frac{U(r)}{n^2(r)}\,. \tag{21}$$

We note that relation (21) is completely analogous to the well-known Fick's law for an anisotropic medium. In the case of an isotropic medium considered here, tensor $M_\varepsilon(r)$ differs from the diagonal tensor with components $M_\varepsilon^{ii} = D(x)n^2(x)$ only within a small ε neighborhood of the boundary.

Unfortunately, we could not find a rigorous mathematical proof of the convergence of the proposed approximation to discontinuous solutions. This convergence is to be verified by computational experiments.

If we introduce a new quantity $N = \frac{U}{n^2} = \int_{4\pi} I/n^2 d\Omega$, the record of the diffusion equation (14) can be future simplified

$$-\nabla M_\varepsilon \nabla N + EN = EU_p(T(x))\,, \tag{22}$$

where new coefficients M_ε and E of the new diffusion equation are related to old coefficients by relations

$$M_\varepsilon = D_\varepsilon n^2\,, \tag{23}$$

$$E = kn^2\,. \tag{24}$$

It is not difficult to verify that only the quantities M_ε and E enter into the usual boundary conditions of the diffusion equation. After solving equation (22), the divergence of radiation flux $w(x)$ which enters as the source into the equation of total energy balance [19, 20], can easily be calculated by the formula:

$$w(x) = \nabla q = E\big(U_p(T(x)) - N\big)\,. \tag{25}$$

Moreover, as measured in experiments, the coefficient of diffuse reflection or the emissivity of a plane layer [25] also depends only on the coefficients M_ε and E:

$$1 - R = \varepsilon = 4n^2\sqrt{kD}th\big(H\sqrt{k/D}\big) = 4\sqrt{EM}th\big(H\sqrt{E/M}\big)\,, \tag{26}$$

where H is the layer thickness. Thus, if optical properties of a material are measured in an experiment, then to calculate the energy transfer by radiation in the diffusion limit [25], one can completely eliminate the refractive index.

The same is obviously true in the multidimensional case. This means that in the diffusion limit for calculating energy transfer by radiation, known software packages developed for a constant refractive index can be used simply by redefining the coefficients of the radiation diffusion equation.

In one-dimensional case, the proposed approach has been successfully used in numerical modeling of laser ablation of porous heat shielding [21]. The work took into account significant changes in the refractive index due to softening and shrinkage of the material.

5. Discussion

From the equations (22) and (25), it is easy to see that the radiative heat transfer operator acting on the Planck radiation energy density and calculating the divergence of the radiant flux is self-adjoint operator. This is equivalent to conservativeness of the corresponding equations, that is, the validity of the energy conservation law. This is an important argument in favor of the validity of the obtained radiation diffusion equation in a medium with a variable refractive index.

When describing radiation transfer using equation (1), we actually divide a system into subsystems - radiation and matter, which, interacting, exchange energy and momentum with each other. For simplicity, we first consider modeling highly rarefied medium consisting of macroscopic particles (e.g., micron glass beads) in vacuum. When a photon is absorbed (or emitted) by a ball, the photon transfers (takes away) its energy $h\nu$ and momentum $h\nu/c$ from the radiation to the matter. The intensity of this process is obviously proportional to the absorption cross section $\sigma_{a\nu}$ of the particles or to the absorption coefficient k_ν, $k_\nu = \sigma_{a\nu}C$ where C is the particle concentration.

Since the scattering proceeds without changing the frequency (as is usually assumed), there is no exchange of energy between radiation and matter, but only momentum exchange occurs. It is easy to calculate that during scattering the photon transfers the momentum to the ball in proportion to the transport scattering cross section $\sigma_{tr\nu} = \sigma_{s\nu}(1 - \bar{\mu}_\nu)$ or $\beta_\nu = \sigma_{s\nu}C$. In the expression for the radiation diffusion coefficient (17) absorption and scattering enter in such a way that the diffusion coefficient can be naturally interpreted as a coefficient describing the total (due to absorption and scattering) momentum exchange between radiation and matter. Similarly, the absorption coefficient can naturally be interpreted as a coefficient describing the energy exchange between radiation and matter.

In a dense refractive medium, the situation is not so obvious, since the definition of the momentum of a photon (or a wave packet) in a medium is the subject of the old Abraham-Minkowski controversy [22, 23]. Using symmetry considerations, it can be shown that in a dense medium the coefficients M_ε and E, as before, can be interpreted as the coefficients describing the exchange of momentum and energy between radiation and matter.

In conclusion, it is useful to try to derive the diffusion equation (22) directly from the fluctuation electrodynamics [26, 27], bypassing the transport equation. This theory is free from the limitations of geometric optics and, perhaps, may circumvent the above problems.

6. Conclusions

It is well known that light in a medium with a variable refractive index can propagate in a very bizarre manner, generating caustics and other sophisticated objects [12]. Even more surprising that in the radiation diffusion limit a very simple equation (22) arises, for the solution of which the refractive index does not appear to be needed at all. The coefficients of this equation describe the exchange of momentum and energy between radiation and matter.

The equation (22) allows one to take into account the reflection and scattering at internal boundaries uniformly by introducing a delta function in the radiation diffusion coefficient. Radiation diffusion equation (22) makes it possible to calculate heat exchange by radiation in a medium with a variable refractive index by known software packages without additional computational costs.

We started from the definition of intensity through the photon distribution function as $I = h\nu(c/n)$. We have obtained an equation (22) that can naturally be interpreted as the photon diffusion equation. The quantity $N = U = \frac{U}{n^2} = \int_{4\pi} I/n^2 d\Omega$ is proportional to the number of photons.

On the other hand, it is natural to require that in a medium is conserved a radiation energy density, which is proportional to $n^2 \partial n(\nu)\nu/\partial\nu$ [28], where ν is the radiation frequency. In our case, the radiation energy density obviously is not conserved. Apparently, this does not lead to significant problems. However, in our opinion, the choice of radiation transport equation and diffusion equation in a medium with a variable refractive index remains open question and requires further research.

References

[1] Chandrasekhar S 1960 *Radiative Transfer* (Dover)
[2] Sobolev V V 1975 *Light Scattering in Planetary Atmospheres* (Pergamon)
[3] Pomraning G C 1973 *The Equations of Radiation Hydrodynamics* (Pergamon, Oxford, UK)
[4] Bekefi G 1966 *Radiation Processes in Plasmas* (Wiley)
[5] Wang P, Arridge R S and Jiang M 2014 *Phys. Rev.* A **90** 023803
[6] Shendeleva M L 2004 *J. Opt. Soc. Am.* A **21** 2464–8
[7] Tualle J M and Tinet E 2003 *Opt. Commun.* **228** 33–8
[8] Bal G 2006 *J. Opt. Soc. Am.* A **23** 1639–44
[9] Ferwerda H A 1999 *J. Optics* A **1** L1
[10] Marti-Lopez L, Bouza-Dominguez J, Hebden J C, Arridge S R, and Martinez-Celorio R A, 2003 *J. Opt. Soc. Am.* A **20** 2046-56
[11] Premaratne M, Premaratne E, and Lowery A J 2005 *Opt. Express* **13** 389–99
[12] Soloviev V Y 2016 *J. Opt. Soc. Am.* A **33** **3** 383-90
[13] Marti-Lopez L, Bouza-Dominguez J, Martinez-Celorio R A, and Hebden J C 2006 *Opt. Commun.* **266** 44–9
[14] Tualle J M 2008 *Opt. Commun.* **281** 3631-5
[15] Khan T and Jiang H 2003 *J. Opt. Pure Appl. Opt.* **5** 137–41
[16] Marti-Lopez L, Bouza-Dominguez J and Hebden J C 2004 *Opt. Commun.* **242** 23–43
[17] Born M and Wolf E 1975 *Principles of Optics* (Pergamon)
[18] Van de Hulst H C 1981 *Light Scattering by Small Particles* (Dover)
[19] Howell J R, Menguc M P, Siegel R 2015 *Thermal Radiation Heat Transfer* (CRC Press, 6 ed.)
[20] Modest M F 2013 *Radiative Heat Transfer* (Academic Press, 3 ed.)
[21] Galaktionov A V, Stepanov S V 1990 *High Temp.* **28** 105–11
[22] Pfeifer R N C, Nieminen T A, Heckenberg N R, and Rubinsztein-Dunlop H 2007 *Rev. Mod. Phys.* **79** 1197
[23] Spirichev Yu A 2018 *Phys.-Usp.* **61** 303
[24] Hale J K 1980 *Ordinary Differential Equations* (Malabar: Krieger Publishing Company, 2 ed).
[25] Stepanov S V, Berkovskii M A 1985 *High Temp.* **23** **2** 288–96
[26] Levin M L, Polevoy V G, Pytov S V 1980 *J. Exp. Theor. Phys.* **52** **6** 1054
[27] Bimonte G, Emig T, Kardar M and Kruger M 2017 *Annu. Rev. Condens. Matter Phys.* **8** 119
[28] Landau L D, Lifshitz E M and Pitaevskii L P 1984 *Electrodynamics of Continuous Media* (Butterworth – Heinemann, Vol. 8 (2 ed.))

IOP Conf. Series: Journal of Physics: Conf. Series **1112** (2018) 012019 doi:10.1088/1742-6596/1112/1/012019

Three-electrodes open discharge in low-pressure deuterium: transition from the overvoltage regime into low-voltage one

Yu S Akishev[1,2], G I Aponin[1], A A Balakirev[1], V B Karalnik[1], A V Petryakov[1] and N I Trushkin[1]

[1] SRC RF TRINITI, 108840, Moscow, Troitsk, Pushkovykh street, 12, Russia
[2] NRNU MEPhI, 115409, Moscow, Kashirskoe shosse, 31, Russia

*akishev@triniti.ru

Abstract. Strongly overvoltage open discharges in a narrow gap (2-3 mm) between the solid cathode and the grid anode are used for generation of the pulsed high-current electron beams with the energy up to several tens of keV. Strongly overvoltage (SO) regime is unstable and the discharge tends to transit into low-voltage (LV) regime with a high-current. We studied this transition by example of the three-electrodes open discharge in D_2 at low pressure (about 0.5-2 Torr) being powered by stepwise voltage with amplitude up to 25 kV. The existence of the auxiliary third electrode allows one to get more stable operation of the open discharge. The physical properties of the SO and LV regimes and transition between them have been explored in detail with the usage of the fast multi-frame camera synchronized with the current and voltage of the open discharge.

1. Introduction

Electron beams with energy up to several tens of keV can be obtained not only with the vacuum systems but with the usage of gas discharges as well. In the latter case it is necessary to create very high electric fields in the discharge gap. The magnitude of the electric field strength has to exceed appreciably the critical value after which so-called run-away electrons appear in the discharge [1-4].In the most cases, strongly overvoltage open discharges in a narrow gap (2-3 mm) between the solid cathode and the grid anode are used for generation of the pulsed high-current electron beams with the energy pointed above [5-10]. The low pressure discharge in a narrow gap exposed to high voltage is characterized as a discharge existing under overvoltage regime. In our case [6] we have created strongly overvoltage regime with the magnitude of the reduced electric field $E/N > 10^5$ Td. The breakdown process of the strongly overvoltage open discharge (SOOD) cannot be described by the concept of Townsend electron avalanches. A reason is that any electron being appeared in the gap immediately becomes as run-away electron and quickly leaves the gap without the ionizing collision vitally needed for the development of the avalanche. Respectively, the breakdown voltage of SOOD at the given product Pd (P is a gas pressure, d is the length of the inter-electrode gap) does not coincide with Paschen curves described on the base of the Townsend electron avalanches concept: $\alpha_i d \approx \ln(1/\gamma)$, where α_i is the first Townsend coefficient characterizing the dependence of ionization rate on the E/N; γ is the second Townsend coefficient characterizing the electron emissive processes on the cathode. The above statement is illustrated with figure 1 by the example of H_2. Figure 1 shows the dependence of the ionization rate for electrons with increase in the E/N magnitude. The dashed curve is calculated in the approximation neglecting the non-local effects. The solid curve in figure 1 takes into account the transition of more and more electrons into run-away regime under strong electric fields. The appearance of run-away electrons leads to the diminishing of the ionization rate under huge E/N. However, strongly overvoltage (SO) regime is unstable and the discharge tends to transit quickly into low-voltage (LV) regime with a high electric current but without high-energy electron beam. We studied this transition by example of the three-electrodes open discharge in D_2 at low pressure (about 0.5-2 Torr) being powered by stepwise voltage with amplitude up to 25 kV. The existence of the auxiliary third electrode allows one to get more stable operation of the open discharge in the SO regime. The low-current auxiliary discharge between the cathode and the distant anode played important role, namely, it provided preionization of the gas gap for the primary discharge and its fast ignition when the high-voltage pulse was applied.

The physical properties of the SO and LV regimes and transition between them have been explored in detail with the usage of the fast multi-frame camera synchronized with the current and voltage of the discharge. The obtained results give more insight into physics of the overvoltage open discharge generating the high-current pulsed e-beams with energy up to 25 keV.

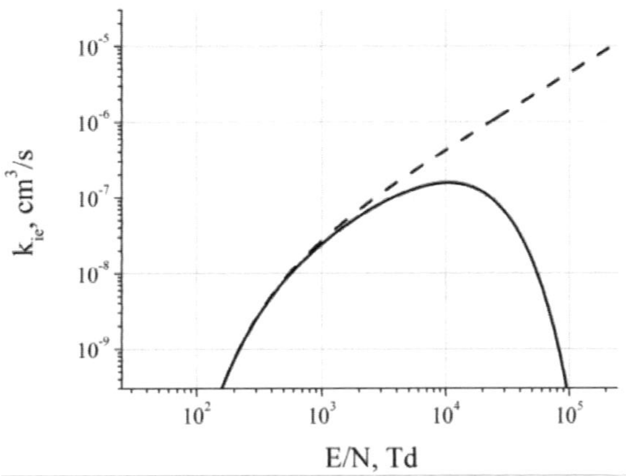

Figure 1 The ionization rate for electrons in H_2 with increase in the E/N. The dashed and solid curves were calculated without and with taking into account the diminishing of the ionization rate with huge E/N.

2. Experimental setup

General scheme of the setup used for generation and investigation of the overvoltage open discharge in D_2 at low pressure with three-electrode system is shown in figure 2.

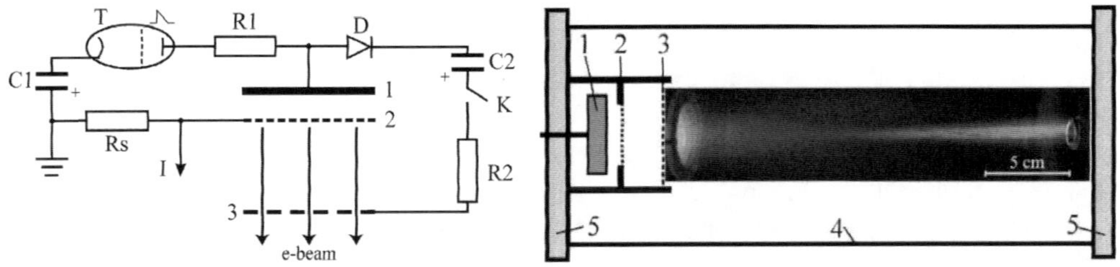

Figure 2 a) The electrical scheme of the three-electrode system generating a high-current e-beam with energy up to 25 keV. 1 is the solid cathode; 2 is the grid anode of the main discharge; 3 is the grid anode of the auxiliary discharge; T is the thyratron switching the capacitor C1 to the electrode gap; $R_1 = 44$ Ohm, $R_2 = 1$ MOhm, $R_s = 0.024$ Ohm; $C_1 = 12.5$ nF, $C_2 = 100$ μF.

b) The sketch of quartz tube with the electrode system generating HV e-beam (the scale is not kept): 1 - the cathode, 2 - the main anode, 3 - the auxiliary anode, 4 –the transparent quartz tube of 24 cm in length, 5 –the flange with a quartz window.

The inter-electrode gap between the solid cathode and meshed anode is 3 mm. The discharge system has been mounted inside the quartz tube which can be pumped down to pressure $P = 10^{-2}$Torr and after that is filled with deuterium up to the needed pressure. The applied high voltage was measured by HV divider PINTEK HVP-39 (1000:1, 40 kV, 200 MHz). The discharge current was measured by a low-inductive shunt with resistance of 50Ohm. All electrical signals were recorded by the digital oscilloscopes such as Tektronix TDS 520, Tektronix DPO2024 and Tektronix TDS 2012 with bandwidth of 500, 200 and 100 MHz. The discharge images were taken by multi-frame fast camera (equipped with the intensifier) with the exposure time down to 50 ns.

3. Experimental results and discussion

The typical current and voltage waveforms of the pulsed open discharge in deuterium at pressure $P =$ 2 Torr is shown in figure 3a. We have performed the experiments with the three-electrode system in the wide range of the applied voltage U from 7 kV to 25 kV. Based on the obtained results, the Current-Voltage characteristic of this discharge in the SO regime at $P = 2$ Torr is presented in figure 3b. Here the magnitudes of I and U were taken at the moments preceding the transition of the discharge from the strong overvoltage regime into the low-voltage regime. This characteristic can be approximated by expression $I(A) \approx (U - U_0)^3/40$, where $U_0 = 5$ kV. Note that the higher applied voltage and gas pressure, the shorter duration of the SO regime (figure 3c, 3d).

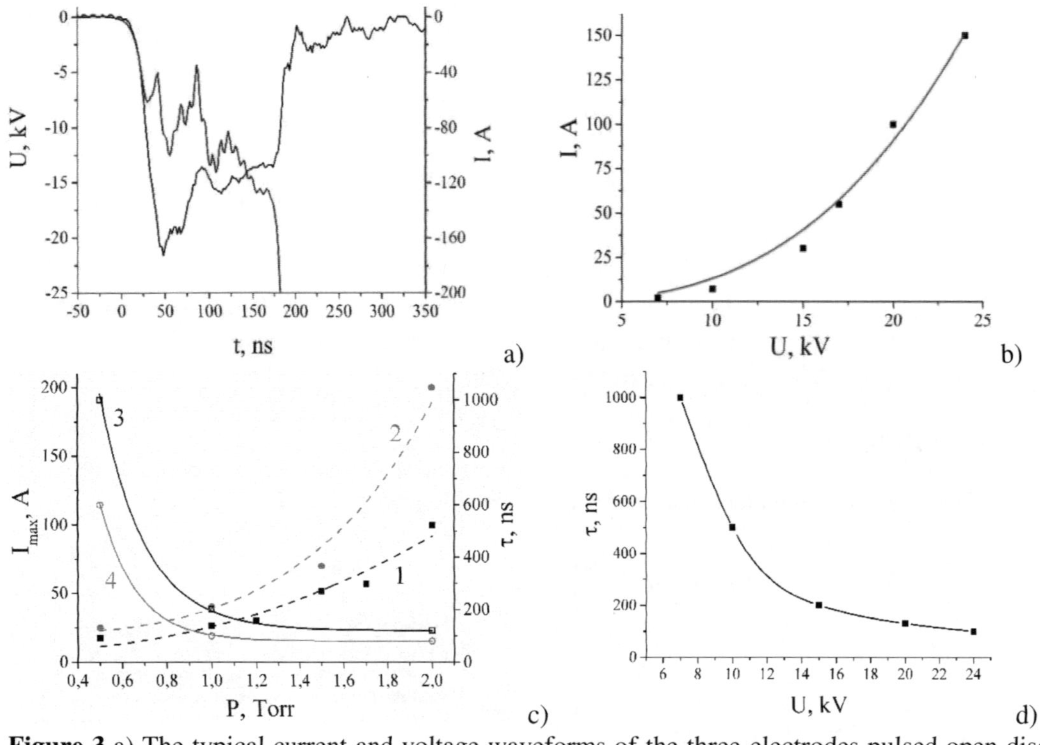

Figure 3 a) The typical current and voltage waveforms of the three-electrodes pulsed open discharge in deuterium. $P = 2$ Torr; b) The Current-Voltage characteristic of the open discharge in the diffuse overvoltage regime at the moments preceding the transition into low-voltage regime. The solid curve is the approximation $I(A) \approx (U - U_0)^3/40$, where $U_0 = 5$ kV. $P = 2$ Torr; c) The dependence of maximum current Imax of SOOD in D_2 before its transition into the low-voltage mode (dashed curves 1 and 2) and duration τ of the high-voltage mode (solid curves 3 and 4) on gas pressure P, $U = 20$ kV (curves 1 and 3), $U = 24$ kV (curves 2 and 4). $P = 2$ Torr. d) The dependence of duration τ of the high-voltage mode on applied voltage. $P = 2$ Torr. The auxiliary low-current discharge between the cathode and meshed anode 3 is powered with the voltage $U^* = 1$ kV.

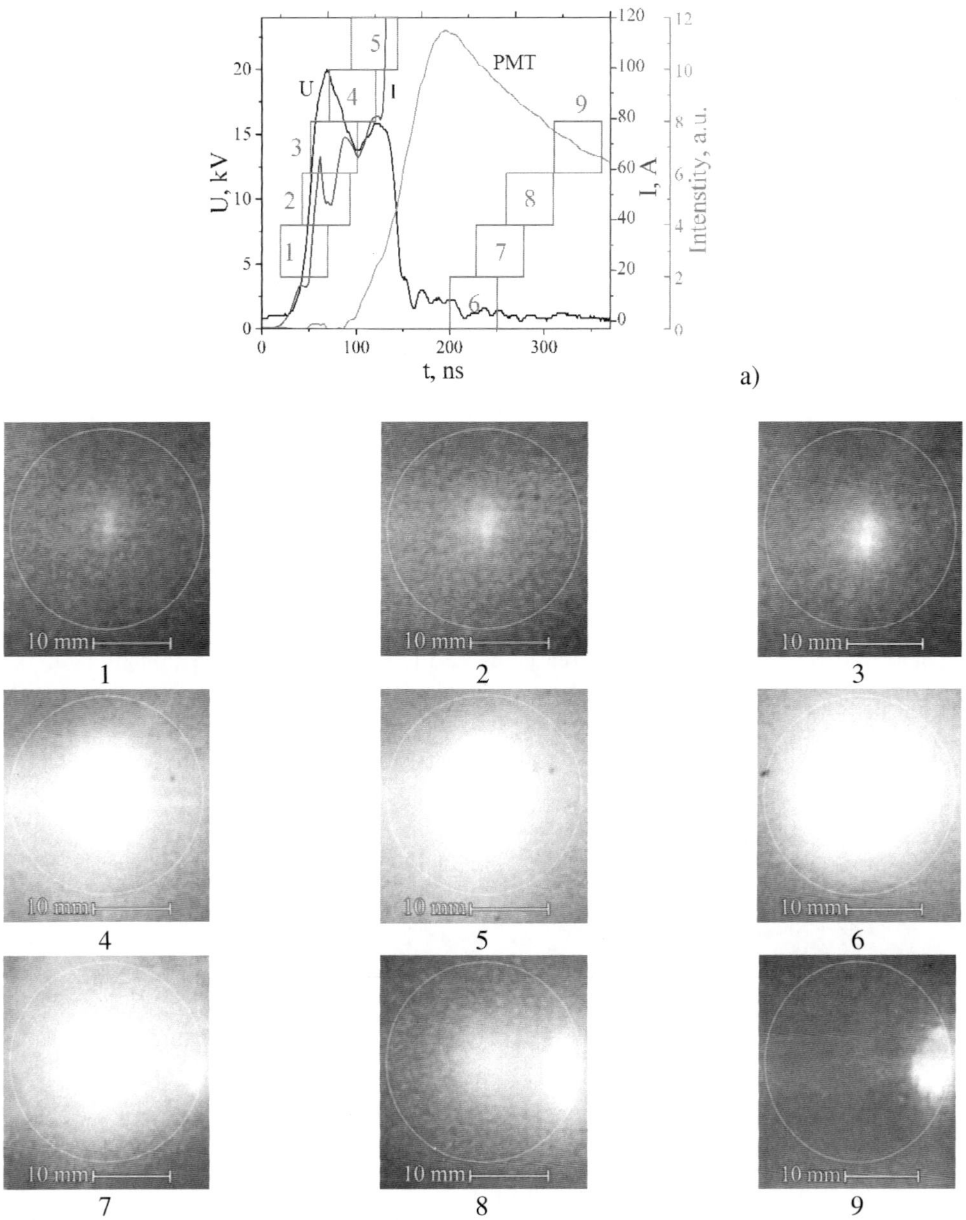

Figure 4. The current-voltage waveforms of the SOOD in deuterium (a) and set of the discharge images (b). The white circle in each shot depicts the boundary of the cathode. P = 2 Torr, U = 20 kV, U^* = 1 kV; the enumerated time windows marked in the current-voltage waveform correspond to the same moments at which the enumerated images (1-9) were taken; the exposure time of each shot is 50 ns.

As it is mentioned above, the higher applied voltage, the shorter duration of the overvoltage regime which generates the high-energy e-beam. In fact, namely this SO regime is the most of interest from the practical point of view. Therefore there is an insistent need in studying of the mechanism of instability of this regime. The experimental data showing in detail the spatial-temporal evolution of the open discharge up to its transition into the low-voltage regime with the high-current cathode spot(s) are presented in figure 4. In this figure, the set of the open discharge images taken by a fast multi-frame camera was correlated with the current-voltage waveform of the discharge.This set of the obtained data can be very useful for the development of the physical mechanisms responsible for the transition of the strongly overvoltage open discharge into the low-voltage regime.

Close examination of all shots in figure 4 leads to the conclusion that, first, the SO discharge does not constrict but contrariwise increases his cross-section and brightness up to the transition into the LV regime and, second, the bright cathode current spot(s) happen after the $SO \rightarrow LV$ transition and therefore they cannot be a reason initiating this transition. A reason is that the cathode spots in itself do not shunt the gap - they can only initiate the streamers (thin current filaments) which can propagate from the cathode towards the anode.

In principle, after the arrival of streamers to the anode, the gap can be shunted because the streamer quickly transforms itself into the high-conductive spark and the overvoltage regime transits into the low-voltage regime. However, the propagation of streamers towards the anode requires the existence in the gap of the intensive direct ionization of neutrals by the electron impact. Unfortunately, the extremely high electric field at the cathode in the overvoltage regime (E/N reaches of huge magnitude of 10^5 Td and even more) leads to that the electrons have become the run-away electrons which have the extremely low efficiency of the impact ionization. This is a reason why the propagation of streamers towards the anode is impossible in the overvoltage regime.

Based on the mentioned above, we can state that, in our opinion, the mechanism responsible for transition of the overvoltage regime into low-voltage mode is not connected with the cathode spots and anode directed streamers. One of the reasons of the observed instabilityof the SO discharge can be a strong gas heating by ions at the cathode leading to switching on the thermal ionization at this place. This ionization provides the dramatic increase in the discharge current and transition of the open discharge into low-voltage regime. Our numerical calculations verified this hypothesis (figure 5). Besides, numerical calculations revealed an essential role of the plasma created by e-beam behind the grid anode in the sustaining the strongly overvoltage open discharge.

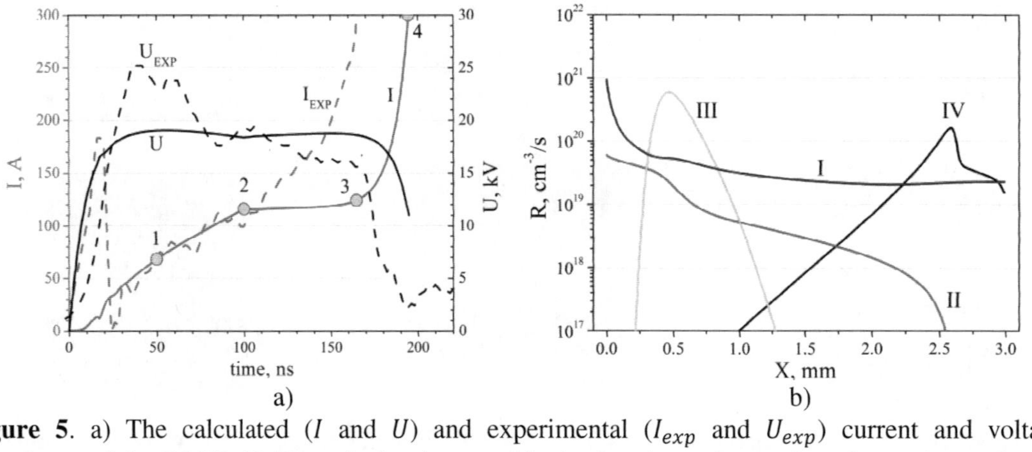

Figure 5. a) The calculated (I and U) and experimental (I_{exp} and U_{exp}) current and voltage waveforms of the $SOOD$. b) The calculated rates of ionization due to impact by e-beam (curve I), fast ions (curve II), slow electrons (curve IV) and the rate of thermal ionization (curve III) for the time moment (4). $U_0 = 25$ kV, D2, $P = 2$ Torr.

One may see in this figure that fast increase in the discharge current is associated with instant "switching-on" the thermal ionization. At this moment, the local gas temperature near the cathode is equal to 6560 K.

4. Conclusion

Our experimental investigation of the pulsed open discharge has led to the conclusion that the strong gas heating by ions at the cathode can be a reason promoting the transition of the open discharge into low-voltage regime. The numerical calculations taking into account the thermal ionization verified this

IOP Conf. Series: Journal of Physics: Conf. Series **1112** (2018) 012019 doi:10.1088/1742-6596/1112/1/012019

hypothesis. Besides, numerical calculations revealed also a crucial role of the plasma created by e-beam behind the grid anode in the sustaining the strongly overvoltage open discharge.

Acknowledgement
This work has been carried out thanks to full financial support by the Russian Science Foundation (Grant № 16-12-10458).

References
[1] Koval N N, Oks E M, Schanin P M, Kreindel Y E and Gavrilov N V 1992 *Nucl. Instrum. Methods Phys. Res.* **A321** 41728
[2] Tarasenko V F and Yakovlenko S I 2004 *Phys.-Usp.* **47** 887
[3] Babich L P 2005 *Phys.-Usp.* **48** 1015
[4] Akishev Yu S, Karal'nik V B, Medvedev M A, Petryakov A V, Trushkin N I and Shafikov A G 2017 *J. Phys.: Conf. Ser.* **830** 012015
[5] Bokhan P A and Sorokin A R 1985 *Tech. Phys.* **55** 88
[6] Akishev Yu S, Karal'nik V B, Medvedev M A, Petryakov A V, Trushkin N I and Shafikov A G 2017 *J. Phys.: Conf. Ser.* **927** 012024
[7] Korolev Y D, Frants O B, Geyman V G, Landl N V, Ivashov R V, Shemyakin I A, Bischoff R E and Frank K 2005 *IEEE Trans. Plasma Sci.* **33** 1648
[8] Akishev Yu S, Karal'nik V B, Petryakov A V, Starostin A N, Trushkin N I and Filippov A V 2016 *Plasma Phys. Rep.* **42** 14
[9] Klimenko K A and Korolev Yu D, 1990 *Tech. Phys.* **60** 138
[10] Akishev Yu S, Karal'nik V B, Petryakov A V, Starostin A N, Trushkin N I and Filippov A V 2017 *J. Exp. Theor. Phys.* **124** 231

IOP Conf. Series: Journal of Physics: Conf. Series **1112** (2018) 012020 doi:10.1088/1742-6596/1112/1/012020

Numerical simulation of argon flow structure in plasma vortex reactor

A I Klimov[1;2], S E Kurushina[1], N E Molevich[1;3], D P Porfiriev[1;3] and I P Zavershinskii[1*]

[1] Samara National Research University, Samara, 443086, Russia
[2] Joint Institute for High Temperatures RAS, Moscow, 125412, Russia
[3] Lebedev Physical Institute, Samara, 443011, Russia

* ipzav63@mail.ru

Abstract. Numerical simulation of the turbulent three-dimensional swirling flow in experimentally tested plasma vortex reactor were carried out to ascertain that earlier limited modelling properly represents flow phenomena and to provide further insights into use of such apparatus. Exact experimental geometry, electrode location that maximizes energy utilization in accordance with previous research, model heat source, and reduced mixture composition (pure Ar) were used. A qualitative agreement between the results of the calculations and the experimental data for pure argon has been obtained. Structure of counterflow zone, crucial for energy extraction, and corresponding temperature field are presented.

1. Introduction

Recently, the problem of creating efficient energy sources has become increasingly noticeable. One of the most promising technologies is the use of plasma vortex reactors (PVRs), which can be both heat and hydrogen source.

Based on the experimental works [1-5] PVR was designed and elaborated. The hydrogen and heat production in the developed setup are based on the hydration reaction of erosion cathode clusters or metal nanopowder introduced into the reaction zone by water vapor, stimulated by plasma formations in a swirling flow. The scheme of this reactor and experimental data are given in [6]. As a working gas, a mixture of argon and water vapor was used. Vortex flow is used for the separation of light hydrogen on the reactor axis, the stability of an electric discharge and the thermal insulation of the walls from the hot plasma flow. It was shown that there is a long recirculation zone located between the anode and the swirler. Measured gas temperature of this heterogeneous plasmoid is about $T = 2000 \div 5000$ K.

The self-consistent simulation of the flow structure is complicated and requires joint modeling of the gas-dynamic and discharge subsystems. Therefore in [7], as a starting step, a simulation of the three-dimensional unsteady viscous turbulent vortex flow of pure argon in a model duct is carried out for different configurations of electrode system. The spatial distributions of the power of the source were chosen phenomenologically with the use of the data [6]. It was verified that the recirculation zone forms in such system and its characteristics strongly depend on the heat source power and the shape of electrodes and their location. The numerical modelling has demonstrated that it is possible to configure electrodes in such a way as to ensure energy transfer mainly downstream from the interelectrode zone.

Otherwise, apparatus suffers from several deficiencies. Among them are a strong overheating of the swirler and fastener elements and undesirable transfer of hydrogen in upstream direction. It was shown that the optimal electrode configuration may be accomplished by the combination of thin anode and pipe-like cathode through which areas of different pressures are connected. A qualitative agreement between the results of the calculations and the experimental data [6] has been obtained.

However, the assumptions made in [6] limit the use of investigation results only to general demonstration of proposed idea. Therefore, a numerical simulation of the unsteady viscous turbulent swirl argon flow structure in the PVR duct, corresponding to the experimental setup in detail, with an optimal configuration of the electrodes obtained in [6] was carried out here.

2. Mathematical modelling

2.1. Numerical model geometry and governing equations

The schematic duct geometry is sketched in figure 1. Governing parameters for the electrode system are the shape, distance between them and the distance from the nozzle. The heat source is located between cathode (at left) and anode. The outlet of the duct has the form of a convergent nozzle. The nickel anode was thin, the nickel cathode was a hollow thin cylinder. The pipe-like cathode has an external diameter of 20 mm and a thickness of 1 mm. The cathode is located at the outlet of the duct, figure 1. The interelectrode distance is 60 mm remains constant, the length of the quartz glass duct is 407 mm.

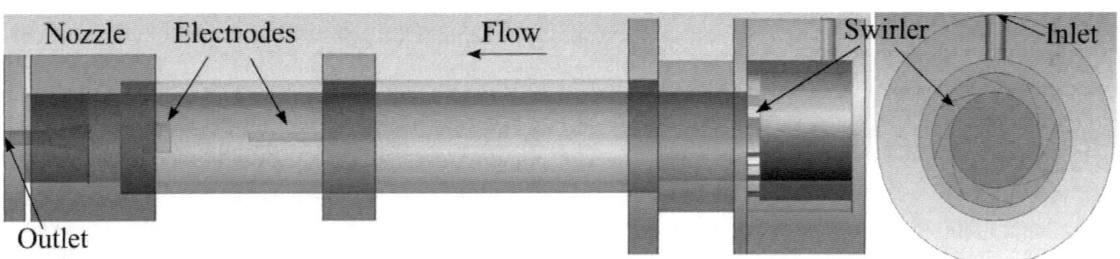

Figure 1 A schematic diagram of the geometry of experimental setup used for numerical simulation

Standard unsteady Reynolds averaged Navier-Stokes (URANS) equations were used to describe the flow:

$$\frac{\partial \rho}{\partial t}+\frac{\partial(\rho v_i)}{\partial x_i}=0,$$
$$\frac{\partial(\rho v_i)}{\partial t}+\frac{\partial(\rho v_i v_j)}{\partial x_i}=-\frac{\partial P}{\partial x_i}+\frac{\partial}{\partial x_i}\left[\mu\left(\frac{\partial v_j}{\partial x_i}+\frac{\partial v_i}{\partial x_j}-\frac{2}{3}\delta_{ij}\frac{\partial v_k}{\partial x_k}\right)\right]+\frac{\partial}{\partial x_i}\left[-\rho\overline{v'_i v'_j}\right], \quad (1)$$
$$\frac{\partial(\rho E)}{\partial t}+\frac{\partial[v_i(\rho E+P)]}{\partial x_i}=\frac{\partial}{\partial x_j}\left[\left(\kappa+\frac{c_P\mu_t}{Pr_t}\right)\frac{\partial T}{\partial x_j}+v_i(\tau_{ij})_{eff}\right]+N(\vec{x}),\ E=h-\frac{P}{\rho}+\frac{v^2}{2},\ P=\frac{\rho T}{M},$$

where $(\tau_{ij})_{eff}=\mu_{eff}\left(\frac{\partial v_j}{\partial x_i}+\frac{\partial v_i}{\partial x_j}\right)-\frac{2}{3}\mu_{eff}\frac{\partial v_k}{\partial x_k}\delta_{ij}$ is the deviatoric stress tensor; $\left[-\rho\overline{v'_i v'_j}\right]$ are the Reynolds stresses which must be modeled to close the set of equations, $v_i, v'_i, \rho, T\ P, E,$ and h are the mean and fluctuating velocity components, density, temperature, pressure, total energy, and enthalpy, respectively; N is the energy source, μ, μ_t, μ_{eff} are the molecular, turbulent, and effective viscosity coefficients, respectively; c_p is the molar specific heat capacity at constant pressure; κ is the thermal conductivity coefficient; and Pr_t is the turbulent Prandtl number.

The model heat source was a homogeneous ellipsoid of revolution located between the anode and the cathode. The small axis of the ellipse equals to 1 cm. In what follows, we investigate the problem for pure Ar close to the experimental conditions [6], corresponding to the total mass flow rate $m_t = 1 - 2$ g/sec, the heat source power $N = 50 - 100$ W and the static gas pressure $P = 10^5$ Pa.

Initial values of pressure and overall temperature were set to 10^5 Pa and 300 K, respectively. The initial value of the input velocity is calculated automatically by the mass flow rate, the inlet diameter and the gas density.

The no-slip velocity conditions were imposed along the tube, nozzle, electrode surfaces etc. Outer surfaces of all internal construction elements were kept at ambient temperature of 300 K. For these surfaces we use the "coupled" boundary condition without additional heat generation on the dividing interior wall. All internal boundaries separating the gas flow with electrodes, a glass tube, a nozzle and all internal boundaries between the solid elements of the setup are thermally conductive.

Boundary conditions for the inlet zone are the mass-flow-inlet type. At the duct inlet fixed mass flow rate condition was used. The maximum inlet turbulent intensity is set to 5%. Variation of the inlet turbulent intensity in the range of moderate values of 1-5% did not affect the flow structure discernibly. The boundary conditions for the outlet zone are of the pressure-outlet type with gauge pressure equals to zero. Therefore, at the outlet, we imposed the boundary conditions with the static pressure equal to the atmosphere pressure.

2.2. Numerical procedure

The numerical simulation of the non-stationary 3D turbulent swirling flow was made using the ANSYS FLUENT 15.0 program package, which solves system (1) using the finite volume method. In our simulation, the Spalart-Allmaras model was used. A detailed analysis of the use of the turbulence models in similar problems is given in [8].

For spatial discretization of density, momentum, energy and turbulent quantities, a second-order upwind scheme is applied. The use of higher-order schemes has not changed the flow parameters considerably. The diffusion terms are central-differenced and second-order accurate. We interpolate the pressure values at the faces using the PRESTO! scheme because of the strong swirl nature of the flow.

For transient terms, we used the fully implicit scheme of the second-order accuracy. Different pressure-velocity coupling schemes were tested and gave equal results. So, we used the SIMPLE scheme as the least resource intensive. The convergence was obtained when the residual reached 10^{-6} for the energy equation and 10^{-4} for the continuity equation, the momentum equation, and the equations for turbulent quantities. The geometric modelling of the setup and the construction of the grid were carried out using the ANSYS Workbench package. The computational grid consisted of about $3 \cdot 10^6$ hexahedral cells, except for a small region, where the cells were tetrahedral (figure 2). The skewness metric has an average value of 0.16, while the orthogonal quality metric equals 0.94. The grid refinement did not considerably alter the gasdynamic parameters.

Figure 2 The horizontal cross sections of grid region with mixed cells structure

Calculations with different values of time step from $1\cdot10^{-4}$ sec to $1\cdot10^{-6}$ sec were carried out. Finally, the time step was set equal to $5\cdot10^{-5}$ sec since using lower values led only to an increase in the computational time.

3. Results of numerical simulation

To a great extent, the structure of the flow is determined by the paraxial counterflow zone, typical for strongly swirled flows [8,9]. Formation of recirculation zone is shown in figure 3. Note that figure 4 demonstrates only counterflow regions. During short initial period of time, the zone of the maximum counterflow velocity is concentrated in the region directly behind the cathode. Then it moves quickly enough to the opposite part of the setup ($t = 1$ sec), and completely forms at $t \approx 5$ seconds. After that, no fundamental changes in the shape of the region and velocity values occur. Thus, a quasi-stationary flow velocity field is formed in the duct.

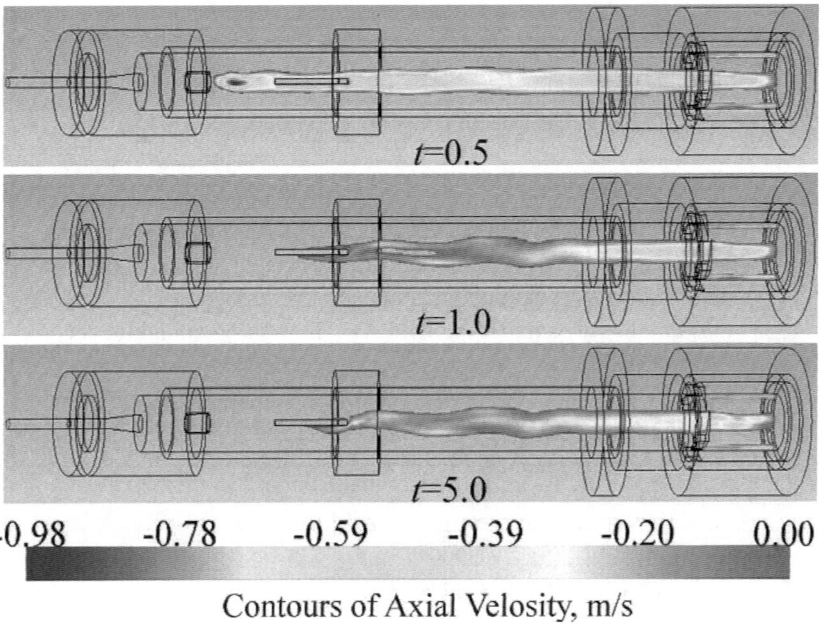

Figure 3 Counterflow zone structure in a fixed cross section obtained at different time moments. The mass flow rate $m_t = 2$ g/sec, the heat source power $N = 100$ W, the static pure argon pressure $P = 10^5$ Pa

Therefore, a flow structure is characterized by the presence of a long paraxial counterflow zone with relatively low flow velocities. Use of solid electrodes leads to the formation of a counterflow with appreciably large velocity magnitudes, which in turn leads to the formation of a heat flux directed toward the swirler and its heating, as noted in [6].

A similar flow structure was described in [8], where analogous swirler and alike paraxial heat source were used, but in combination with a fully open duct. The numerical simulation is also demonstrated that as well as in [8] that a precessing vortex core (PVC) appears. The obtained PVC is the left-handed co-rotated bending single-vortex structure.

The simulation showed that combination of thin anode and pipe-like cathode leads to desirable flow structure when the pressure gradient along the symmetry axis draws out the hot gas from the interelectrode area. As a result, the formed flow structure leads to the propagation of a heat flux mainly downstream, figure 3. It should be noted that figure 4 demonstrates only regions heated to a temperature above 400 K. Comparison with previous investigation [7] did not reveal fundamental differences in flow structure because of additional heat transfer. Due to heating of electrodes and losses through construction elements, there is decrease in output heat power of up to 20%. Although being significant, it leads only to some scaling which is manifested through lower flow temperature, at least at stages far from "saturation" when temperature of subsidiary parts becomes steady. Such process of warming-up takes much more time to complete (~30 min, [6]) than typical gasdynamics phenomena so the whole behavior of the reactor can be considered as quasi-stationary.

Thus, the considered localization and form of the electrodes allows to avoid a noticeable heating of the setup structural elements excepting the massive nickel cathode, which is heated to a temperature close to the melting point and, less, the input of the nozzle. It leads to a decrease of PVR coefficient of energy performance (COP) in comparison with the model system [6]. However, the losses in the case considered here are the smallest among the configurations described in [6].

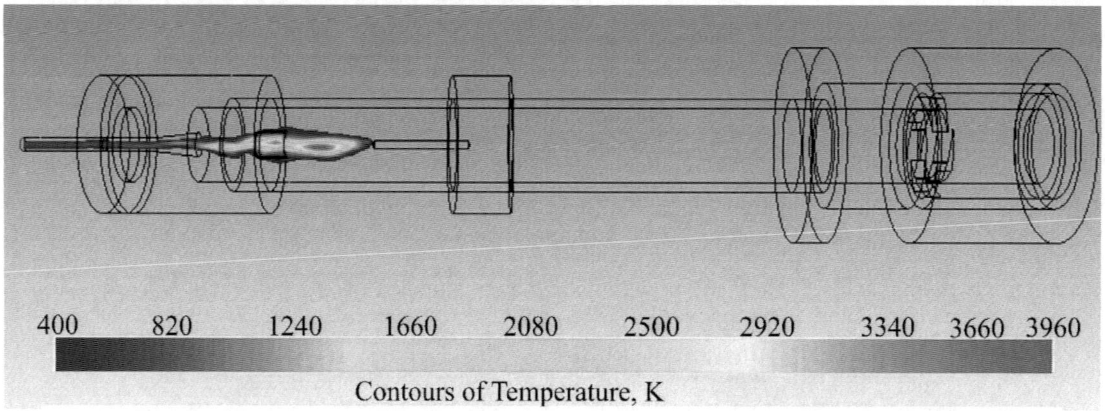

Figure 4 Gas temperature field at fixed time moment t = 100 sec. The conditions correspond to those shown in figure 3

The presented results demonstrate a qualitative agreement with the results of the experimental data. Indeed, the control of the flow mean temperature measured by thermocouple grid TP-K01 and CENTER 304 thermometer showed that the output mean temperature of argon in the presence of two solid electrodes was $T_g \sim 350$ K, whereas in the case of a combination of a thin anode and a hollow cathode located at the duct outlet, it was $T_g \sim 700 - 800$ K. These results are corresponded to the total mass flow rate $m_t = 2$ g/sec, the heat source power N=100 W and the static pure argon pressure $P = 10^5$ Pa, used in numerical simulations.

4. Conclusion

The numerical modelling has demonstrated that it is possible to configure electrodes in such a way as to ensure energy transfer mainly downstream from the interelectrode zone. It may be accomplished by the pipe-like electrode through which areas of different pressures are connected. Pressure forces draw out the hot gas towards the exit nozzle. Thereby, it can be assumed that the considered setup model is optimal for creating a compact vortex heat generator. A qualitative agreement between the results of the calculations and the experimental data for pure argon [6,7] has been obtained for the exact experimental setup.

However, the construction of effective heat and hydrogen generator remains simulation of the flow and discharge structure for real mixtures of metal, H_2O and Ar, taking into account the plasma-chemical kinetics of the discharge mixture components as well as thermal characteristics of the experimental set up. It should be noted in this connection that optimal condition for heat generation and hydrogen generation can differ from each other. Therefore, the next step is the self-consistent simulation of the problem.

Acknowledgments
The study was supported in part by the Ministry of Education and Science of Russia under the public contract with educational and research institutions within the project 3.1158.2017/4.6 and by RFBR under grant 16-41-630591.

References

[1] Yiguang J and Wenting S 2015 Plasma assisted combustion: Dynamics and chemistry *Prog. Energy Combust. Sci.* **48** 21
[2] Jasiński M, Dors M, Nowakowska H and Mizeraczyk J 2008 Hydrogen production via methane reforming using various microwave plasma sources. *Chem. Listy* **102** 1332–9
[3] Klimov A, Bitiurin V, Tolkunov B, Chinnov V, Godin S, Efimov A, Kutuzov D and Polyakov L 2011 Hydrogen Plasma Flow Creation for MHD Power Generation *AIAA Pap.* 2011-3285
[4] Bitiurin V A, Klimov A I, Korshunov O V and Chinnov V F Kinetic 2014 Model of Aluminum Oxidation by Water Vapor in Heterogeneous Plasma: Gas-Phase Kinetics *High Temp.* **52** 657–63
[5] Bitiurin V A, Klimov A I, Korshunov O V and Chinnov V F 2015 Kinetic Model of Al Oxidation by Water Vapor in Heterogeneous Plasma: Heterophase Kinetics *High Temp.* **53** 21–6
[6] Belov N K, Zavershinskii I P, Klimov A I, Molevich N E, Porfiriev D P and Tolkunov B N. 2018 High effective heterogeneous plasma vortex reactor for production of heat energy and hydrogen *J. Phys: Conf. Ser.* **980** 012040
[7] Kazanskii P N, Klimov A I, Molevich N E, Porfiriev D P and Zavershinskii I P 2018 Numerical simulation of an argon swirling flow in the presence of a DC discharge *J. Phys: Conf. Ser.* **980** 012010.
[8] Gorbunova A, Klimov A, Molevich N, Moralev I, Porfiriev D, Sugak S and Zavershinskii I. 2016 Precessing vortex core in a swirling wake with heat release *Int. J. Heat and Fluid Flow* **59** 100–8
[9] Syred N 2006 A review of oscillation mechanisms and the role of the precessing vortex core (PVC) in swirl combustion systems *Prog. Energy. Combust. Sci.* **32** 93–161

IOP Conf. Series: Journal of Physics: Conf. Series **1112** (2018) 012021 doi:10.1088/1742-6596/1112/1/012021

Thunderstorm activity influence mechanism on lower ionosphere

N V Ardelyan [1], V L Bychkov [1*], G V Golubkov [2], M G Golubkov [2] and K V Kosmachevskii [1]

[1] Moscow State University, Moscow, Leninskie gory, 119991, Russia
[2] Semenov ICP RAS, Moscow, Kosygin str. 4, 119991, Russia

*bychvl@gmail.com

Abstract The influence of a charged cloud the electric field on chemical processes in the air at heights of 95-100 km has been studied. Numerical simulation of the plasma at these altitudes is performed, the concentrations of main charged particles are determined under the conditions of the electric field created by the unipolar charge of the cloud. It is shown that the electric fields of the clouds after the discharges of linear lightning affect the electron concentration at these altitudes.

1. Introduction

The study of electric fields influence on the ionosphere, in particular of thunderstorm clouds, is of considerable interest for global satellite positioning systems GPS/GLONASS. Since in normal geomagnetic conditions, the average electron concentration in the altitude layer 95 - 100 km of the order of 10^3 cm^{-3}, then in this case the characteristic delay time of the signal is of the order of 100 ns, and the usual horizontal positioning error is several meters [1,2]. The situation drastically changes when the periods of increased activity of the Sun are formed and strong magnetic storms appear in the Earth's ionosphere. In these cases, the operation of satellite navigation systems is violated and leads to significant errors of the order of hundreds of meters, and sometimes the satellite signal disappears at all [3]. This is a fundamental scientific and technical problem, which has not been yet solved.

In recent years, mainly with a help of ballistic missiles it has been established that the strongest distortions of satellite signals occur in the ionospheric layer 80-110 km [4]. If the concentration of slow electrons at the upper 110-km boundary is of the order of 10^6 cm^{-3} [5], in the 90-100-km layer that we are examining, it corresponds to a value of the order of 10^5 cm^{-3}, then the GPS/GLONASS satellite systems experience noticeable perturbations [6]. In this paper, an analysis is made of the increase in the electron concentration as the result of thunderstorm activity. This phenomenon has not been studied earlier.

Strong interest to influence of electric fields on the ionosphere is connected with creation of artificial perturbations of *D*-layer of the ionosphere by powerful short wave radiation such as transmitting complexes Sura and Poisk [7,8] and by investigations in the field of non-linear propagation of radio waves in the ionosphere [9].

Interest to ionization processes at high altitudes is caused by a possibility of ionization processes in the air under thunderstorm conditions when the appearance of atmospheric discharges such as sprites, and jets takes place. As follows from [7], it is the thunderstorm phenomena that lead to their realization.

The implementation mechanism is as follows. When a cloud strikes the lightning on the ground, the charges of the same sign are leaking, while the charges of the other sign remain on the cloud. The magnitude of charges on the cloud reaches 5 C, which create an electric field above the cloud. This electric field falls off slowly with altitude, and the density of neutrals exponentially fast, so at heights of 70-100 km the conditions necessary for the implementation of sprites are realized. It was shown in [8] that simultaneously with the realization of the conditions for the appearance of streamer discharges, avalanche ionization occurs, leading to a significant increase in the electron concentration of the electron and gas temperatures, which depends on the magnitude of the external electric field that can be realized by the experimental devices, clouds, and mesospheric processes.

When lightning fronts move at altitudes of 2-10 km in the mesosphere at altitudes of 90-100 km, conditions can be created for the formation of a front of ionized air with a width of ~ 1%, larger than the width of the cloud front at the Earth. In this case, the lifetime of such an ionized region will be equal to the time of neutralization of the cloud by tropospheric ions, which can reach hundreds of seconds.

The purpose of this work is: to determine the possible electron concentration of the plasma on the basis of plasma chemical kinetics studying at altitudes of 90-100 km in a thunderstorm field.

2. Formation of the charged region in the lower ionosphere

For analysis, we use a well-known expression for the dependence of the quasi-stationary electric field over the height [10] h

$$E = \frac{zQ}{\pi\varepsilon_0 h^3}\left[1 + \left(\frac{h_i}{2h_i - h}\right)^3\right], \quad (1)$$

here E is the value of the electric field at the required altitude, h_i is the height of the ionosphere, which is assumed to be an ideal conductor (in the calculations $h_i = 120$ km), z is the height of the charge Q over an ideally conducting surface (5 km, $z \ll h$). This formula describes the electric field strength over a charged cloud, taking into account the polarization effect in the ionosphere.

Estimates obtained on the basis of (1) give the following electric field value of 1.1 V/m at the altitude of 95 km and 0.9 V/m at the altitude of 100 km for a charge of the cloud equal to 5 C located at the altitude of 5 km. With this respect, the problem is similar to those considered in [7]. Consequently, the presence of charges on the clouds which can occupy several tens km areas [11] and the maximal charge areas, according to figure 3.4 from [11] can be of several km, lead to a sharp increase in the electron concentration. In the case of the cloud system, when the electric field strengths of individual clouds add up in the mesosphere the effective charge effect can be much greater at the mesoscale level.

To study the chemical kinetics of plasma in air at altitudes of 90-100 km, as in our works [8-10], as the starting point in the analysis of the dry chemical plasma-chemistry, we chose a system of chemical reactions corresponding to the works [12-16]. This model includes 26 components (neutral, positive and negative ions, electrons and excited particles) and more than 180 plasma-chemical reactions with their participation. In this case, the electron energy distribution function was considered to correspond to a glow discharge in air, the rate constants of electron-molecular reactions were calculated on the basis of the Boltzmann equation with cross sections from [17]. The rates of excitation, dissociation, and ionization of molecules by fast particles are taken from [18].

Along with the balance equations for atomic particles, an equation for the temperature of electrons and gas was solved. In the equation for the electron temperature we took into account: rotational and elastic excitation of molecules, the effects of electron heating in the recombination of electrons and ions [15], the cooling of electrons during the triple attachment of an electron to an oxygen molecule; ionization and excitation of molecules in collisions of electrons with air-molecules. Heating of electrons in the external electric field of the cloud was taken into account. We also considered heating of electrons during the injection of fast electrons of the background into the region below the vibrational excitation threshold [14,19] as it is usually applied in problems connected with degradation of energy of fast particles. In the equation for electron temperature we accounted cooling of electrons in elastic and inelastic collisions with O_2 and N_2 molecules.

IOP Conf. Series: Journal of Physics: Conf. Series **1112** (2018) 012021 doi:10.1088/1742-6596/1112/1/012021

In the equation for the gas temperature T we included terms corresponding to heating as a result of the relaxation of the vibrational and electronic degrees of freedom of molecules, heating and cooling of the gas in chemical and plasma-chemical reactions (heating by electrons in collisions is Joule heating). Reactions involving positive ions $O+, O2+, O4+, O6+, O8+, N+, NO+, N2+$, negative ions O^-, O_2^-, O_3^-, atoms N, O, and molecules O_2, O_3, N_2, N_2O, NO were taken into account in the simulation.

Excited vibrational states of nitrogen N_2 (v), excited electronic states of nitrogen N_2 (b), N_2 (z), excited vibrational states of oxygen O_2 (v), excited electronic states of oxygen $O_2(\,^1\Delta_g)$ were also accounted. The recombination reactions of each positive ion with each negative ion are taken into account. The data on these processes were determined on the basis of the Flannery approach [15], taking into account the gas temperature and pressure. In the calculations, it was assumed that de-excitation of vibrational and rotational states occurs mainly in collisions with molecules. The model also takes into account the processes of charge exchange of ions [20]. The data on excitation and ionization of the background in the ionosphere was taken from [21], where we used the total excitation and ionization rates at the considered altitude. The power W inputted into the gas by the fast particles of the background is usually determined by us in units of eV/(cm^{3}sec), and the excitation rate of the molecules Q is connected with it by the relation [14,18] $Q = W/Ui$, where Ui is the ionization cost, in particular in the air, $Ui = 31.6$ eV.

In the analysis, the background values of the components were calculated first for a complete set of components at given values of the excitation and ionization costs as functions of the altitude and the temperature, taken from [21, 22].

They served as initial values for calculating the kinetics of charged, neutral, and excited particles in the plasma. This corresponded to the model establishment of plasma-chemical equilibrium at a given altitude at a given pressure and temperature.

The plasma in the calculations is represented as an ideal gas with an adiabatic exponent of 1.4, which restricts the applicability of the model only to the case of a weakly ionized gas, when the concentrations of charged particles are much smaller than the concentration of diatomic components of the air.

In this study and other ones on a similar subject, we use a model described by the ODE system of plasma chemical kinetics, taking into account the energy equations of gas and electrons. The model makes it possible to find the time dependences for the concentrations of the components of a weakly ionized plasma and temperatures. In the context of the purposes of the simulation, it is not necessary to consider more complex models that take into account the motion of the plasma and the dependence of the concentrations on the spatial coordinates, since we are interested in the most common fundamental aspects - the composition of the plasma, its concentration, the influence of external factors such as electric field, effect of fast particle fluxes and others. Spatial distributions are significant in more specific applied problems of plasma interaction with bodies, as well as in the application of developed fundamental concepts and methods for analyzing thermal inhomogeneities arising in the ionosphere. In addition, the spatial-temporal modeling, taking into account the fairly complete plasma chemical description used by us, including dozens of components and hundreds of reactions, requires a dramatic increase in computational resources and a fundamental development of the computational methods used.

From the point of view of taking into account the motion, the model includes two approximations: the approximation of constant density and the approach of constant pressure. In the first case, the gas does not have time to move in the volume under consideration during the times under consideration, in the second case times are so large (or linear dimensions are so small) that the gas pressure is equalized by the gas dynamic motion. When considering processes in a gas, motion can occur as a result of change in pressure during heating or cooling. Thus, in the absence of heating or cooling, these models are equivalent. As can be seen from the results of numerical simulation presented below, the processes of plasma modification that we are interested in under the influence of an electric field occur at a very slightly varying gas temperature, and therefore they are adequately described in any of the two indicated approximations. The thermal instability observed at the final stage, accompanied by strong heating and an increase in the degree of ionization, is described qualitatively within the framework of the model, and it should be noted here that calculations in the constant-pressure approximation yield qualitatively the same results. A more accurate simulation of the stage of thermal instability and subsequent processes goes beyond the scope of this paper and requires both the development of a description of plasma chemical reactions at relatively high temperatures and the development of the model in the direction of non-fulfillment of weak ionization approximations and a slight change in the partial composition of the gas as a whole.

3. Calculation results

The calculations were performed over a wide range of excitation rates by the external electric fields. The data from [17, 18] and presented in the Table were chosen as the basis. The values of $E = 1.0, 1.5$ and 2.0 V/m obtained by formula (1) at heights of 95 and 100 km were taken as the background electric field. The following notation is used in the figures below: e - electron concentration, $Op - O^+$, $O2p - O_2^+$, $O4p - O_4^+$, $O6p - O_6^+$, $O8p - O_8^+$, $Np - N^+$, $NOp - NO^+$, $N2p - N_2^+$, $O2pN2 - O_2^+N_2^+$, $Om - O^-$, $O2 - O_2^-$, $O3m - O_3^-$, $O2 - O_2$, $O3 - O_3$, $N2 - N_2$, $N2O - N_2O$, $N2v - N_2(v)$, $N2b - N_2(B^3\Pi_g)$, $N2z - N_2(X^1\Sigma_g^+)$, $O2v - O_2(v)$, $O2d - O_2(\,^1\Delta_g)$.

Table Parameters of atmosphere at heights 95 and 100 km

h, km	N, cm^{-3}[21]	T_{gas}, K [21]	N_e, cm^{-3}, December [21]	Q, (cm^{3}s)$^{-1}$ [21]
95	2.93 10^{13}	188	2 10^4	100
100	1.93 10^{13}	195	5 10^4	5.0 10^2

Figures 1a-c show the plasma component concentrations at the altitude of 95 km. Figure 2 shows the plasma component concentrations at the altitude of 100 km. Symbols T correspond to gas temperature, which rises at the stage of thermal instability taking place under long influence of electric field.

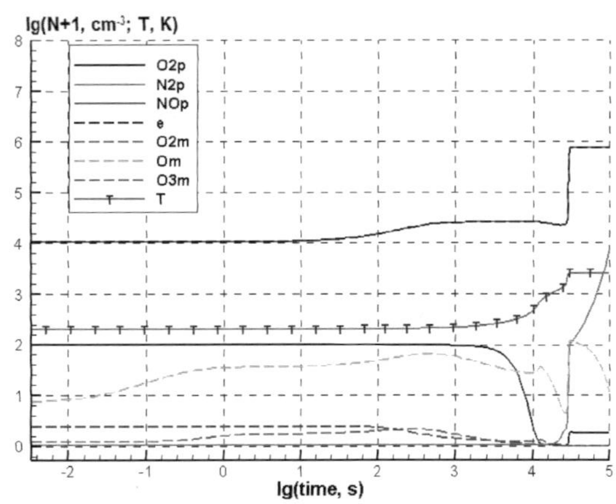

Figure1a Main charged particles in air at the altitude of 95 km at the external electric field of 1.0 V/cm, initial E/N =3.14·10^{-14} V·cm^2 (symbols T correspond to lg(T,K)).

An investigation in such a wide range of gas excitation by the external electric field was caused by the need to elucidate the behavior of the main components excited by the external source. In this case, we present data over a wide range of excitation times. Therefore, the results obtained are also useful in analyzing the effect of not only charged clouds, but also artificial sources of the electric field.

It can be seen from the calculations that even under the conditions of excitation with one cloud charge at the altitude of 95 km, the electron concentration increases by a factor of 1.5-2 in the time 10^2-10^3 seconds. At long times, the input of energy by a discharge leads to the heating of the gas and the development of thermal instability accompanied by a sharp increase in the electron concentration. With an increase in the electric field by a factor of 1.5, the electron concentration increases in 10^2 seconds by an order of magnitude with respect to the background concentration of 10^4 cm^{-3} and reaches the value of $2.5 \cdot 10^5$ cm^{-3} during the next 10^2 seconds. With an increase in the electric field up to 2.0 V/m, the electron concentration rises to 10^5 cm^{-3} during 6 seconds, it is followed by a rapid increase in the electron concentration accompanied by the gas heating.

As the field increases, along with the positive NO^+ ion, the role of the O_2^+ ion becomes significant. Also, at a maximum field of 2 V/cm considered, the role of the negative O^- ion becomes significant at times greater than 0.1 s.

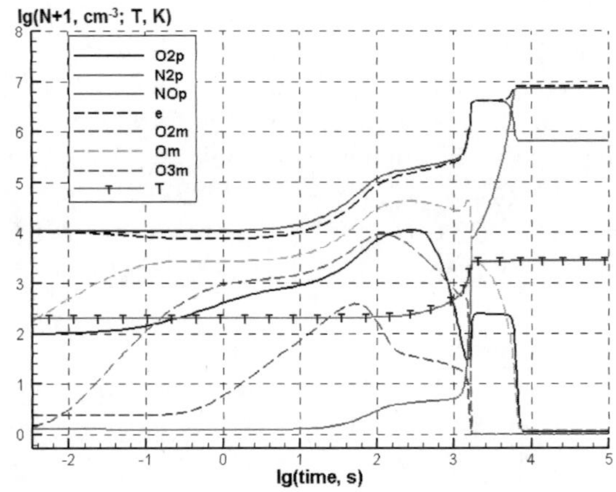

Figure 1b. Main charged particles in air at the altitude of 95 km at the external electric field of 1.5 V/cm, initial $E/N = 4.7 \cdot 10^{-14}$ V·cm^2, (symbols T corresponds to lg(T,K)).

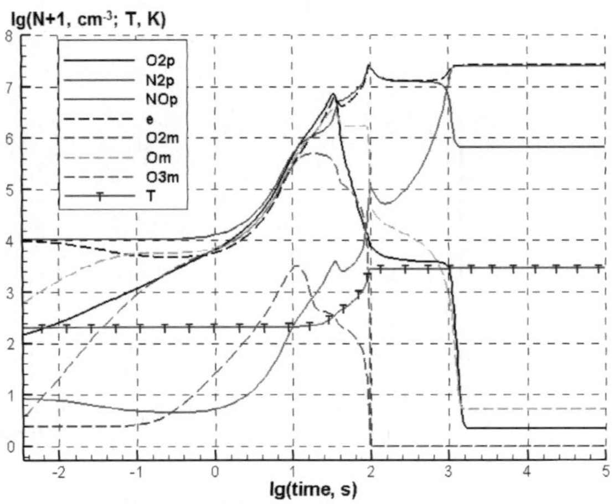

Figure 1c. Main charged particles in air at the altitude of 95 km at the external electric field of 2.0 V/cm, initial $E/N = 6.3 \cdot 10^{-14}$ V ·cm^2 (symbols T correspond to lg(T,K)).

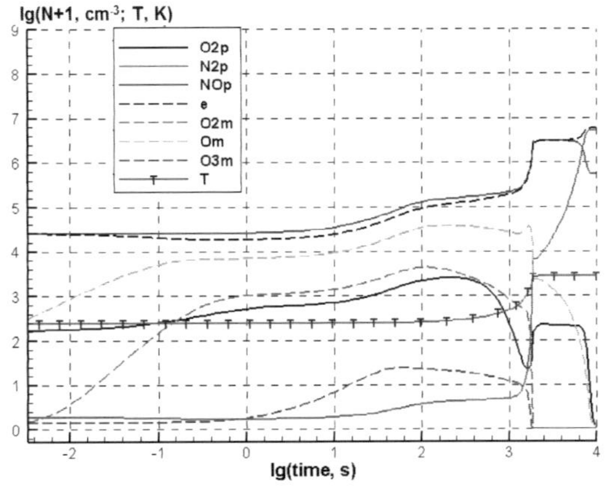

Figure 2a Main charged particles in air at the altitude of 100 km at the external electric field of 1.0 V/cm, initial E/N =4.8·10^{-14} V·cm^2 (symbols T corresponds to lg (T,K)).

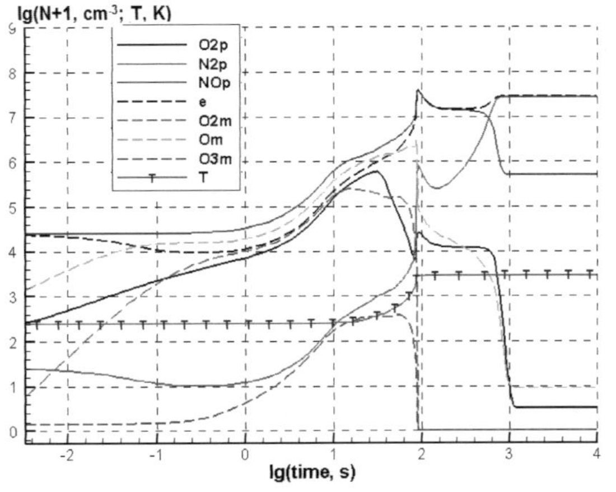

Figure 2b Main charged particles in air at the altitude of 100 km at the external electric field of 1.5 V/cm, initial E/N =7.1·10^{-14} V·cm^2 (symbols T corresponds to lg (T,K)).

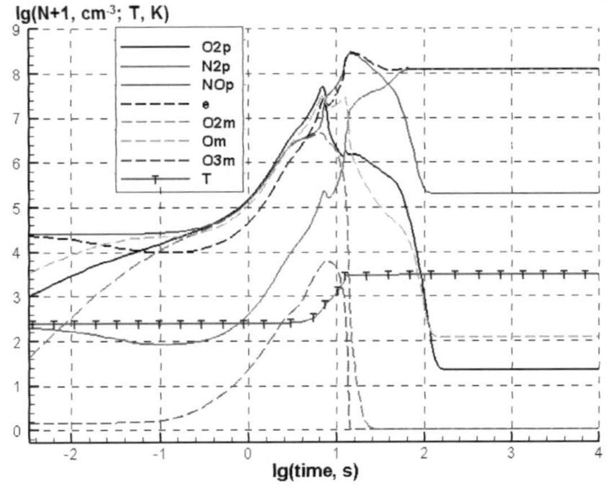

Figure 2c Main charged particles in air at the altitude of 100 km at the external electric field of 2.0 V/cm, initial E/N =9.6·10^{-14} V·cm^2 (symbols T corresponds to lg (T,K)).

Calculations for 100 km show that even at $E = 1.0$ V/m, the electron concentration reaches 10^5cm^{-3} during the time of 10^2 seconds, and during $1.5 \cdot 10^3$ seconds the plasma transits into a thermal instability regime with avalanche growth of the electron concentration. At $E = 1.5$ V/m, the electron concentration increases to 10^5 cm^{-3} in 6 seconds. At long times from 10 to 100 seconds, the input of energy by a discharge leads to the heating of the gas and the development of thermal instability accompanied by a sharp increase in the electron density to. With an increase in the electric field to $E = 2$ V/m, the electron concentration increases to 10^5 cm^{-3} in 1.5 seconds and reaches the value of $3.0 \cdot 10^7$ cm^{-3} during the next 6 seconds.

The ion composition changes with an increase of the electric field (within 100 seconds) from the main NO^+ ions to the mixture of O_2^+ and NO^+ ions, at that the negative ion O^- begins to play an important role in the balance of particles at this altitude. Under the influence of fields of 1.5 V/cm and 2 V/cm, the plasma becomes electronegative in the intermediate stage of plasma formation, at times from 0.1 s. up to 10 s.

Considering the general features of the ionization process under the influence of an electric field at altitudes of 95-100 km, we should note a significant increase in the concentration of charged particles for a relatively long time, taking place without heating the gas. Heating occurs later, with the development of thermal instability. The description of this stage within the framework of the model is of a qualitative nature and is of interest to us only in terms of fixing the transition of the cold plasma to a heated state with an increased degree of ionization occurring at sufficiently large fields.

It should be noted that after rapid heating of the plasma under the influence of thermal instability, one observes a substantial dissociation of oxygen and nitrogen molecules. However, at the growth times of the electron concentration, the concentrations of the atoms remain smaller than the concentrations of the O_2 and N_2 molecules, so we can assume that the parameters of the model (the adiabatic exponent) do not change in the calculations.

With the growth of the electric field, takes place a substantial dissociation of oxygen and nitrogen molecules. However, at the calculated growth times of the electron concentration before the stage of thermal instability, the concentrations of the atoms remain smaller than the concentrations of the O_2 and N_2 molecules, so the parameters of our model (the adiabatic exponent) do not change in the calculations at the stage of plasma formation of interest to us.

Conclusion

Based on the plasma-chemical model, the composition of air plasma at heights of 95-100 km in a wide range of air excitation by the electric field of clouds or some other external sources of the electric field was analyzed. It showed that in the stationary conditions at fields $E = 1.5 - 2$ V/m comparable and somewhat larger than the electric fields created by individual clouds, sharp ionization can occur. With the formation of electrons with a concentration of up to 10^5 cm^{-3} and above, which can affect the functioning of electronic instruments at mesospheric heights. This effect will be manifested much greater at passing of thunderstorm fronts above the ground, when a separate effect of clouds merge into the total effect of the front. The same impact can also have artificial sources of a strong field, the actions of which have also to be taken into account.

The increase in the electron concentration by two orders of magnitude as a result of thunderstorm activity calculated in this paper should lead to rather strong distortions in the operation of satellite positioning systems. Since previously, this phenomenon has not been systematically studied it requires further research.

Acknowledgements

This work was carried out in the framework of State Assignment of Russian Federal Agency of Scientific Organizations (project 0082–2018–0001, registration code AAAA–A17–117112240026–5).

References

[1] Golubkov G V, Manzhelii M I and Karpov I V 2011 *Khim. Fiz.* **30** 55

IOP Conf. Series: Journal of Physics: Conf. Series **1112** (2018) 012021 doi:10.1088/1742-6596/1112/1/012021

[2] Golubkov G V, Golubkov M G and Manzhelii M I 2012 *Khim. Fiz.* **31** 31
[3] Golubkov G V, Manzhelii M I and Karpov I V 2013 *Khim. Fiz.* **32** 64
[4] Golubkov G V, Golubkov M G and Manzhelii M I 2013 *Dokl. Akad. Nauk* **452** 510
[5] Golubkov G V, Golubkov M G and Manzhelii M I 2012 *Khim. Fiz.* **33** 64
[6] Golubkov G V, Golubkov M G Manzhelii M I and Karpov I V 2014 *The Atmosphere and Ionosphere. Elementary Processes, Monitoring, and Ball Lighting* ed Bychkov V L Golubkov G V et. al. (Springer International Publishing)
[7] Belikovich V V, Benediktov E A, Vyakhirev V D 1995 *Geomagn. Aeron.* **35** 107
[8] Bakhmetieva N V Grigoriev G I Tolmacheva A V 2018 *Atmosphere Ionosphere Safety* eds. Karpov I V, Borshevkina O P et. al. (Kant Baltic federal university publishers) vol. **1** p. 20
[9] Gurevich A V Svartsburg A B 1973 *Non-linear theory of radio waves propagation in ionosphere* (Moscow: Nauka)
[10] Raizer Yu P, Milikh G M, Shneider M N and Novakovski S V 1998 *J. Phys.* D: *Appl Phys* **31** 325
[11] MacGorman R D, Rust W D 1998 *The electrical nature of storms* (Oxford University press, Oxford)
[12] Ardelyan N V, Bychkov V L, Kochetov I V and Kosmachevskii KV 2014 *The Atmosphere and Ionosphere. Elementary Processes, Monitoring, and Ball Lighting* ed Bychkov V L Golubkov G V et. al. (Springer International Publishing) p 69
[13] Ardelyan N V, Bychkov V L, Volkov S A, Kosmachevskii K V and Kochetov I V 2015 *Khim. Fiz.* **34** 63
[14] Ardelyan N V, Bychkov V L, **Bychkov D V and Kosmachevskii K V** 2013 Electron-Beam and Non-Self-Maintained Driven Plasmas for PAC. *Plasma assisted combustion, gasification and pollution control,* vol.1, ed. I.B. Matveev. (Outskirts press. Denver. Colorado) p 183
[15] Kossyi I, Kostinsky A, Matveyev A and Silakov V 1992 *Plasma Sources Sci.Technol* **1** 207
[16] Akishev Yu S, Deriugin A A, Kochetov I V, Napartovich A P, Trushkin N I 1996 *Fiz. Plazmy* **20** 585
[17] Yousfi M, Azzi N, Segur Pet al. 1987 *Electron-molecule collision cross sections and Electronswarm parameters in some atmospheric gases (N_2, O_2, and H_2O)* (Center de Physique Atomique de Toulouse & Instituto di Electtrotecnica ed Electronica. Universita di Padova) 1
[18] Konovalov V P Son E E 1987 *Plasma Chemistry* ed. Prof. B M Smirnov (Moscow: Energoatomizdat) vol. 14 p. 194
[19] Bychkov V L, Eletskii A V 1985 *Plasma Chemistry* ed. prof. B M Smirnov (Moscow: Energoatomizdat) vol. 12 p. 119
[20] Virin L I, Dzhagatspanyan R V, Karachevtsev G V, Potapov V G, Tal'roze V L 1979 *Ion-molecule reactions in gases* (Moscow: Nauka publishers)
[21] Brasseur G B and Solomon S 1984 *Chemistry and physics of the Stratosphere and Mesosphere* (Dordrecht. Holland. D. Reidal publishing company)
[22] Sedunov Y S 1991 *Atmosphere. Reference book* (Leningrad)

IOP Conf. Series: Journal of Physics: Conf. Series **1112** (2018) 012022 doi:10.1088/1742-6596/1112/1/012022

Production of long-lived heterogeneous structures using plasma

V I Abakumov[1], V L Bychkov[1*], A R Bikmukhametova[1], V A Chernikov[1], G V Golubkov[2], D A Safronenkov[1] and S V Zaitsev[1]
[1]Lomonosov MSU, Faculty of Physics, Leninskie Gory, Moscow 119991, Russia
[2]Semenov ICP RAS, Moscow, Kosygin str. 4, 119991, Russia

*bychvl@gmail.com

Abstract Experiments in the air were carried out using capillary and corona discharges in order to obtain long-lived heterogeneous structures up to 2 mm and greater. Electron microscopy photo of one of them obtained in the capillary discharge included a core of a vaporous substance and a quasi-solid shell. Carried out experiments with corona discharges over tap and distilled water, alcohol, glycerin and their mixtures have shown appearance of liquid jets and columns which can be caused by different hydrodynamic instabilities. In complex liquids there is a possibility of creation of structures with long lifetime.

1. Introduction

Investigation of large (up to 2 cm and greater in diameter) luminescent heterogeneous structures, the so-called long-lived heterogeneous structures (LLGS), is of significant interest for low-temperature plasma physics, both in terms of determining conditions and thermodynamics of such structures created in non-equilibrium conditions [1-4]. They represent interest from a point of view of their possible applications as combustion activators in plasma aerodynamics and plasma assisted combustion at injection of LLGS created in plasma into fuel mixtures [5], or creation of corona discharges in combustors for creation of small particles of a fuel in result of developed instabilities and applied as combustion activators.

Already during the last twenty years, studies have been carried out [4,6-8] showing that in a plasma created by erosive plasma generators, large enough long-lived luminous formations with a diameter of up to 8 millimeters and a lifetime of up to 6 seconds appear when plasma is exposed to samples of metals and polymeric materials. These formations have a complex composition and differ in structure from the drops of molten metal. A sequential study of the multiple production of such formations and their structure has not been carried out yet, therefore, studies of the possibility of formation of a flow of such objects are of considerable interest.

Because of possible applications in plasma aerodynamics present investigations were also devoted to formulation and undertaking of experiments with corona discharges realized over surfaces of different liquids. Investigations when a surface of one of the electrodes is covered by a liquid are of interest for hydrocarbon fuel activation, search of undesirable liquids elimination in engines, problems of ecology and disinfection [9]. They are of a fundamental interest, since the corona discharge allows to obtain new information on phenomena connected with development of electro-hydrodynamic instabilities in new, earlier unrealized conditions [10-12]. Such investigations are developing now, that is a reason for undertaking of these ones. In this work we were concentrated on electric-hydrodynamic features of corona discharges which are practically uninvestigated yet. However, there is limited

IOP Conf. Series: Journal of Physics: Conf. Series **1112** (2018) 012022 doi:10.1088/1742-6596/1112/1/012022

information about any discharges over surfaces of liquids [9] and it is obtained mainly for electrolytes or water. From the application point of view it is interesting to know how a surface of a liquid fuel acts under an impact of gas discharges. Therefore our work is devoted to investigations of electric-hydrodynamic (EHD) effects caused by corona discharges over liquid. They are continuation of our experiments [10-12].

For investigations, we used two types of plasma generators - the capillary discharge [1-4], which usually is applied for interaction of high temperature plasma in air with a metallic or a dielectric surface [1-4].

Another type of generators is a corona discharge over liquids [10-12]. Interaction of plasma with a surface of a liquid leads to creation of charged structures with a long lifetime over their surface. The purpose of this paper is a continuation of previous investigation for creation of long-lived heterogeneous structures in plasma.

2. Experiments on generation of two-phase objects

Experiments on formation of heterogeneous luminescent formations - LLGS with a thin shell under action of an erosive plasma generator plasma jet on various materials, including pure metals, metallic alloys and metals covered with a film of alcohol or glycerin, were performed. Under heterogeneous or two phase objects we mean the objects consisting of vapor and a shell (cover) of quasi-liquid or quasi solid material. In plasma experiments such objects can have additional plasma cover and in this case they represent three-phase objects.

A scheme of the experimental device is shown in Figure 1. It is the same as we used in [3,4,8]. It represents a capillary plasma generator with plasma forming dielectric plate between two electrodes, which works in pulse modes with the pulse time ~10 ms. The energy input during the pulse is about 200 J.

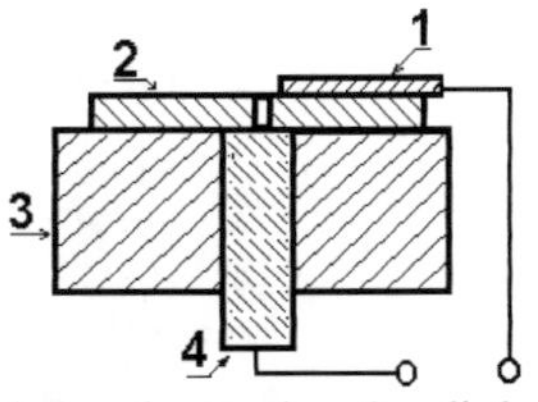

Figure 1 Capillary plasma generator: 1,4 - electrodes, 2 - dielectric plate with a capillary (discharge chamber), 3 – a base of the plasma generator of Plexiglas.

In Figure 2 in the left photo one can see an appearance of the set up during generation of the erosion plasma. Experiments were carried out to obtain LLPF based on the alloy of tin and lead, pure tin, pure lead and pure copper using the erosion plasma generator with energy input of about 190 J during an action a time of ~ 10 ms. A lifetime of objects with a shell and a vapor nucleus reaches 4 seconds.

We found that when the plasma generator's jet interacts with materials moistened by various liquids, LLPF number and size change. In particular, glycerol leads to increase the amount and size of LLPF, while alcohol, on the contrary, reduces these characteristics.

In Figure 2 (in the right photo) typical trajectories of LLPF obtained in experiments are presented. Figure 3 shows traces of the molten droplets and tracks of explosive LLPF. One can see in Figure 3 traces in the form of stars. Stars reflect explosion and destruction of LLPF. According to [4] it is the cover (shell) presence that allows to explain LLPF explosion when it falls on a substrate by an existence of a condensed vapor under the cover. Because of the objects fragility we studied various conditions and ways of LLPF shell conserving.

In Figure 2 one of the experimental setup arrangement variants is shown. The plasma generator capillary was directed upwards. We tried to conserve the undamaged shells by catching LLPF in various liquids (alcohol, glycerin, water).

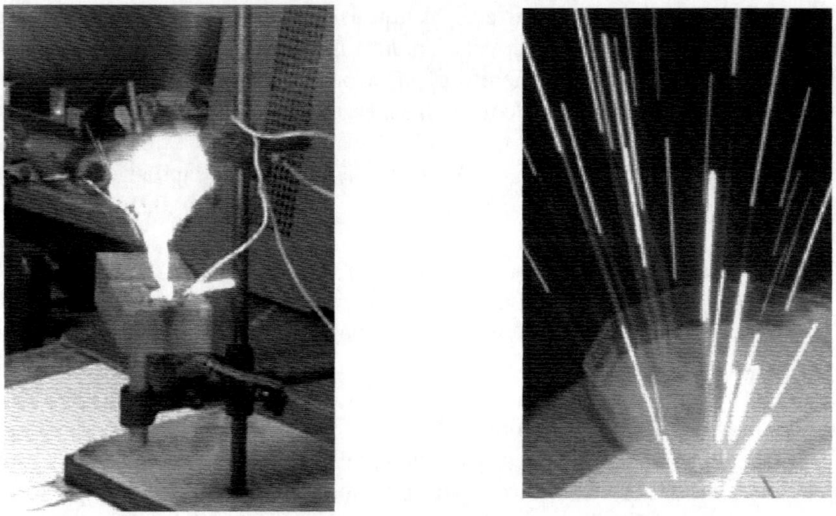

Figure 2 Left photo: plasma generator: capillary directed upwards. Right photo: typical paths of LLPF.

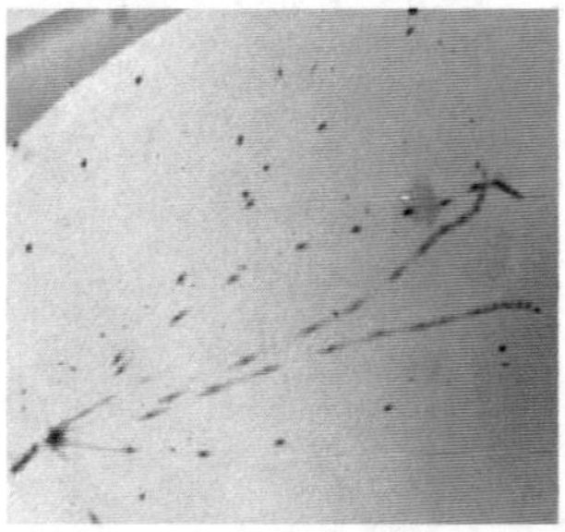

Figure 3 LLPF tracks on the paper.

In Figure 4 in the left photo the LLPF (resulting cover with a ball core inside it) is presented. In Figure 5 in right photo one can see the shell and the "core" of LLPF (of the solder POC-61) dropped out of it.

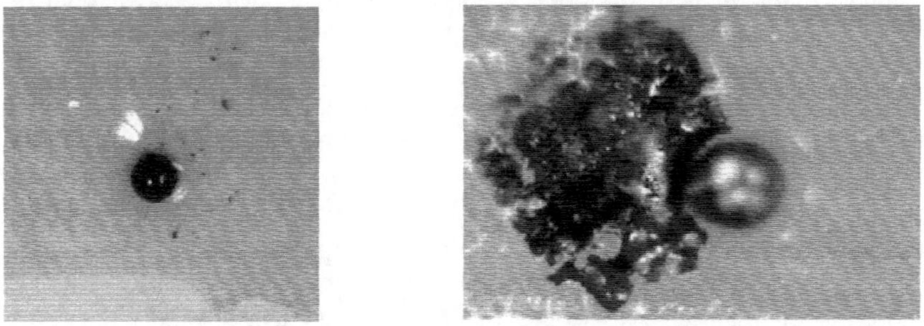

Figure 4 Left photo: the envelope of the LLPF. Right photo: the shell and "core" of LLPF (of~600 μm diameter).

It was possible to keep the whole shells for a week, after which they disintegrated. Oil and glycerin were used to save the undamaged shells.

IOP Conf. Series: Journal of Physics: Conf. Series **1112** (2018) 012022 doi:10.1088/1742-6596/1112/1/012022

Investigations of spheres obtained in plasma

In these experiments, in order to obtain individual spheres for research, we used a syringe without a needle so that, the LLPF should be preserved without damaging the shell. After a shot, plasma formations fell on a paper leaving traces in a form of tracks or stars, as shown in Figure 3. We did not succeed in retaining of LLPF in such experiments. Therefore, we conducted experiments, replacing paper for various liquids, namely: water, glycerin and alcohol.

Because of a high density of the liquid most of the LLPF was destroyed when it came into contact with the surface of glycerin, while contacting the alcohol surface, the LLPF was retained and drowned, but the most fragile formations were destroyed by interacting with the bottom of the cuvette. In the case of water, some samples were drowned, and some were broken when struck against the surface of the liquid. In all three cases, strong formations could be analyzed.

Predominantly water was the liquid into which the formations fell after the shot, because on the surface of it the greatest amount of LLPF with the shell was retained. From the surface of the water with a syringe without a needle, samples were carefully collected and placed in individual containers with water and oil. To continue research without contaminating the cuvette, and also to learn how the storage conditions affect the lifetime of the casing. As it turned out, in water the formations are stored on average 3 days, and in oil 5 days, individual samples can have shell lifetime up to 7 days in both cases.

After the destruction of the shell of the LLPF, the "core" fell out. The composition of the obtained spheres includes Al_2O_3, Pb, Sn. The presence of Sn and Pb in the analysis of the nuclei is explained by the fact that the sample on which the jet of the plasma generator acts, and this wire from the solder POC-61 consists of 59-61% of tin and 39-41% of lead.

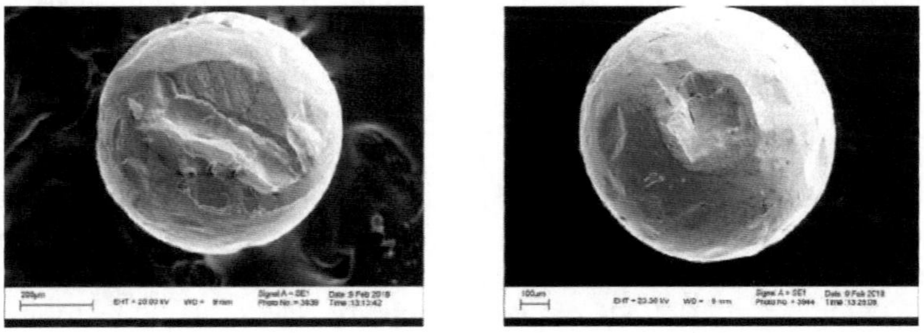

Figure 5 Cores of the colored LLPF: red-to the left, blue –to the right.

Al_2O_3 is present in the analysis of samples, because of the sputtering from the electrode. The presence of carbon on one of the samples is explained by that during the shot a part of the nozzle, which is made of plexiglass $(C_5O_2H_8)_n$, is burnt. A photograph of two nuclei were obtained using an atomic force microscope, are shown in Figure 5. They differ in colors and sizes (because of different additions of small components). A Photograph of LLPF shells were obtained using an atomic force microscope, see Figure 6. The thickness of the shell can be estimated as 5-10 microns.

Also, experiments were carried out in which we tried to preserve the whole shells by catching LLPF in various liquids (alcohol, glycerin, water). A spectral analysis of the resulting formations was carried out, and their linear dimensions, density, mass, and heat of evaporation were estimated.

The linear size of each "core" was about several hundred microns. For each of the formations, the volume was estimated. On average, the result was tenths of a cubic millimeter. With a help of spectral analysis, the content of tin, lead, and Al_2O_3 was determined. To determine the density, we used the obtained volume values, tabulated values of the density for tin, lead, and Al_2O_3, as well as the percentage of these metals and the oxide of the relative whole sample. On average, the density of "nuclei" was 5.13 g/cm^3, and the average value of the mass of the nuclei was 1.7 mg.

An estimate was also made of the specific heat of vaporization of the obtained nuclear alloy by means of tabulated values of the specific heat of vaporization. The minimum values stored in each "core" of the LLPF energy were 2 J, the values per unit mass of an average of 1139000 J/kg. It shows that such LLPF have high-energy content and can be considered as activators for combustion.

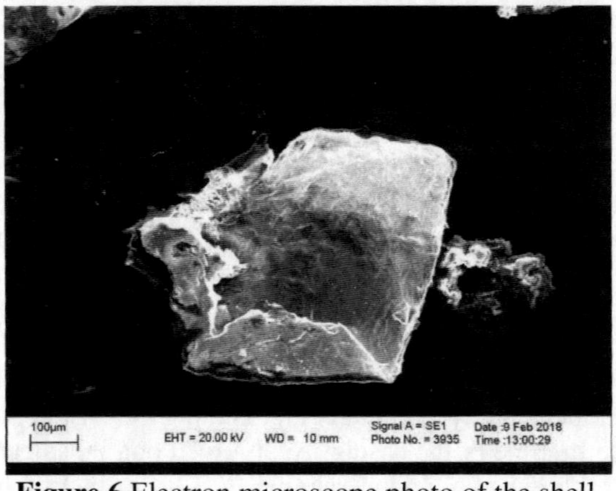

Figure 6 Electron microscope photo of the shell.

3. Structures in corona discharge

Corona discharges are typically obtained by applying of a high potential between a small diameter needle, charged positively or negatively and plate or coaxial cylinder [9-13] charged oppositely. The principle scheme of the experimental device is represented in Figure 7. It consists of a ditch filled with a liquid (water, alcohol, kerosene, etc), and electrical circuit. A current in a discharge was measured by a milliamperemeter A1, a voltage was measured by a chain consisting of resistance R_1 and a milliamperemeter A_2. Experiments have been carried out with a help of different cavities: rectangular – dielectric, cylindrical – dielectric and steel. Sizes of the rectangular cavity were 70×20×15 mm (length, width and depth). Characteristics of cylindrical cavities were ∅125 mm - diameter, 10 mm - depth and 185 mm - diameter, 20 mm- depth, respectively. An investigating liquid was poured into a cavity, a negative or positive electrode was located directly over the liquid, a distance between the electrode and the liquid could be varied in a range (1- 30) mm. The needle diameter was 3 mm, a radius of a tip was 0,4 mm. An electrode of another polarity was located directly in the liquid. (The metallic cavity was used as the electrode). It was connected with a feeding source through a hermetic unit. In this case, the liquid becomes the second electrode.

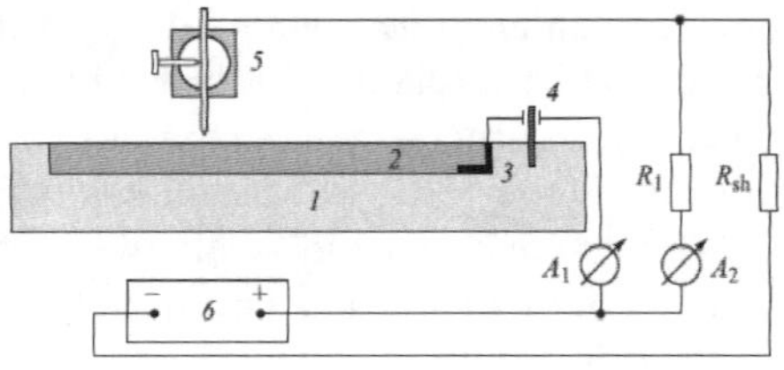

Figure 7 Principle scheme of the experimental device. 1-Ditch, 2- Liquid (water, alcohol, kerosene), 3-Anode, 4- Post, 5- Cathode, 6-Power supply

A high voltage generator was used as the feeding source, it allowed to measure a voltage on the electrodes from 2 kV to 25 kV with a step of 1 kV. Typical range of the voltage was 5-25 kV. Appearance of the spark discharge was accompanied with a drop of the voltage at the capacity, which was detected by the voltmeter. In this device we used a ballast resistance, which value was 510 MΩ. Typical values of a current were 1-50 μ A.

IOP Conf. Series: Journal of Physics: Conf. Series **1112** (2018) 012022 doi:10.1088/1742-6596/1112/1/012022

Experiments with discharges over tap water

Our experiments have shown that application of negative and positive corona discharges over water surfaces in plastic or metallic cavities lead to appearance of a rotating funnel, it was evidently caused by the ion wind.

Vortices over surfaces of liquids appeared in case of non-central position of the upper electrode with respect to the cavity or in rectangular cavities.

Experiments with alcohol

In case of alcohol in the plastic or metallic cavity, corona discharges lead to appearances of different electric hydrodynamic effects over the surface of the liquid.

Funnels appeared at both polarities of the upper electrode. In Figure 8 (left photo) we present a photo of such a phenomenon. In case of the positive upper electrode a rotating funnel usually appeared (it was evidently caused by the ion wind) initially. This process finishes by an appearance of liquid columns, see Figure 8 (right photo).

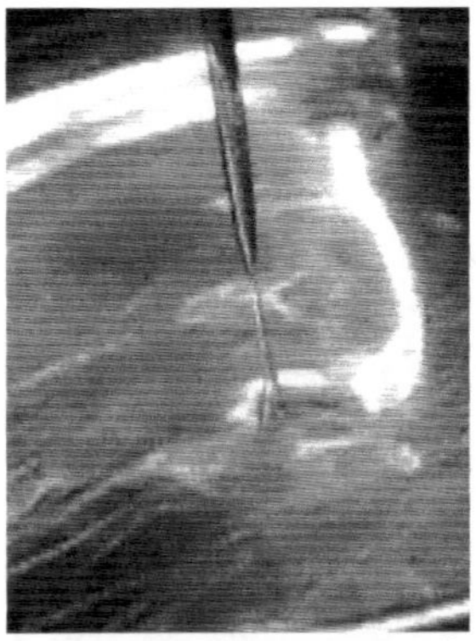

Figure 8 Left photo: appearance of a cavity over the alcohol surface, metallic cavity, a distance between the upper electrode and a surface is 5 mm. Positive corona. Right photo: Appearance of a conical column over the alcohol surface, metallic cavity, a distance between the upper electrode and a surface of the liquid is 7 mm. Positive corona.

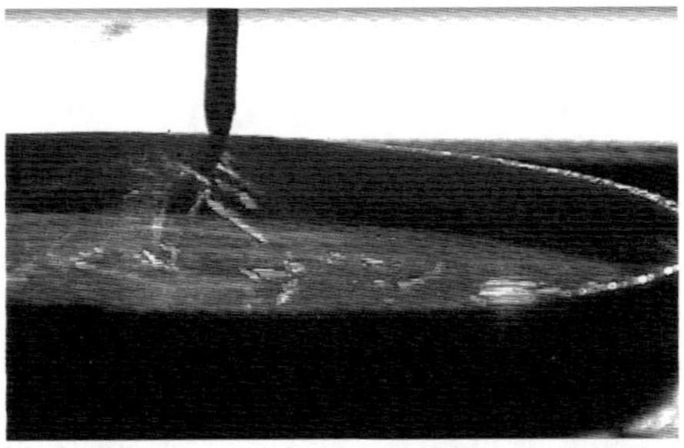
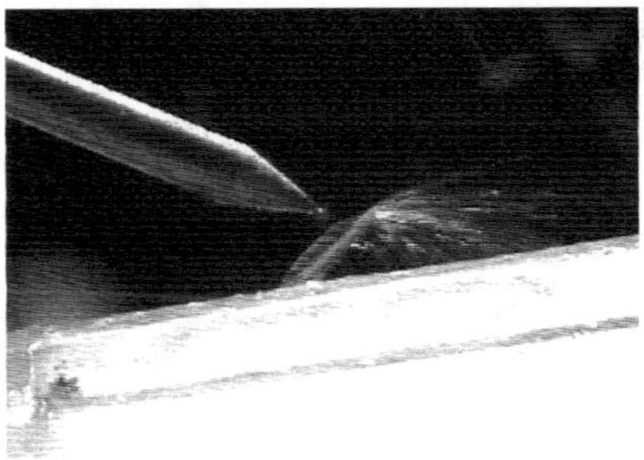

Figure 9 Appearance of jets and drops over the alcohol surface, dielectric cavities, a distance between the upper electrode and a surface of the liquid is 7 mm (left photo) cylindric cavity, 17 kV and 6 mm (right photo) rectangular cavity, 20 kV. Negative corona.

The development of the process finally finishes by appearance of the electric arc. Usually this takes place at high applied voltage (15-25 kV). These columns are either standing in the funnel center or on a side of a funnel. Liquid columns have usually a widening at the tip. Sometimes a violet-bleu glow on the tip appears. It speaks about forming of the breakdown field near the tips. At voltage about 15-25 kV [10-13] these columns often transform into bursting jets. In Figure 9 one can see such jets consisting of drops. Images were obtained with a help of a digital video camera Canon Digital IXUS 70 with frame duration 33 ms.

Namely creation of such charged drops can be used for applications for vaporization of liquid fuel in combustors [4].

Similar phenomena are known in EHD [14-17]. They are connected with a development of Rayleigh, Taylor, Tonks and Frenkel electrostatic instabilities on surfaces of charged drops and jets when Coulomb forces of accumulated charges become greater than the surface tension force.

Experiments with Al_2O_3 powder

Experiments with the Al_2O_3 powder on a liquid surface are of particular interest. The powder particles were of 30 μm in diameter. When the applied voltage exceeds a value of 7 kV vortices very intensively mixing of the liquid appears.

Figure 10 Column has a length of 7mm and a diameter 1 mm a distance Between the upper electrode and a surface is 5 mm. Positive corona.

Experiments with dry and wet clay covered with aluminum powder were conducted also. Some of them demonstrated interesting results, such as arise of a clay column (see Figure 10) covered with metallic powder that provides a possibility to deeper understand the structure of columns. This composite column widens at its head as if some charge tried to separate it into drops. So, one can suppose a manifestation of two electro hydrodynamic instabilities in this case: the first one - the appearance of the column and the second one– Rayleigh instability of charged object electrostatic separation. Since we have realized instabilities then it was difficult to formulate an experiment where such columns were well observed.

4. Conclusions.

Experimental studies on creation of long-lived heterogeneous structures (LLGS) in air have been carried out.

It was possible to experimentally obtain LLGS of the alloy of lead and tin. The composition of the shell and the core of formations are significantly different; this means that boiling can occur inside the LLGS leading to the separation of the composition into a light part-the shell and the heavy part-the core. This result is new and it allows to consider LLGS as a new object for investigations in low temperature plasma. As a result of the work, it was possible to obtain photographs of the core and the shell of plasma formations, to estimate specific energy values of four LLGS cores. The obtained values of specific heat of vaporization of LLGS are close to those of explosives [4]. This fact allows to consider creation of such LLGS structures for addition to combustors for activation of a fuel.

Creation of corona discharges has been realized over tap and distilled water, alcohol, glycerin and their mixtures. Conducted experiments and their analysis show that the corona discharge over liquids can cause different hydrodynamic phenomena including appearance of the ion wind and different hydrodynamic instabilities. Funnels appear both in positive and negative coronas. Appearance of liquid jets and columns is a result of the Tonks - Frenkel and Rayleigh instabilities in conditions of positive and negative corona. This phenomenon depends on a value of electric field strength and is better realized in liquids with low density and surface tension. Separation of these jets and columns in drops are connected with instabilities of charged drops, which can also be considered as LLGS. These experiments help to clarify modes of discharges realization and their connection with appearance of funnels on the surfaces of liquids and columns in case of alcohol. Application of corona discharge for creation of charged particles and structures in plasma aerodynamics or plasma assisted combustion require additional investigations.

Acknowledgements
This work was carried out in the framework of State Assignment of Russian Federal Agency of Scientific Organizations (project 0082–2018–0001, registration code AAAA–A17–117112240026–5).

References

[1] Avramenko R F Bakhtin B I Nikolaeva V I, et al 1990 *Zhur. Tekh. Fiz.* **60 12** 73.
[2] Avramenko R F Bakhtin B I Nikolaeva V I, Poskacheeva L P 1994 *Ball lightning in the laboratory* ed. Avramenko R.F., Bychkov V.L., Klimov A.I., Sinkevich O.A. (Moscow: Khimiya Publishers) 15
[3] Ershov A P Rozanov V V Sysoev N N et al 1994 *Physical hydrodynamics* ***4*** *Preprint of phys. dep. of MSU Moscow MSU 8/1994*
[4] Bychkov V L Chernikov VA Osokin A A et al 2015 *IEEE Trans. Plasma Sci.* **43** 4043.
[5] Pokhil P F Belyaev A F Frolov Y V Logachev V S Korotkov A I 1972 *Combustion of powdered metals in active media* (Moscow: Nauka publishers)
[6] Emelin S E Semenov V S Bychkov V L et al 1997 *Tech. Phys.* **42** 269
[7] Emelin S Bychkov V Astafiev A Kovshik A Pirozersky A 2012 *IEEE Trans. Plasma Sci.* **40** 3162.
[8] Bychkov V L Bychkov A V Timofeev I B 2004 *Tech. Phys.* **49** 128.
[9] *Encyclopedia of low temperature plasma* 2009 *(vol.3. Applied Plasma Chemistry)* ed. Lebedev Yu A, Plate N A, Fortov V E. (Moscow: Yahys-K publishers)
[10] Aleksandrov A F Bychkov V L Bychkov D V et al 2011 *Moscow Univ. Phys. Bull.* **66** 390
[11] Bychkov V L Chernikov V A Volkov S A et al 2011 *IEEE Trans. Plasma Sci.* **39** 2640
[12] Bychkov V L Chernikov V A Volkov S A et al 2011 *IEEE Trans. Plasma Sci.* **39** 2643
[13] Landau L D Lifshitz E M 1982 *Electrodynamics of continuous media* (Moscow: Nauka Publishers)
[14] Saranin V A 2002 *Equilibrium of liquids and its stability* (Moscow: Institut Kompiuternykh Issledovanii Publishers)
[15] Grigor'ev A I Sinkevich O A 1984 *Zhur. Tekh. Fiz.* **54 7** 1276.

IOP Conf. Series: Journal of Physics: Conf. Series **1112** (2018) 012023 doi:10.1088/1742-6596/1112/1/012023

Using triple Langmuir probe method for study high-speed plasma flow created by a pulsed capillary discharge

A S Pashchina* and R E Karmatsky

Joint Institute for High Temperatures, Russian Academy of Sciences, Moscow, Russia

*fgrach@mail.ru

Abstract. The results of triple Langmuir probe method diagnostics of plasma created by a pulsed discharge in a capillary with an ablating wall and injected into a rarefied atmosphere ($p < 1$ Torr) are presented. The possibility of obtaining high-speed plasma flows in the wide range of electron density $n_e = 10^{11} - 10^{15}$ cm^{-3} and the temperature $T_e = 1 - 5$ eV by varying the discharge power input in the range of $N = 15 - 500$ kW is shown. The effects indicating the violation of approximation of the infinitely thin probe layer while reducing the electron density below $n_e < 10^{13}$ cm^{-3} and requiring account in the theoretical model of a triple probe are found. The results of measurements are in good quantitative agreement with the results of diagnostics of plasma flows similar in parameters.

1. Introduction

A pulsed capillary discharge is known as a relatively simple and efficient method of obtaining plasma over a wide range of pressures, temperatures, and charged particles concentrations [1–3]. It was shown in our previous work [4] that such discharge can be successfully used to produce quasistationary high-speed plasma flows in a rarefied atmosphere that have good prospects in magnetoplasma aerodynamics applications, including problems of laboratory modeling of physico-chemical and gas-dynamic effects and experimental study of bodies interaction processes with hypersonic gas-plasma flows. The problem of plasma diagnostics takes an important place in these studies, which, taking into account the plasma nonequilibrium, high flow rates, and the rapidity of the processes, represent a difficult experimental task. Among the whole variety of diagnostics methods and tools available in the experimental arsenal of plasma physics, the electro-probe methods [5] are the most suitable and convenient under these conditions. Locality of the measurements, as well as the absence of necessity for prior knowledge of the plasma thermodynamic state, are the main advantages of electro-probe methods.

Multi-probe methods offer an advantage over single and double Langmuir probes, allowing simultaneous measurement of plasma parameters without the need for voltage sweeps that can be limited in pulsed plasmas. The triple probe method, first proposed by Chen S.L. and Sekiguchi T. [6,7], is one of such tools, allowing measurement of electron density and temperature in real time. Triple probe method is characterized by high speed, limited by the ion plasma frequency that provides wide opportunities to study fast processes, such as transient processes, low-frequency plasma oscillations and waves, plasma shock waves, etc. Multi-probe methods are successfully used for diagnostics the non-stationary high-speed plasma flows [8,9], obtained in similar conditions considering in this paper.

At the same time, in spite of the relative simplicity, the use of multi-probe methods in plasma-aerodynamic experiments requires certain accuracy when performing measurements. One of the important requirements for conducting such measurements is the spatial uniformity of plasma in the region of probe location, assumed the identity of plasma parameters near each electrode. This requirement can be easily satisfied in a motionless plasma or in a low-speed plasma flows, $v \ll c$ (c is the speed of sound). It is much more difficult to fulfill this condition in supersonic flows $v \geq c$ because the strong perturbation of the flow in the region of probe location. In this case, the collecting electrodes of the probe are located behind the shock wave, where the pressure, density, concentration and temperature of the charged particles, as well as the spatial distribution of these parameters, differ significantly from those in the unperturbed flow. The accuracy and reliability of the obtained results, as well as the very possibility of using multi-probe methods under these conditions, depend largely on how much the flow inhomogeneity influences the plasma parameters near each of the electrodes.

The aim of the work is the experimental study of high-speed plasma flows, created by a pulsed capillary discharge and injected into the atmosphere of rarefied gas, by the triple Langmuir probe method, as well as the assessment of applicability of this method under the experimental conditions, including the estimation of the flux spatial inhomogeneity and its influence on the accuracy of the obtained results.

2. Laboratory installation and research methods

The capillary arrester, whose detailed description and parameters are given in [10], is used to create high-speed plasma flow. Perspex is used as a capillary wall material (chemical formula $C_5H_8O_2$). The depth and the initial diameter of a capillary are respectively $h = 5$ mm and $d = 1$ mm. A capacitive storage device with series-connected inductor is used as the discharge power supply. The algorithm of power supply approximately corresponds to the sine half-wave. The following parameters and conditions are typical for our experiments: the stored energy of a power supply $Q = 90 - 640$ J, the peak discharge current $I_m = 200 - 2500$ A, the discharge pulse duration $\tau_d = 2$ ms (a "short-pulse" mode of discharge) and $\tau_d = 18$ ms (a «long-pulse» mode of discharge).

Experimental investigations were carried out in a vacuum chamber 1, inside which a capillary arrester 2 and a triple probe 3 were placed (figure 1). The residual pressure in the chamber is $p_\infty \approx 0.1$ Torr. The probe is placed in the central zone of the plasma jet at a distance of about 10 cm from the capillary outlet. The probe consists of three identical electrodes made of tungsten wire with a diameter of $2r_p = 0.23$ mm. The length of each electrode is $l_p = 5$ mm, and the distance between their axes is $s = 1$ mm. Inside the vacuum chamber, the probe is placed in such a way that the electrodes axes would be oriented along the direction of the plasma flow.

The so-called "current-mode" operation of triple probe [8,9] is used in experiments (figure 2) that makes it possible to minimize the influence of electromagnetic interference on the measured signal. One of the electrodes in the probe circuit (P_1) is selected as the reference one, and the negative bias voltage is applied to the other two electrodes (P_2, P_3). The bias voltages (their values are equal to $V_{12} = -8.5$ V and $V_{13} = -40$ V in our experiments, see also figure 1) is chosen from the condition of location of the operating points of both probes on ion saturation branch of probe characteristic. Each circuit probe contains low-inductance metal film resistors $R1, R2$ and $R3$, which are used as shunts to measure the probe currents I_1, I_2, I_3.

The use of conductive-type shunts to measure probe currents leads to the necessity of grounding the reference electrode P_1. In this case, to ensure galvanic isolation between the probe and the arrester's power supply, the electrodes of the latter one are not grounded. It is obvious that such circuit modification leads to the fact that the probe system ceases to be isolated, since the potential difference between the plasma and the probe' electrodes causes the additional leakage of ion current to the ground. This circumstance, in general, leads to a change of the probe layer parameters and, as a consequence of the probe characteristic, and must be taken into account in calculating the plasma parameters (electron temperature and density).

IOP Conf. Series: Journal of Physics: Conf. Series **1112** (2018) 012023 doi:10.1088/1742-6596/1112/1/012023

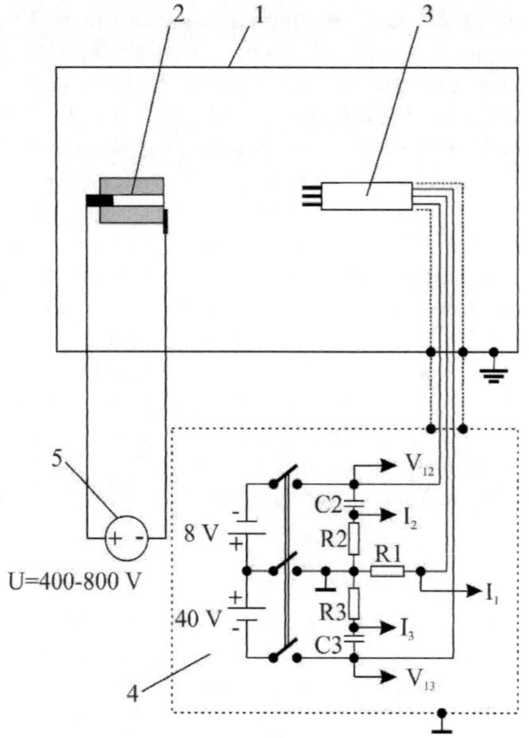

Figure 1 Principal scheme of laboratory installation: (1) vacuum chamber, (2) capillary arrester, (3) triple Langmuir probe, (4) probe measurement unit, (5) discharge power supply.

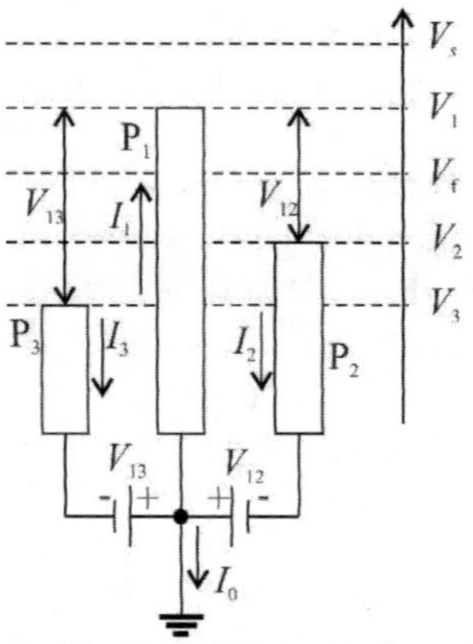

Figure 2 Electric diagram of a triple (P_1, P_2, P_3) Langmuir probe in a "current-mode" operation.

All elements of the probe measurement unit (see figure 1) are mounted inside the metal casing, and shielded cable is used to enter the probe into a vacuum chamber. The measures taken, in conjunction with "current-mode" operation of the probe circuit, made it possible to minimize the influence of electrical interference on the measuring equipment. The probe currents and voltages between the reference and negative electrodes recorded during the experiments were used for calculation the electron density and temperature.

3. Calculation of plasma parameters

The multi-probe methods are based on the theory of the Langmuir probe in collisionless plasma [11], which completely determines the requirements for the range of the pressure, concentration of charged

IOP Conf. Series: Journal of Physics: Conf. Series **1112** (2018) 012023 doi:10.1088/1742-6596/1112/1/012023

particles, and also probe dimensions in accordance with the limits of applicability of the theory. To estimate the applicability of the Langmuir probe theory with respect to the gas pressure, we will proceed from the fact that the characteristic radius of the probe is $r_p \sim 0.1$ mm, the mean free paths of electrons and ions are of the order of $\lambda_e \sim 10^{-1}/p$ [cm] and $\lambda_i \sim 10^{-2}/p$ [cm] (the pressure p is expressed in Torr). Then the conditions $r_p \ll \lambda_e$, λ_i and $r_D \ll \lambda_e, \lambda_i$ are satisfied for $p < 1$ Torr ($r_D = \sqrt{\varepsilon_0 k T_e / e^2 n_i}$ – Debye length).

In the range of electron density $n_e = 10^{13} - 10^{15}$ cm^{-3} and temperature $T_e = 0.5 - 10$ eV, the ratio of the probe radius to the Debye length is $r_p/r_D > 100$, so the case of an infinitely thin layer is realized. This makes it possible to simplify the procedure for calculating the plasma parameters by using the approximation of the independence of the ion saturation current density from the probe potential, i.e. $J_i(\phi_{s1}) \approx J_i(\phi_{s2}) \approx J_i(\phi_{s3}) = J_i$ (here ϕ_{sn} – the potential of n-th probe electrode relative to the plasma potential). The ion current density to the probe in this case is determined by the Bohm relation [6]: $J_i = \exp\left(-\frac{1}{2}\right) e n_e \sqrt{\frac{kT_e}{M}}$.

If $n_e < 10^{13}$ cm^{-3}, then the size of the probe layer increases, and the condition of an infinitely thin layer ceases to be satisfied. The violation of this condition can also be caused by the leakage of the ion current to the ground. In this case, to calculate the density of the ion saturation current, that depend on the probe potential, it is necessary to use the modified relations, for example, obtained by Peterson and Talbot [12] on the basis of Laframboise data [13], which are valid in the range of moderate values of the probe layer thickness $5 < r_p/r_D < 100$. In this paper, in order to simplify the calculation procedure, we used the assumption of an infinitely thin layer.

An important condition for the applicability of the Langmuir theory is the small thickness of the probe layer in comparison with the distance between the probes, i.e. $(r_s - r_p) \ll s$. It can be seen that this condition is satisfied throughout the whole range of expected values of electron temperatures and densities.

Taking into account above approximations, the initial system of equations for the probe currents can be written in the form

$$I_1 = -SJ_{e0} exp(-\phi_{s1}) + SJ_i$$
$$I_2 = -SJ_{e0} exp(-\phi_{s2}) + SJ_i$$
$$I_3 = -SJ_{e0} exp(-\phi_{s3}) + SJ_i$$

The solution of the system, taking into account the leakage of the ion current by introducing into the Kirchhoff law the corresponding term I_0 (i.e., $I_0 + I_1 + I_2 + I_3 = 0$), leads to the relations for determining the electron temperature

$$\frac{-I_1 + I_2}{-I_1 + I_3} = \frac{1 - exp(-\phi_{12})}{1 - exp(-\phi_{13})}$$

the ion saturation current density

$$J_i = \frac{1}{S} \frac{I_3 - I_2 \exp(-\phi_{32})}{1 - \exp(-\phi_{32})},$$

and the electron density

$$n_e = \frac{1}{S} \frac{I_3 - I_2 \exp(-\phi_{32})}{1 - \exp(-\phi_{32})} \frac{1}{0.6 e \sqrt{\frac{kT_e}{M}}}.$$

Here $M = \mu m_p$ – the ion mass, $m_p = 1.6726219 \times 10^{-27}$ kg – the proton mass, μ – the effective molecular weight of plasma ($\mu = 6.67$ for Perspex), $S = \pi d l$ – the surface area of the probe electrode.

4. Results of experiments

The oscillogramms of currents I_1, I_2, I_3 onto the probe electrodes and also of the leakage current I_0 for the two discharge modes ("long-pulse" and "short-pulse") are shown in figure 3. The change in currents starts synchronously for all electrodes with a certain time delay relative to the beginning of the discharge. The delay time is determined by the speed of the front of the plasma jet and depends on the discharge power. The velocity of the jet front, calculated from the delay time, is approximately of

300-400 m/s for the "short pulse" and of 40-60 m/s for the "long-pulse" discharge mode. We did not find explicit signs of the plasma homogeneity violation in the region of the probe location: the floating potentials measured for this purpose showed the same values for all three electrodes regardless of the discharge power. The change in the discharge power is accompanied by a corresponding change in the ion current density to the electrodes, whose magnitude varies from 0.1 mA/mm^2 to 10 A/mm^2 for the "long-pulse" and "short-pulse" discharge modes, respectively. It is noteworthy that in the "short-pulse" discharge mode (the discharge power $N > 100$ kW and electron density $n_e > 10^{14}$ cm^{-3}), the condition of an isolated system for a triple probe is performed with good accuracy: the leakage current does not exceed 10% of the probes currents. This condition violates in the “long-pulse” discharge mode (the discharge power and electron density are less than $N < 20$ kW and $n_e < 10^{12}$ cm^{-3}, respectively), in which the leakage current becomes comparable with probes currents.

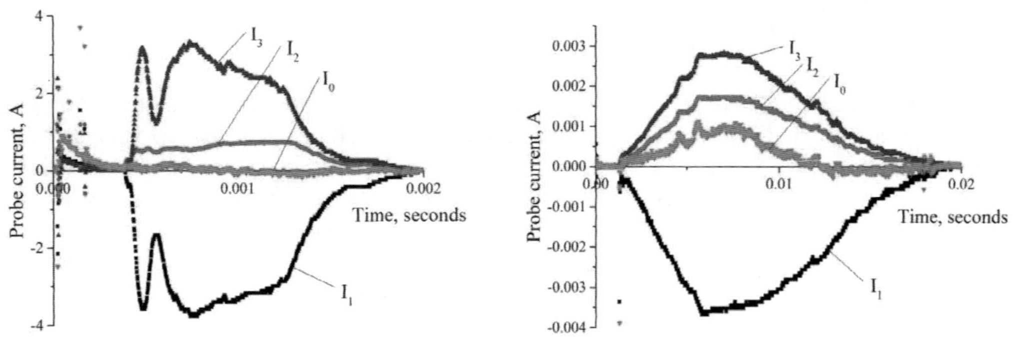

a) b)

Figure 3 Typical oscillogramms of currents in triple probe circuit for (a) the “short-pulse” and (b) “long-pulse” discharge modes. The voltage and energy of the power supply are $U = 600$ V and $Q =$ 360 J, respectively.

The results of calculating the electron densities and temperatures are presented in figure 4. The time course of the electron density in both modes monitors the discharge power algorithm and reaches the maximum approximately in the middle of the discharge pulse. Thus, if the electron density in the "short-pulse" and "long-pulse" modes differs by more than three orders of magnitude (its typical value in the "short-pulse" mode is $n_e \sim 10^{14} - 10^{15}$ cm^{-3}, and in the "long-pulse" mode - $n_e \sim 10^{11} - 10^{12}$ cm^{-3}), then the discharge power in these modes changes only by an order of magnitude (from $N = 150$ kW in "short pulse" mode to $N = 15$ kW in "long-pulse" mode). The observed feature is related to the Joule heating of a section of a plasma jet by remote currents. These currents travel far beyond the capillary and heat a sufficiently long section of the plasma jet (up to some hundred calibers relatively to the capillary diameter). The size of the zone of remote currents and, correspondingly, the length of the heated section is the greater, the higher the discharge power. Therefore, in the "short-pulse" discharge mode, the remote currents heat up a substantially longer section of the plasma jet than in the "long-pulse" mode. So, in the "short-pulse" mode, the spatial position of the probe corresponds to the section of the jet heated by the remote currents, and in the "long-pulse" mode - to the section of relaxed plasma. This conclusion is indirectly confirmed by the results of the electron temperature estimates, which value in the "short-pulse" mode ($T_e = 4 - 6$ eV) more than twice exceeds that for the "long-pulse" mode ($T_e = 1.5 - 3$ eV).

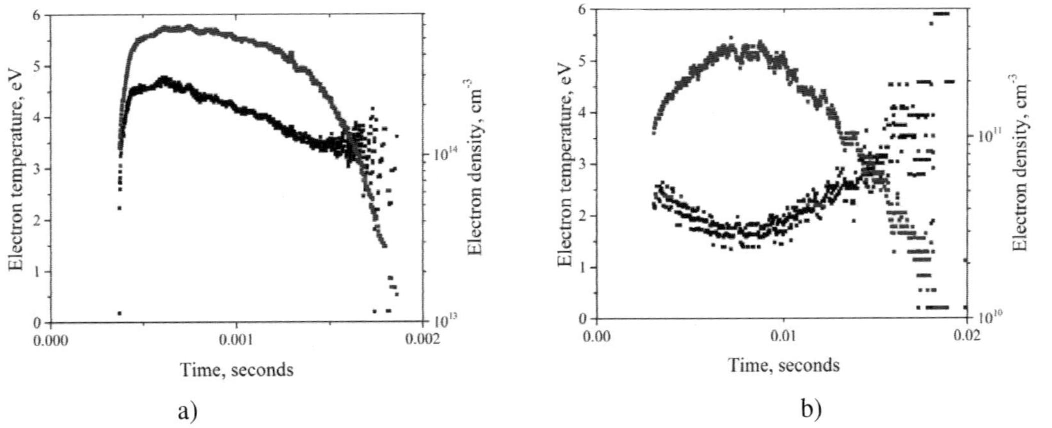

Figure 4 Temporal dependences for the electron densities and temperatures for (a) the “short-pulse” and (b) “long-pulse” discharge modes. The discharge parameters are the same as in figure 3.

Attention is drawn the different temporal trend of electron temperature observed in two discharge modes: in-phase with the discharge power in the "short-pulse" mode, and the antiphase one in the "long-pulse" mode. In our opinion, the main reason for this difference is the violation of the approximation of an infinitely thin probe layer because the low electron density $n_e < 10^{13}$ cm^{-3} and the significant leakage of the ion current observed in the "long-pulse" discharge mode. These factors, obviously, should be taken into account in the theoretical model of the triple probe, which is planned in the future.

5. Conclusion

Thus, even the first experiments showed the principal possibility of using the triple Langmuir probe method for diagnostics the high-speed plasma flows created by a pulsed capillary discharge and injected into a rarefied atmosphere. It was shown experimentally that such plasma flows can be obtained in the range of electron densities $n_e = 10^{11} - 10^{15}$ cm^{-3} and temperatures $T_e \sim 1 - 5$ eV by varying the discharge power in the range $N = 15 - 500$ kW. A change in the temporal trend of electron temperature is observed when the electron density decreases below $n_e < 10^{13}$ cm^{-3}, which is caused by a violation of the approximation of an infinitely thin probe layer, as well as by the increasing role of leakage of ion current under these conditions. The role of these factors must be accounted in developing the theoretical models of triple probe.

In general, the results of probe measurements are in good quantitative agreement with the results obtained by various authors who used the triple Langmuir probe method to study plasma with close parameters. This fact, as well as the absence of obvious signs of the spatial uniformity violation of the plasma in the probe localization region, is the encouraging arguments of the reliability of obtained results and the applicability of this method for diagnostics the capillary discharge plasma injected into a low-pressure atmosphere. In the future, it would be advisable to perform measurements and compare the results of diagnostics obtained in open and isolated probe systems, which will help to determine the influence of ion leakage current on the accuracy of the measurements. The preparation and approbation of the quadruple Langmuir probe method [14], which will expand the set of measured parameters by including such important parameter as the plasma flow velocity, should be the important step in future research.

References

[1] Min'ko L Ya 1970 *Obtaining and study the pulsed plasma flows* (Minsk: Nauka i tekhnika) (in Russian)

IOP Conf. Series: Journal of Physics: Conf. Series **1112** (2018) 012023 doi:10.1088/1742-6596/1112/1/012023

[2] Ogurtsova N N, Podmoshensky I V and Shelemina V M 1974 *Teplofiz. Vys. Temp.* **6** 48–53 (in Russian)

[3] Pashchina A S, Efimov A V and Chinnov V F 2017 Optical investigations of multicomponent plasma of capillary discharge. Supersonic outflow regime *High Temp.* **55** 650–64

[4] Pashchina A S, Karmatsky R E and Klimov A I 2017 *Tech. Phys. Lett.* **43** 1033–6

[5] Lochte-Holtgreven W 1968 *Plasma Diagnostics* Ed. North-Holland (Amsterdam; New York:Wiley–Interscience)

[6] Chen S L and Sekiguchi T 1965 *J. Appl. Phys.* **36** 2363–75

[7] Chen S L 1971 *J. Appl. Phys.* **42** 406–12

[8] Tilley D L, Kelly A J and Jahn R G 1990 The Application of the Triple Probe Method to MPD Thruster Plumes (AIAA-90-2667) *21st International Electric Propulsion Conference* (Orlando, Florida, USA)

[9] Eckman R, Byrne L, Gatsonis N a. and Pencil E J 2001 *J. Propuls. Power* **17** 762–71

[10] Pashchina A S, Efimov A V and Chinnov V F 2016 *High Temp.* **54** 488–502

[11] Mott-Smith H M and Langmuir I 1926 *Phys. Rev.* **28** 727–63

[12] Peterson E W and Talbot L 1970 *AIAA J.* **8** 2215–9

[13] Laframboise J G 1966 *Theory of Spherical and Cylindrical Langmuir Probes in a Collisionless, Maxwellian Plasma at Rest (UTIAS REPORT NO. 100)*

[14] Johnson B H and Murphree D L 1969 *AIAA J.* **7** 2028–31

IOP Conf. Series: Journal of Physics: Conf. Series **1112** (2018) 012024 doi:10.1088/1742-6596/1112/1/012024

Thermal Energy Release and Hydrogen Production in Swirl Heterogeneous Plasma-Chemical Reactor

A Klimov [1*], P Kazanskii [1], N Belov [1], B Tolkunov [1], I Zavershinskii [2] and N Molevich[2]

[1]Joint Institute for High Temperatures RAS, Moscow, 125412, Russia
[2] Samara National Research University, Samara, 443086, Russia

*klimov.anatoly@gmail.com

Abstract. Basic researches in field of plasma –assisted combustion and aluminum-hydrogen energetics have been carried out in JIHT RAS during last 10 years. Aluminium -hydrogen plasma vortex reactor (PVR) was designed, manufactured and tested in this Institute. We use this PVR with gas testing mixture argon- water steam to obtain of pure hydrogen and heat power. Plasma-assisted hydration reaction of different metals with water steam was realized in this reactor.

1. Experimental setup

The experimental set up PVR was considered in our previous works in detail [2-7]. Electric pulsed repetitive discharge is used in this reactor. The scheme of the experimental setup PVR is shown in the figure 1. Heterogeneous plasmoid (HP) is created near cathode in the setup PVR. Mean electric power used for HP's creation in this PVR is about of 1÷2 kW. Mean DC current is about of 1÷2 Amp. The tested gas mixture argon + water steam is used in this reactor. Argon mass flow rate equals to $M_{Ar}\sim 1-$ 4 G/s. Water steam mass flow rate equals to $M\sim 1\div 3$ g/s. It was revealed that this HP consists of many tiny erosive charged metal nano-clusters (cathode erosive material) and hydrogen ions [2-7].

Hydrogen mass flow rate and its power budget of one hydrogen molecule production are measured in this work. Standard hydrogen analyser АВП-01 is used for hydrogen mass flow rate measurement, figure 4.

2. Thermal Energy Release in Vortex Heterogeneous Plasmoid

Physical properties, dynamics and structure of a heterogeneous plasmoid (HP) in swirl gas flow have been carried out in our previous works [2-7]. It was revealed that there is considerable energy release in this HP with hydrogen atoms and erosive metal nano-clusters. It is very important to note that there is high specific energy q~ 1KeV/atom in this HP, [4,6]. It was revealed that this energy release is connected with creation of intensive soft X-ray radiation in the HP. The typical quantum energy of this X-radiation is about of 1-10 KeV [4]. HP's parameters and its characteristics were measured by optical spectroscopy, X-ray spectroscopy, ESP method, IMS method and chemical analysis. Electric discharge parameters (current, voltage, power) were measured by Tektronix voltage probe and Tektronix current probe [2-7].

The present work is continuation of the previous ones [2-7]. The accurate measurements of a power balance in the PVR and a hydrogen concentration were realized in this work. We used two well-known calorimetric methods in our experiments, namely: (1) - thermocouple method and (2) - water pool calorimeter method with hot gas bubble barbotage, figure 3. Dry gas mixture (argon+ hydrogen) and

IOP Conf. Series: Journal of Physics: Conf. Series **1112** (2018) 012024 doi:10.1088/1742-6596/1112/1/012024

humid gas mixture (argon + water steam) were used in these experiments. Heat looses in this PVR were measured in the calibration experiment before each calorimetric one. Electric heater (or electric discharge) in a noble gas (argon) was used in these calibration experiments (see below). The typical argon flow rate was 1-3 g/s. The typical water steam flow rate was 1-3 g/s. The typical hydrogen flow rate was 1 mg/s.

Gas flow enthalpy (thermal energy) was measured before PVR's input and behind its nozzle in the first method (see below). Power efficiency *K* of the PVR was estimated by the ratio of energy release/ input electric power. Thermal energy looses in the PVR are taking account in this ratio also. Estimated accuracy of the thermocouple method was 10-15%

Thermal energy of the water in a pool calorimeter was measured in the second method (see below). Power efficiency *K* of the PVR was estimated by the ratio of water thermal energy/ input electric power. Thermal energy looses in the PVR are taking account in this value also. Estimated accuracy of this method was about of 30%.

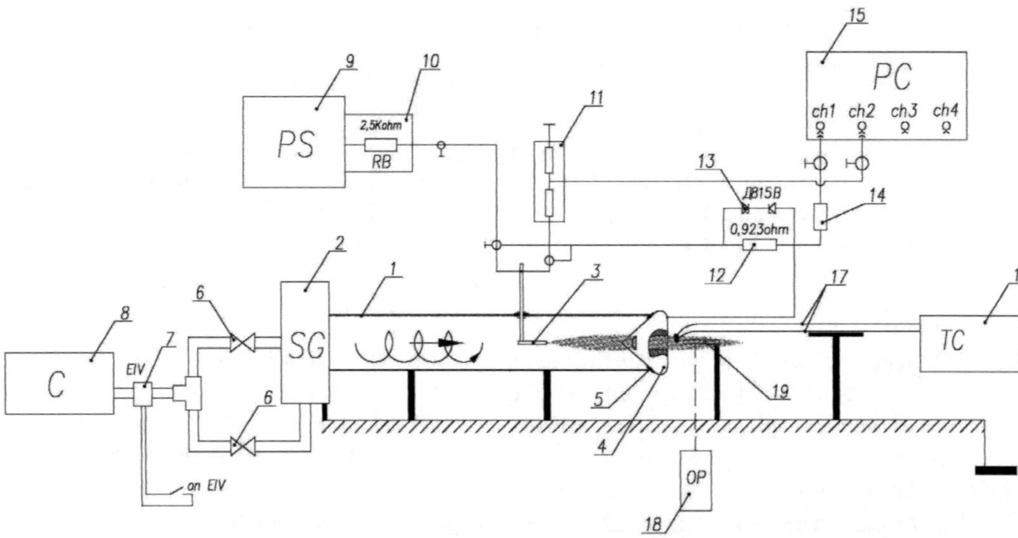

Figure 1 The experimental set up PVR. 1 – quartz tube 60 mm; 2 – swirl generator (SG); 3 – anode; 4 – cathode and nozzle; 5 – rubber; 6 – mechanical valves; 7 – electromagnetic valve (ElV); 8 – gas mixer + steam generator + argon tank, (C); 9 – power supply; 10 – resistor 2,5 K; 11 – resistor divider 1:1000; 12 – current probe; 13 – diodes (Д815В); 15 – oscilloscope TDS 2014B; 16,17 – thermocouples (CENTER 304); 18 – IR pyrometer (MS6560A), 19 – thermocouple holder

2.1. Thermocouple method. Experimental conditions.

The following parameters of an initial tested gas are measured before PVR's mixer, (figure 1 (2)) in this experiment: argon mass flow- G_1 (g/s) and its initial temperature T_1 (°C), water steam mass flow- G_2 (g/s) and its initial temperature T_2 (°C).

The following parameters of tested gas mixture are measured before discharge zone (duct cross section after mixer (2) in this experiment: - gas mixture temperature T_3 (°C).

Input mean electric power N_e (W) and integral electric energy input Q_e (J) to plasma in the PVR are measured in this experiment also.

Mean temperature of output gas mixture $T_6 = (T_{61} + T_{62})/2$ (where T_{61} - axial gas temperature and T_{62} - nozzle wall gas temperature) is measured by two thermocouples.

M_{k1}, M_{k2} - cathode mass before experiment and after it are measured by balance. We used the published values: - Q_c - specific chemical oxidation energy of cathode material and C_{p1}, C_{p2}, - specific heat capacities of argon and water steam in our simulations.

2.2. The typical experimental results on power measurements obtained in the PVR.

Calibration regime. Pure argon is used in the PVR only. Thermal power in the PVR are estimated by the following formula:

$$Q_{1T} = C_{p1}G_1(T_6 - T_1)\,. \tag{1}$$

Calorimetric working regime. Gas mixture water steam + argon are used in this experiment. Thermal power in the PVR are estimated by the following formula:

$$Q_{2T} = (C_{p1}G_1(T_6 - T_1) + C_{p2}G_2(T_6 - T_2))\,.$$

The typical experimental results are shown in the table 1.

Table 1

M_{Ar}, g/s	M_{H2O}, g/s	T_1, °C	T_6, °C	T_3, °C	N_e, W	Q_T, W
Pure argon. Calibration regime.						
1.80	0.0	25.5	202.4		248	159.2
Water steam+ argon. Calorimetric working regime.						
1.80	1.43	25.5	377.3	85.3	624	1098

where initial water steam temperature $T_2 = 102$ °C, measured cathode material erosion rate less than 1 mg/s

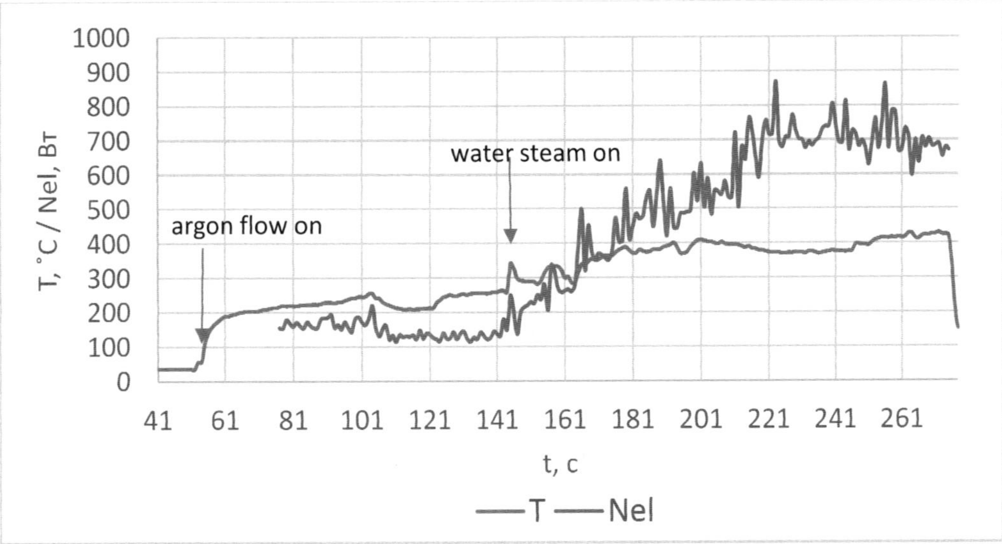

Figure 2 The typical signals of the time evolutions of the output axial gas mixture temperature T_{61} and electric power N_e input in plasma

The closed experimental results were obtained in second water pool calorimeter method, see figure 3. The value of thermal power was measured by following formula:

$$Q_{2T} = C_{pH2O}\, M_{H2O}\, (T_2 - T_0)/\tau = 1000 - 1200 \text{ W} \tag{3},$$

where M_{H2O} – water mass in the calorimeter, $Q_{H2O} = C_{pH2O}\, M_{H2O}\,(T_2 - T_0)$, T_0 - initial water temperature, final water temperature - T_2, τ - experimental time.

$$Q_{1T} = C_{pH2O}\, M_{H2O}\,(T_{2L} - T_0)/\tau = 300 - 340 \text{ W}, \qquad (4),$$

where calibration coefficient K_L obtained in the calorimeter with pure argon flow (without water steam). These results were obtained at value N_e~200-500W.

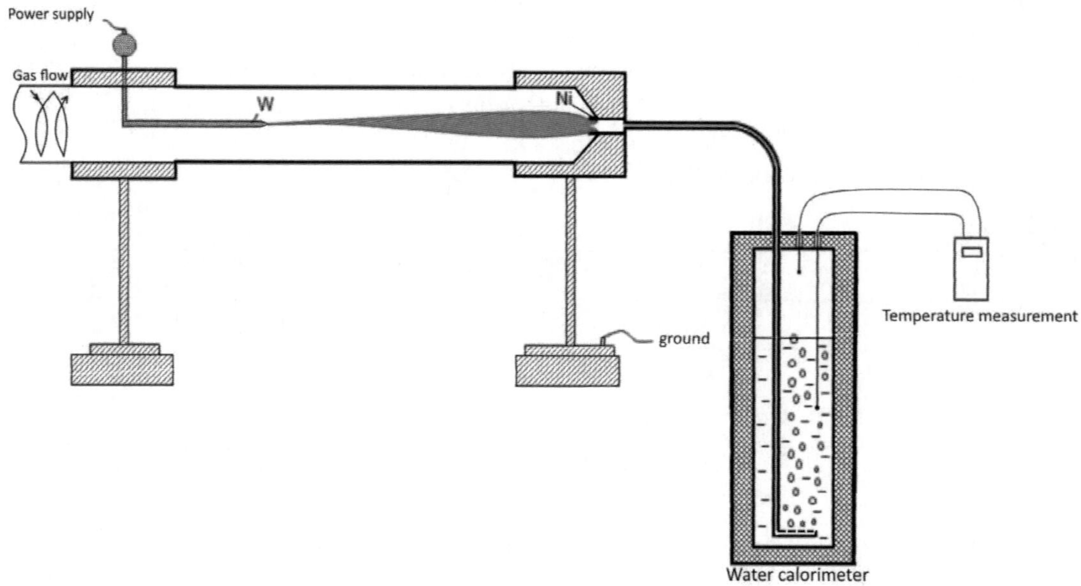

Figure 3 Scheme of the PVR with water calorimeter. W-tungsten anode, Ni- Nikole cathode, 1-PVR, 2-water bath, 3- thermocouple with recorder CENTER 306

Parametric study of PVR's design and its operation modes at different electrode configuration and electrode materials were obtained in this work also. It was revealed that the optimal electrode configuration is realized at its location near PVR's nozzle. Numerical simulation results obtained in Samara State University prove this conclusion [7].

It was revealed in these experiments also:

- Considerable gas temperature decrease in a recombination heterogeneous plasma jet behind nozzle. The typical decrement value is about of 200°C/m;

- Plasma jet gas temperature is increased at initial static pressure increase in PVR's testing chamber.

These two important results may be explained by relaxation and phase processes in the recombined heterogeneous plasma probably. This question will be studied in our future experiments in detail.

3. Measurement of hydrogen gas concentration

The scheme of hydrogen concentration measurement is shown in the figure 4. Tested gas goes from the PVR's nozzle to water bath (4) with storage tank (5). Residual water steam in this gas flow is cooled and condensed by cold water liquid. Then dry gas is stored in the tank (5). Then this gas goes to gas analyser (9) with the help of a vacuum pump (7). Residual gas flow (H_2+Ar+O_2) behind sensor (7) is burned by an igniter (8). Electric signal from gas analyser's sensor is recorded by PC.

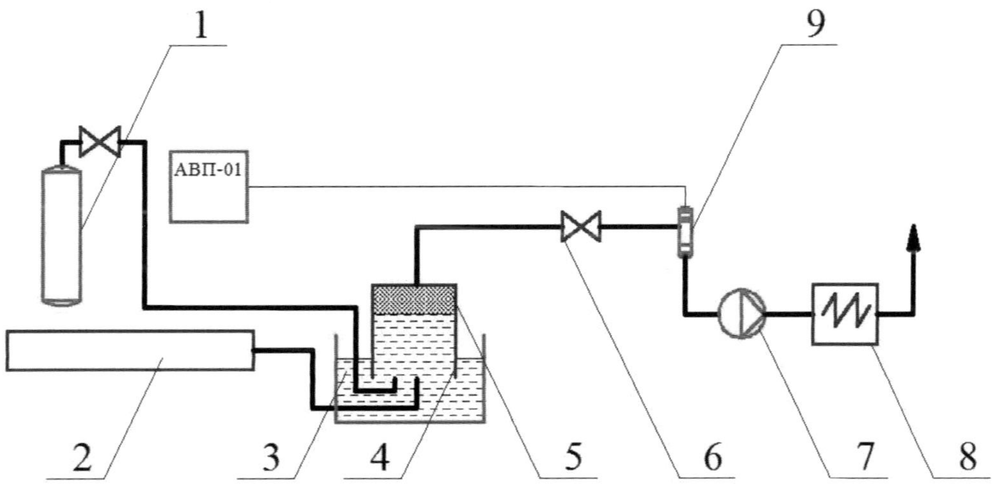

Figure 4 Scheme of hydrogen concentration measurement. 1 – balloon with hydrogen for calibration of gas analyser, 2 – setup PVR, 3 – water bath, 4 – gas tank, 5 – gas mixture, 6 – valve, 7 – vacuum pump, 8- igniter, 9 – gas analyser sensor.

It was measured that the typical hydrogen concentration is about of ~ 20 ÷ 30 % in output gas mixture. Note that estimated maximal hydrogen concentration should be [H_2]~33 % at the total water steam dissociation in hot plasma region. Therefore, definite part of hydrogen (3-13%) is absorbed by metal clusters probably. In a result metal hydride is created in the PVR. It was measured that maximal H_2 mass flow rate creation by the HP in the PVR is about of 10^{-3} g/s at water steam flow rate 1 g/s.

4. Conclusions

1. It is measured considerable thermal energy release in the PVR according formulas (1) and (2), see section 2.2. The closed experimental results were obtained in second water pool calorimeter method by formulas (3) and (4). The typical value of a thermal power release in this reactor is about of ~1kW. It is measured that the typical mass flow rate of an erosive cathode material is about of 1mG/s. Therefore, specific energy release of "nano-cluster fuel" is about of 1 KeV/metal atom.
2. Hydrogen production is realized in the setup PVR by a heterogeneous plasmoid. The typical value of a hydrogen mass flow rate is about of 1 mg/s at water steam dissociation in the discharge zone.
3. According our opinion high-energy chemical reactions with internal electrons are responsible for obtained extra power release in the setup PVR. Experimental results on analysis of the soft X-ray radiation spectra and the optical spectra recorded in this plasmoid prove this conclusion [4].

Acknowledgments
This work was fulfilled by the financial support of RAS Government Education Department (Project 3.1158.2017/4.6).

References

[1] Sheidlin A E 2007 *Aluminum-Hydrogen Energetics* (Moscow: JIHT RAS) p.250 (in Russian)
[2] Klimov A, Bityurin V, Chinnov V, et.al, Plasma-Assisted Reaction of Heterogeneous Mixture in

IOP Conf. Series: Journal of Physics: Conf. Series **1112** (2018) 012024 doi:10.1088/1742-6596/1112/1/012024

High-Speed Swirl Flow *Proc. Of the X International Workshop on Magneto-plasma aerodynics, Moscow, April 2011*, p.294-301

[3] Klimov A, Bityurin V, Grigorenko A et.al 2009 Plasma Assisted Combustion of Heterogeneous Fuel in High-Speed Airflow *AIAA Pap.* AIAA-2009-1411 11

[4] A Klimov, A Grigorenko, N Evstigneev, et.al. 2016 High-Energetic Nano-Cluster Plasmoid and Its Soft X-radiation *Condens. Matter Nucl. Sci., Proc. Int. Conf. Cold Fusion 19th* 1–10

[5] Klimov A, Tolkunov B, Evstigneev N, et.al. 2015 Patent RF 2 554 512 C1, 27.06.2015, (in Russian)

[6] N K Belov, I P Zavershinskii, A I Klimov, et.al 2018 High effective heterogeneous plasma vortex reactor for production of heat energy and hydrogen *J. Phys.: Conf. Ser.* **980** 012040

[7] P N Kazanskii, A I Klimov, N E Molevich, et.al 2018 Numerical simulation of an argon swirling flow in the presence of a DC discharge *J. Phys.: Conf. Ser.* **980** 012010

IOP Conf. Series: Journal of Physics: Conf. Series **1112** (2018) 012025 doi:10.1088/1742-6596/1112/1/012025

A semi-analytical model for the study of the evolution of a microwave plasma channel

V G Brovkin* and P V Vedenin

Joint Institute for High Temperatures, Russian Academy of Sciences, Izhorskaya 13 Bd 2, Moscow 127412, Russia

*brovkin47@mail.ru

Abstract. A simplified electrodynamic (non-electrostatic) approach is developed to study the evolution of a single microwave (MW) plasma channel (plasmoid) forming in the above breakdown electric field of linearly polarized MW beam (or beams). The results of numerical calculations in air within the pressure range $P = 60 - 140$ Torr based on this approach coupled with a sufficiently complete system of plasma chemical reactions, diffusion and photoionization are presented. Quasi-stationary values of the main MW plasmoid characteristics are estimated. The estimated values of the plasma density, the length of the plasmoid and its dipole moment agree well with the experimental results.

1. Introduction

In the problems of plasma aerodynamics and plasma-assisted combustion, a MW channel is considered as an object transforming the energy of the electromagnetic radiation into the gas energy [1,2]. The plasma channel is formed as a result of the development of a freely localized (away from surfaces) non equilibrium MW discharge in the above breakdown electric field of a linearly polarized wave beam (or beams) at high pressure under the condition $\nu \gg \omega$ (ν is an effective collision frequency of electrons with molecules, ω is the electromagnetic wave angular frequency). The investigation of the system MW radiation- plasmoid-gas requires a self-consistent solution of Maxwell's equations, plasmachemical kinetics equations and gasdynamics equations. The characteristic time scale at which the gas density changes noticeably is longer than the channel formation time. This fact makes it possible to use the following simplified scheme of the investigation.

The stage of a plasmoid formation under the condition of constant gas density. Three main phases of the plasma channel formation in the above breakdown electric field are revealed both theoretically [3] and experimentally [4]. (i) The development of an electron avalanche in an incident electric field. (ii) The MW channel-streamer elongation in mutually opposite directions in parallel to the incident electric field up to the length comparable to the wavelength, λ, of the incident radiation. (iii) At the quasi-stationary phase, the dimensions of the plasmoid change very slightly [4], the plasma density, the electric field inside a plasmoid, the total charge and current as well as the absorbed power reach quasi stationary levels [3]. It should be emphasized that the plasmoid accumulates its energy at the quasi-stationary phase.

The gasdynamic stage. At this stage, the accumulated energy can be used as an input parameter in the equations of gasdynamics.

Here, we present a semi-analytical self-consistent model which makes it possible to investigate the space-time evolution of a plasmoid under the condition of constant gas density (first stage). The model

IOP Conf. Series: Journal of Physics: Conf. Series **1112** (2018) 012025 doi:10.1088/1742-6596/1112/1/012025

is based on a simplifying assumption that allows the research to be carried out within one-dimensional approximation. The main feature of the model is the analytical expression for the electric field on the axis of the plasma channel. The model contains one undefined parameter - the characteristic radius of the streamer - that can be estimated on the basis of the discharge photos. The model has been applied to a freely localized self-sustained microwave discharge developing in the form of a single thin plasma channel in the above breakdown electric field of linearly polarized microwave beams in air within the pressure range of $P = 60 - 140$ Torr. We have estimated quasi-stationary values of the main MW plasmoid characteristics (the plasma density, the absorbed power). The estimated values of the plasma density, the length of the plasmoid and its dipole moment agree well with the experimental results.

2. Model

A plasmoid is in a high-frequency electric field $Re[\mathbf{E}_0(\mathbf{r})\exp(-i\omega t)]$ with components $\mathbf{E}_0 = (0,0,E_0)$. Its longitudinal (along the incident electric field $\mathbf{E}_0$) dimension is less than the wavelength and characteristic transverse dimensions are much less than λ. The coordinate origin is placed at the plasmoid center. We assume that the variations in the amplitude of the electric and magnetic fields, the averaged over a period, $T = 2\pi/\omega$, plasma density, the characteristic dimensions of the plasmoid are insignificant for the cycle T.

The complex amplitude of the total electric field can be written in the integral form [5]

$$\mathbf{E}(\mathbf{r},t) = \mathbf{E}_0(r) + \mathbf{E}_p(\mathbf{r},t)\,, \quad (1)$$

where $\mathbf{E}_p$ is the slowly time varying ($\partial\mathbf{E}_p \ll \omega\mathbf{E}_p$) complex amplitude of the plasma response electric field produced by electron current and space charges

$$\mathbf{E}(\mathbf{r},t) \approx iE_c(t)\sigma_{*c}(t)\mathbf{\Psi}(\mathbf{r},t)\,, \quad (2)$$

$$E_c \approx \frac{E_0}{1-i\sigma_{*c}\Psi_c}\,, \quad (3)$$

$$\Psi_{x,y} = -\frac{k^3}{4\pi}\int dV' f(\mathbf{r}',t)\,G(\kappa)\eta_{x,y}\eta_z(1 + i\kappa^{-1} - 3\kappa^{-2})\,,$$

$$\Psi_z = \frac{k^3}{4\pi}\int dV' f(\mathbf{r}',t)\,G(\kappa)[1 + i\kappa^{-1} - \kappa^{-2} - \eta_z^2(1 + 3i\kappa^{-1} - 3\kappa^{-2})]\,,$$

$E_c, \sigma_{*c}, \Psi_c = E_z, \Psi_z(\mathbf{r} = 0)$, $\sigma_* = \sigma/\omega\varepsilon_0$, $\sigma = e^2N_e/m(\nu - i\omega)$ is the high-frequency complex electrical conductivity, N_e is an averaged over the period T electron density, -e is the electron charge, m is the electron mass, ε_0 is the permittivity of free space, $k = \frac{2\pi}{\lambda}$, $f = \sigma/\sigma_c$, $G = \exp\frac{(i\kappa)}{\kappa}$, $\kappa = k|\mathbf{r} - \mathbf{r}'|$, $\eta = (\mathbf{r} - \mathbf{r}')/|\mathbf{r} - \mathbf{r}'|$. Everywhere below, we will assume the condition $\vartheta < 1$ ($\vartheta = \omega/\nu$) to be met. The results of the experiments [4-9] indicate that a plasma channel is nearly ellipsoid. This allows us to consider the interaction of an electromagnetic wave with a channel, which averaged over a microwave cycle electron density is distributed as follows

$$N_e(\mathbf{r},t) = N_{ec}(t)f(t,\xi(t)), 0 \le \xi \le 1\,, \quad (4)$$

where $\xi = \sqrt{(r/R_{max})^2 + (z/Z_{max})^2}$, $r = \sqrt{x^2 + y^2}$, R_{max} and Z_{max} are the minor and major semiaxis of the prolate spheroid, respectively. Within the scope of our model $Z_{max} \ge \lambda$, $Z_{max} = const$, $R_{max}(t) = Z_{max}/L(t)$. The value L comes from the experiment:

$$L(t) = \frac{Z_{vis}(t)}{R_{vis}(t)}\,, \quad (5)$$

where Z_{vis}, R_{vis} are the visible dimensions of the plasmoid in the photos. We assume that the visible volume is localized inside the spheroid with

$$N_e(r \le R_{vis} z \le Z_{vis}) \ge 10^{12}\ \text{cm}^{-3}. \quad (6)$$

The value $Z_{vis}(t)$ is computed. The assumption (4) transforms the 2D problem into the 1D one. Based on this assumption, we can consider the self-consistent evolution of the electron density and longitudinal electric field, $E_z(z,t)$, along only z-axis.

In the approximation (4) the expression for the form-factor Ψ_z is

$$\Psi_z(z_*,t) = \Phi(z_*,t) + \textstyle\sum_{m=0}\Psi_m(z_*,t)\,, \quad (7)$$

where $z_* = z/Z_{max}$,

IOP Conf. Series: Journal of Physics: Conf. Series **1112** (2018) 012025 doi:10.1088/1742-6596/1112/1/012025

$$\Phi(z_*,t) = -f(z_*,t) + \frac{2\mu^4}{L^2}\int_0^1 d\xi \frac{\xi^2 f(z_*\cdot\xi,t)}{(\mu^2-\xi^2)^2}, \mu = L/\sqrt{L^2-1},$$

$$\Psi_m(z_*,t) = \frac{(kR_{max})^2(ikZ_{max})^m}{m!(m+2)\mu^{m-1}}\int_{-1}^1 d\xi\xi^{m-1}f(\xi,t)\Gamma_m\left(\frac{z_*}{\xi}\right),$$

$$\Gamma_m\left(\frac{z_*}{\xi}\right) = \int_{-1}^1 dx\left[U^{m-1} + \frac{(m-1)(1-x^2)}{2(L^2-1)}U^{m-3}\right], U = \left(x^2 - 2\mu^2\frac{z_*}{\xi} + \frac{\mu^2}{L^2} + \left(\mu\frac{z_*}{\xi}\right)\right)^2,$$

Here we used the representation for the parameter Ψ_z obtained by expanding function G in series in powers of kr'.

The averaged over a MW cycle electron number density in air follows the equation

$$\partial_t N_e = Q_{ion} + Q_{ass} - Q_{rec} - Q_{at} + Q_{det} + Q_{ph} + \nabla\cdot\left(D_{eff}\nabla N_e\right), \quad (8)$$

where $Q_{ion}, Q_{ass}, Q_{rec}, Q_{at}, Q_{det}, Q_{ph}$ are contributions of the impact ionization, associative ionization [10], recombination, attachment, detachment from negative ions and photoionization [11] processes, respectively, D_{eff} is the effective diffusion coefficient. The effective diffusion coefficient is given by [12]

$$D_{eff} = \frac{C_D D_{ee} + D_a}{1+C_D},$$

where D_{ee}, D_a are the free and ambipolar diffusion coefficients, respectively, $C_D = \nu_i\tau_M$, $\tau_M(z,t) = \varepsilon_0/\sigma(z,t)$ is the local Maxwell's relaxation time.

The kinetic block of the model in air includes the system of processes that describe the change in the densities of the main neutral and charged components of the nitrogen–oxygen mixture and the vibrational excitation of the gas. Nine types of positive and negative ions are taken into account: O_2^+, O_4^+, $N_2^+, N_4^+, NO^+, O^-, O_2^-, O_3^-, O_4^-$. The system of ion-molecular reactions [10] was taken as the basis. In addition, the reactions involving the following neutral particles were included in the model: $N_2(X^1\Sigma_g^+), N_2(A^3\Sigma_u^+), N_2(B^3\Pi_g), N_2(C^3\Pi_u), N_2(a'^1\Sigma_u^+), N(\,^4S), N(\,^2D), N(\,^2P), NO, O_2(X^3\Sigma_g)$, $O_2(a^1\Delta_g), O(\,^3P), O(1D), O(1S)$ [10, 13]. In the conditions under consideration, the particles are assumed to be excited mainly by electron impact from the ground electronic state.

Using this model we considered the main features of the development of a plasmoid in air within the ranges of discharge parameters $P = 60 - 140$Torr, $E_{eff} = (1.1 \div 1/3)E_{br}$, $E_{eff} = |E_0|/\sqrt{2(1+\vartheta^2)}$, E_{br} is the breakdown electric field, $\frac{E_{br}}{P} = 40$V·cm^{-1}·Torr^{-1} and obtained the quasi stationary values of its basic characteristics.

Numerical results=0.05

The initial stage of the development lasts as long as a plasma cloud is transparent to an incident electromagnetic radiation. The duration of the initial stage, t_0, for $P > 50$ Torr, $\frac{E_{eff}}{P} \geq$ 45 V·cm^{-1}·Torr^{-1}, and the corresponding maximum electron density at the cloud center, $N_{ec0}(t_0)$, are given by [5]

$$t_0 \approx \frac{1}{\nu_1^{(eff)}\left(\frac{E_{eff}}{P}\right)}\ln\left(\frac{0.5}{\sigma_{*0}(t=0)n_z(t_0)}\right), N_{ec0} \approx \frac{5\cdot10^{11}P[\text{Torr}]}{\lambda[\text{cm}]}, \quad (9)$$

where $\nu_i^{eff} = \nu_i - \nu_a$, ν_a is the attachment frequency, $n_z = \mu\ln\left(L+\sqrt{L^2-1}\right)/(L^2-1)$ is the depolarization factor in the z direction ($n_z = 1/3$ for a sphere). We assume that by the end of the initial stage the profile $f(\xi, t_0)$ is Gaussian with the radius r_0. A rough, order-of-magnitude estimation of this radius from above is

$$r_0 < \sqrt{D_e t_0}\,. \quad (10)$$

Running ahead, we will note that the streamer dynamics and its basic quantitative characteristics, at the final stage of the evolution, are virtually independent of the choice of the parameter r_0 .

The photos of the streamer evolution demonstrate that the visible radius is proportional to the visible length, when $Z_{vis} \geq Z_{vis0} \geq 0.05$ cm. On the basis of this result as well as the approximation from [3] in our computations we used the following expression

$$R_{vis} \approx \begin{cases} r_0(Z_{vis}/r_0)^b \,, r_0 \le Z_{vis} \le Z_{vis0} \,, [3] \\ 0.2Z_{vis} \,, Z_{vis} \ge Z_{vis0}, \text{our experiments}, \end{cases} \quad (11)$$

where $b = \ln(0.2Z_{vis0}/r_0) / \ln(Z_{vis0}/r_0)$. Note that the sizes R_{vis}, Z_{vis} depend slightly upon air pressure within the range $P = 60 - 140$ Torr. Under our experimental conditions, a plasma channel develops in the focal region of the standing wave, where the electric field spatial distribution can be approximated by the expression

$$E_{0z}(\mathbf{r}) = Re\left\{\exp(-i\omega t)\left[A_+ \exp\left(ikz - \frac{(y^2+z^2)}{a_f^2}\right) + A_- \exp(-ikz)\right]\right\} \quad (12)$$

with $a_f \approx 0.7\lambda$, $\lambda = 2.3$ cm, $A_- \approx 0.1A_+$.

The simulation results at times $t > t_0$ (the streamer and quasi stationary phases are under the investigation) are presented in figures 1 – 5.

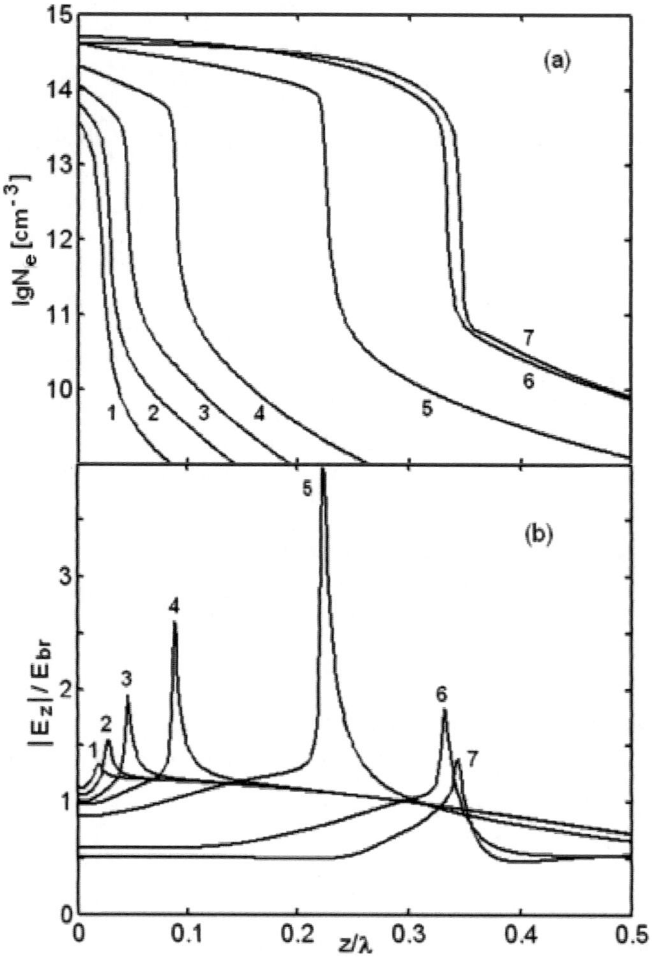

Figure 1. Axial profiles of (a) electron density, (b) axial component of the normalized electric field amplitude plotted from the streamer center (one half of the streamer is represented). Time moments $\tau = 50 - 350$ ns (1-7), step 50 ns. E_{br} is the breakdown level of the electric field. Air $P = 100$ Torr, $E_{eff} = 1.2E_{br}$

Figure 1 shows a typical picture of the evolution of electron density profiles and the amplitude of the electric field along the z axis. As follows from this figure, at $t < 250$ ns, the maximal magnitude of the amplitude of the electric field increases along with the plasma density. In the central region the amplitude of the electric field slowly decreases. With increasing of the plasma density and dimensions of the streamer, at $t > 250$ ns, the role of the electric field produced by the current flowing along the streamer significantly growth. This field tends to compensate the field of the space charges both in the central area and at the tips of the streamer [3]. As a result, the ionization waves are sharply slowed down, when the length of the streamer is about 0.7λ. Later, at $t > 300$ ns, the electrostatic field of the space charges can no longer ensure further noticeable elongation of the plasmoid. Finally, at $t > 350$ ns, the state of an ionization-recombination quasi equilibrium reaches. In this state, the plasma density and the amplitude of the electric field reach their quasi-stationary levels. The dynamics of the contour lines of the electron density $N_e = 10^{11}$ cm^{-3} and $N_e = 10^{13}$ cm^{-3} demonstrates figure 2.

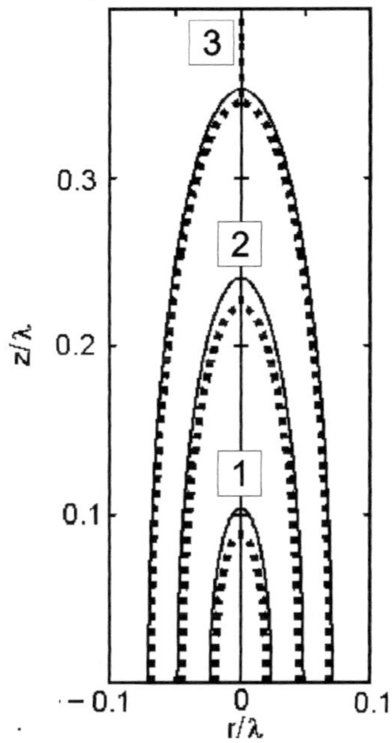

Figure 2 Contour lines of the electron density $N_e = 10^{11}$ cm^{-3} and $N_e = 10^{13}$ cm^{-3} (dots). Time moments are 200 ns (1), 250 ns (2), 350 ns (3).

In order to verify some predictions of our model, we compared the experimental data with the results of numerical calculations. As follows from the experiments, (i) normalized full maximum channel length, $2Z_{vis}^{(max)}/\lambda$, is in the ranges of values $2Z_{vis}^{(max)}/\lambda \approx 0.6 \div 0.8$, and this length is weakly dependent on the pressure and amplitude of the incident field (our study); (ii) the quasi stationary value of the amplitude of the normalized dipole moment of the plasmoid, $\left|d_*^{(qs)}\right|$, $d_* = k^3|d|/4\pi\varepsilon_0 E_0$, $|d|$ is the dipole moment amplitude), is in the range of values $\left|d_*^{(qs)}\right| \approx 0.5 \div 0.7$, within the ranges of discharge parameters $P = 60 - 140$ Torr, $E_{eff} = (1.1 \div 1.3)E_{br}$ [5]; (iii) the value of the electron density estimated in nitrogen by the spectroscopy method (Stark broadening of the hydrogen spectral line H_β) under the experimental conditions $\lambda = 4$ cm, $P = 50$ Torr [6] and $\lambda = 2$ cm, $P = 70$ Torr [14] are $N_e = 10^{14}$ cm^{-3} and $N_e = 0.9 \cdot 10^{14}$ cm^{-3}, respectively.

Figure 3 shows time dependences of the normalized channel length (by the channel length we will mean the distance $2l_{ch}$ between the points at which the electric field amplitude is at a maximum) and amplitude of the normalized dipole moment of the plasmoid at different values of pressure and amplitude of the incident electric field. Figure 4 reflects the dynamics of the averaged over a MW cycle electron density at four axial points in the studied range of discharge parameters. In our opinion, the results of the experiments are in agreement with the predictions of our model.

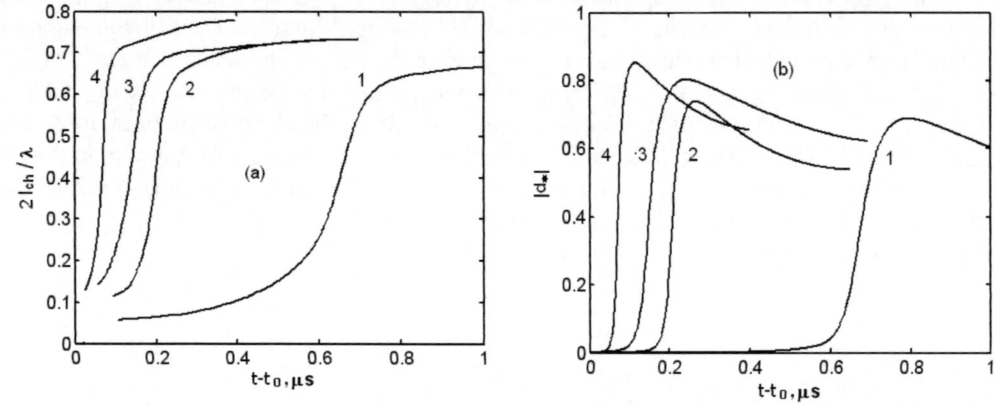

Figure 3 The dynamics of the (a) normalized full plasma channel length, $2l_{ch}/\lambda$, and (b) amplitude of the normalized dipole moment, $|d_*|$, of the plasma channel in air for $E_{eff} = 1.1E_{br}$, (1) P=60 Torr, (2) $P = 140$ Torr; $E_{eff} = 1.3E_{br}$, (3) $P = 60$ Torr, (4) $P = 140$ Torr.

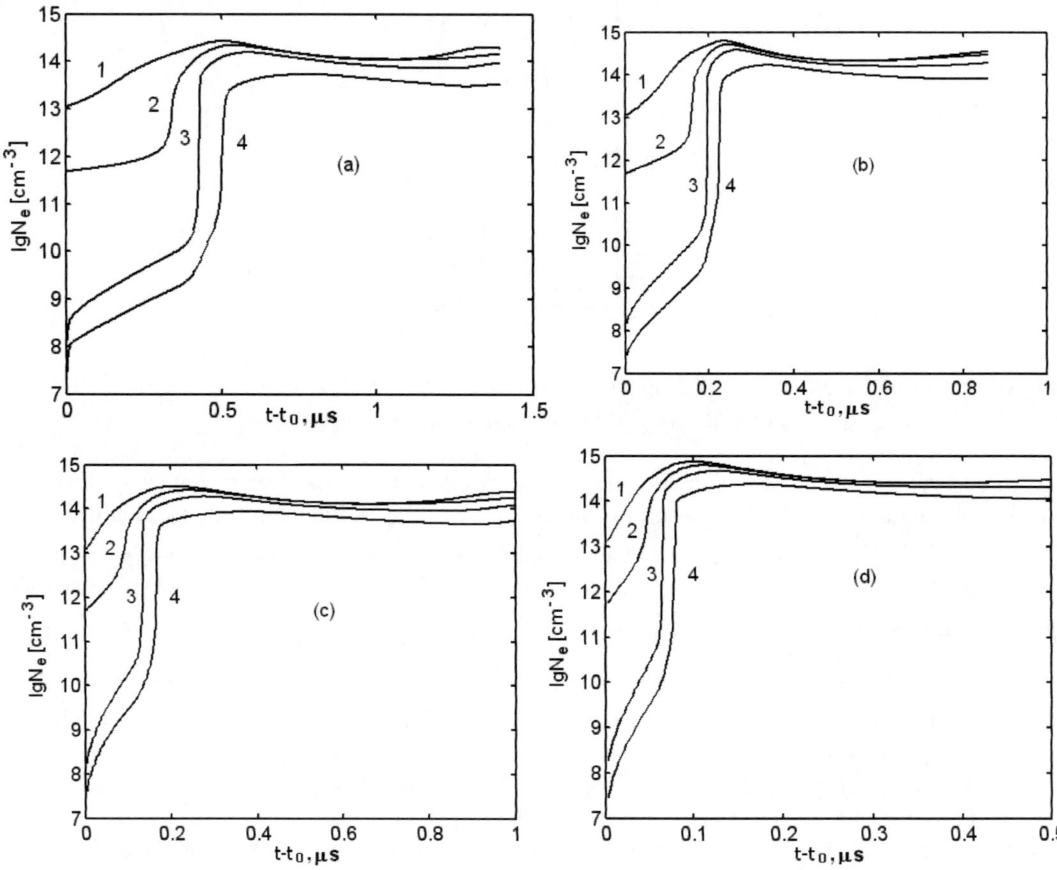

Figure 4 The dynamics of the time-averaged electron density at the points $\frac{z}{\lambda}$ =0 (1), 0.1 (2), 0.2 (3), 0.3 (4) for $E_{eff} = 1.1E_{br}$, (a) $P = 60$ Torr, (b) $P = 140$ Torr; $E_{eff} = 1.3E_{br}$, (c) $P = 60$, Torr, (d) $P = 140$ Torr.

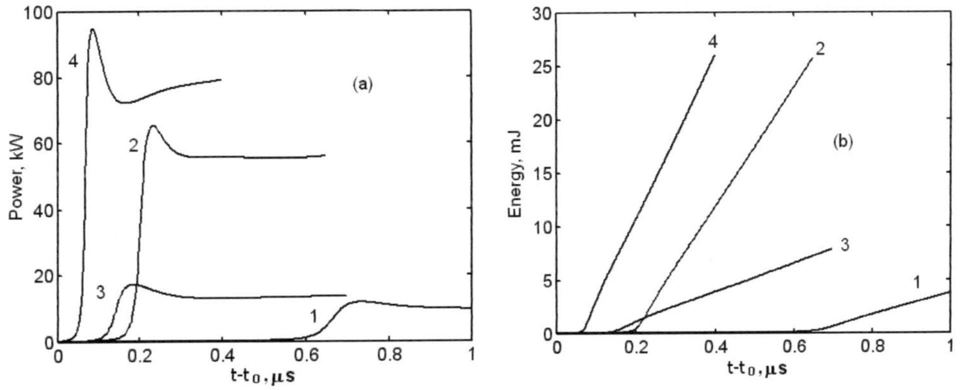

Figure 5 The dynamics of the (a) absorbed power and (b) energy in the conditions of figure 3.

The most important characteristics (for gasdynamic and plasma-assisted combustion applications) of a microwave plasmoid are the absorbed power (Joule losses)

$$W_J = \frac{\int dV\sigma|E|^2}{2(1+\vartheta^2)} = E_{eff}^2 \int dV\sigma \left|\frac{E}{E_0}\right|^2 \qquad (13)$$

and the absorbed energy. As follows from expression (13) $W_J \propto \left(PE_{eff}/E_{br}\right)^2$. Figure 5 shows the evolution of the averaged over a MW cycle absorbed power and energy under the same conditions as in figure 4.

Summary
The semi-analytical self-consistent model is developed to study the space-time evolution of a single MW plasma channel forming in the above breakdown electric field of linearly polarized MW beam (or beams). The model is based on simplifying assumption that transforms the 2D problem into the 1D one. Using the model, we considered the main features of the development of a plasmoid in air within the ranges of discharge parameters $P = 60 - 140$ Torr, $E_{eff} = (1.1 \div 1.3)E_{br}$ and obtained the quasi stationary values of the absorbed power and energy. The quasi stationary values of the plasma density, the length of the plasmoid and its dipole moment agree well with the experimental results.

References

[1] Y F Kolesnichenko, O A Azarova, V G Brovkin, VG Grudnitsky, V A Lashkov, and I C Mashek *AIAA Pap.* 2003-0361
[2] B Michael, A Dogariu, M N Shneider, and R B Miles 2010 *J. Appl. Phys.* **108** 093408
[3] V A Bityrin and P V Vedenin 2010 *Sov. Phys. JETP.* **111** 512
[4] V A Bityurin, V G Brovkin, and P V Vedenin 2012 *Tech. Phys.* **57** 95.
[5] V A Bityurin, V G Brovkin, and P V Vedenin 2017 *J. Phys.* D: *Appl.Phys.* **50** 75201
[6] A L Vikharev, A M Gorbachev, A V Kim, and A L Kolysko 1992 *Sov. J. Plasma Phys.* **18** 554
[7] V. G. Brovkin and Yu. F. Kolesnichenko 1995 *J. Moscow Phys. Soc.* **5** 23
[8] Yu F Kolesnichenko, V G Brovkin, D V Khmara et al. 2003 Fine structure of microwave discharge *Proceedings of the 41st AIAA Aerospace Meeting and Exhibition, 6-9 January 2003, Reno, Nevada* 362
[9] V M Shibkov 1996 *High Temp.* **34** 525
[10] I A Kossyi, A Y Kostinsky, A A Matveev, and V P Silakov 1992 *Plasma Sources Sci. Technol.* **1** 201
[11] M B Zheleznyak, A Kh Mnatsakanyan and G V Naidis 1982 *High Temp.* **20** 357
[12] J P Boeuf, B Chaudhury, and G Q Zhu 2010 *Phys. Rev. Lett.* **104** 015002
[13] J T Herron 1999 *J. Phys. Chem. Ref. Data.* **28** 1453
[14] V V Zlobin, A A Kuzovnikov, and V M Shibkov 1988 *Moscow Univ. Phys. Bull.* **43** 89

IOP Conf. Series: Journal of Physics: Conf. Series **1112** (2018) 012026 doi:10.1088/1742-6596/1112/1/012026

Numerical study of hydrodynamic perturbations caused by filiform spark discharge near wall

E V Dolgov and A A Firsov*

Joint Institute for High Temperatures (JIHT RAS), 125412, Moscow, Russia

*af@jiht.org

Abstract. A numerical simulation of the high-speed jet caused by the decay of the thermal cavern after cylindrical 10 mm long spark discharge located on the wall in ambient air is performed. It is shown that after the shock wave retreats, two counter-flowing jets are formed in a low-density hot gas, moving inside the thermal cavern from the electrodes to the center of the area. A strong distortion of the hot gas region is observed and a cold gas suction into the heat cavern is appeared. At the next phase, the formed hot gas jets collide, and the flow is directed radially from the axis of the initial discharge channel. Finally, the hot low-density gas takes the form of half the toroid, which size increases in time. The central part of the flowfield is occupied with the cold gas. The jet velocity and momentum are simulated at variation of the discharge energy (5-500mJ). The simulation results are considered to be consistent with the experimental data previously published by the authors.

1. Introduction

The spark electrical discharge has been thoroughly investigated [1] and is actively used in various fields of science, including ignition in internal combustion engines [2] and generation of NOx for biomedical applications [3], etc.

Recently, an interest was arisen for a long (5-50mm) pulsed (100ns-5μs) spark discharge due to its ability in generation of gasdynamic perturbations potentially beneficial for a flow control. In several publications, the results of studies on the features of the spark discharge thermal cavern decay were presented: it was shown that large-scale jet flows are formed in the direction perpendicular to the channel due to the spark channel bends during the decay of the long spark discharge [4]. The effect of discharge shape (bending) on pattern of the gasdynamic perturbations was shown by using additional electrodes to distort the shape of the discharge [5], and also by numerical simulation [5, 6]. Later, using an 8-frame IIT framing camera, direct experimental proof of the described effect was obtained [7].

It was proposed to use these specific properties of the spark discharge as a tool that rapidly increases the mixing degree of the injected fuel with an incoming high-speed (including supersonic) flow [8]. It was shown that the long spark discharge arranged transversely to the flow and passing through the core of the stream is localized along the jet of fuel and introduces significant perturbations into the area of fuel jet. This behavior of the discharge was supposed to promote a stretching the interface between the fuel and the oxidizer that might lead to intensification of mixing. However, the method used in the study - schlieren visualization of the flowfield - did not allow exact determination of the interface between the jet and the flow under the influence of the discharge. When fuel is injected from the wall-mounted nozzle into the supersonic flow, the fuel is localized in a near-wall layer and doesn't penetrate far into the flow. In this case, the need to use a discharge that crossed the

IOP Conf. Series: Journal of Physics: Conf. Series **1112** (2018) 012026 doi:10.1088/1742-6596/1112/1/012026

core flow is eliminated. It appears that the use a spark discharge right on the wall might be a more appropriate approach. In paper [7], using numerical simulation it was shown that spark discharge near the surface in the region of fuel injection into a supersonic flow leads to a substantial increase in the interface between the fuel and the oxidizer, thereby accelerating the mixing of the components.

Another direction of the use of gasdynamics caused by a spark discharge is development of plasma actuators for suppression of a transonic buffet - the normal shock wave oscillations that takes place in transonic flow around a wing at high Mach number or high angle of attack [9-12]. The spark discharge acted as a vortex generator in the boundary layer on a flat surface [9-10, 12], and in the work published in paper [11] the spark discharge was combined with protrusions on the wing surface to create a directed transverse action. In paper [9], it was shown that the spark discharge near the surface leads to the formation of a jet of a rather complicated shape: after the shock wave has run away, the thermal cavern is significantly modified in geometry - a gas flows from the ends of the discharge channel to the center. Then the region of the low-density hot gas moves into plane perpendicular to the initial position of the spark, into a wide sector of about 180 degrees in the form of a half of toroid, the center of which takes a gas from the surrounding area. In paper [12], it was shown experimentally that such a discharge on the surface of the wing can lead to decrease the buffet amplitude. By the numerical simulation it was shown an entire suppression of the oscillations of the shock wave upon action by a pulse-periodic spark discharge.

There is of a practical interest to study a very short nanosecond (~ 100ns) discharges due to potential applications for the airflow control, including applied aerodynamics problems (flow around various bodies) and problem of combustion/ignition of lean mixtures. In paper [13], it was shown that the decay of a thermal cavern of a nanosecond spark discharge with a spark gap of 1 mm results in the transfer of active particles into the surrounding volume by a distance of ~ 10 mm from the discharge axis. To use the discharges for aerodynamic applications, it is important to optimize the discharge parameters in terms of the flow control authority, the effect on the airflow pattern depending on the discharge parameters, such as interelectrode distance and energy deposition [14]. In this paper, the major focus is made on the parametric study of structure and dynamics of gas perturbations caused by the long spark discharge.

2. Description of simulation model

Simulation of the gasdynamics caused by the decay of a thermal cavern arising after a breakdown of a long spark discharge was performed in the software package FlowVision 3.10.02. The simulation is based on the solution of the three-dimensional unsteady system of Navier-Stokes equations supplemented by the k-ε model of turbulence. Our previous simulation of spark in ambient air was performed without turbulence model [5], but real applications of spark for flow control or mixing corresponds to the turbulent flow. Therefore, in order to better reproduction the result of this work under turbulent flow conditions the turbulent model was used in current simulation. Gas density was calculated using ideal gas law. For description of transport gas properties tables with dependencies of viscosity, thermal conductivity and specific heat on temperature (up to 30kK) at constant pressure were used [15]. The simulation domain (shown in figure 1) was the one-eighth of the sphere with a radius of 30 mm due to the symmetry of the region: on the planes XY and YZ, the symmetry boundary condition was chosen; on the XZ plane, the adiabatic wall boundary condition with no-slip and a wall function for modeling the boundary layer in the first cell was set. On the remaining convex surface, the "input/out" boundary condition was set with the fixation of the total pressure 101325 Pa and the temperature 273K. Due to the fact that this condition leads to reflection of the shock wave back to the simulation domain when a shock wave passes through the boundary, shortly before the interaction of the shock wave with the boundary of the region, and for some time after it leaves the calculation domain, this boundary condition was replaced by the Riemann nonreflecting boundary. It was done in order to allow the shock wave leave the considered calculated domain without reflection. Unfortunately, this condition in FlowVision can be used with zero gas velocity indication on it only for a short time, otherwise unphysical vortex flows begin to form on the boundary. Therefore, this condition cannot be used during the whole calculation.

IOP Conf. Series: Journal of Physics: Conf. Series **1112** (2018) 012026 doi:10.1088/1742-6596/1112/1/012026

The 10-mm spark discharge was simulated by a volumetric heat source and was placed in the center of the sphere underlying the calculation domain. The shape of the volumetric heat source was a half of cylinder with rounded ends, recessed into the surface (figure 1). The cylinder radius was 0.3 mm, and its length was 5 mm due to the symmetry of the region. The spatial profile of the energy input was not applied, it was accepted that the energy input is uniform in length and radius. The dependency of heat power on time during the volumetric heat source operation was also not taken in the account, the energy input duration was set to 200 ns. Calculations were carried out for energy release into in the discharge of 5, 20, 60, 200 and 500 mJ.

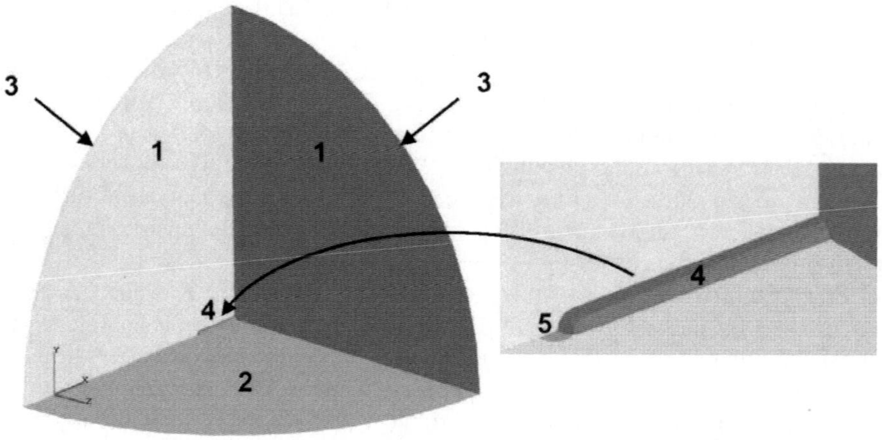

Figure 1 The calculation domain. 1 – symmetry condition, 2 – adiabatic wall condition, 3 – outlet with fixed total pressure, 4 – volumetric heat source, 5 – approximate location of electrode for such shape of discharge

In simulation, a variable time step was used. At the time of the volumetric heat source operation, the time step was 1 ns, and after that the time step was set by the Courant number $CFL = 1$. The calculation was performed on a reconfigurable rectangular grid using local adaptations in the specific area. Before the activation of a volumetric heat source, the grid in discharge region and in the space at a distance of 2.7 mm around the heat source was adapted, resulting in a spatial resolution of 54 cells per 1 mm. Periodic adaptation to the pressure gradient was used to accompany the shock wave by fine mesh, and periodic adaptation of the density field in the interval from 0 to 1.2 kg/m^3 was used to accompany the expanding heat cavern area. Moreover, in both cases, a space with a thickness of 6-20 cells around the indicated regions was additionally adapted to prevent the movement of gasdynamic features of considered flow between adaptations performed once per 5 time steps. Also in this procedure, outside discussed area, smaller cells were united into larger one if a shock wave or a low density gas left the given location of the computational domain. Adaptation made it possible to reduce the number of computational cells and the calculation time while maintaining sufficient spatial resolution in the area of the shock wave and thermal cavern. Periodic use of adaptation also reduced the calculation time, because one calculated step with the adaptation requires substantially more time than the step on which the grid was not rebuilt. Subsequently, with the motion of the shock wave to the boundary of the computational domain, the spatial resolution of the shock wave by the calculated grid was reduced due to the decrease of the influence of the shock wave on the thermal cavern. Starting from 70 μs, the spatial resolution of the computational grid in the region of thermal cavern was reduced to 14 cells per 1 mm, and was retained until the end of the calculation. As a result of the calculation, it was possible to sufficiently resolve by a grid and time step the gasdynamical features of the process under investigation and keep the number of cells at a level no greater than 5e+06, and on average about 5e+05.

3. Results of simulation

The results of the numerical calculation are periodical (in increments of 10 ns to 5 μs) complete data saves of all gasdynamic quantities in volume under consideration, what makes possible after completion of the calculation to carry out a full analysis of gasdynamics of the thermal cavern decay. The gasdynamics of the thermal cavern passes through the following stages (see figure 2). First, within 200 ns, the region occupied by the volumetric heat source is heated, the thermal cavern rapidly expands and a shock wave leaves this region. After the retreat of the shock wave, high-speed jets are formed in the low-density hot gas, moving inside the thermal cavern from the electrodes to the center of the region (center of sphere). The position of one electrode matches to the left end of volumetric heat source in 1, and second one is located symmetrical about the dark blue (YZ) plane. As a result, deformation of the heat cavern occurs and cold gas suction is formed inside the heat cavern. After a while (at 80-180 μs, depending on the energy), the jets collide and a new jet flow arises in the YZ plane, directed outward from the center to the angle about 180°. As a result, the hot low-density gas takes the form of half the toroid, whose size increases in time, and the central region is occupied by the cold gas. This development of the gasdynamics caused by thermal cavern decay at the surface is in good agreement with the results of the experimental studies that presented in paper [6]. Typical experimental schlieren images of spark discharge cavern decay in YZ plane from paper [6] are presented in figure 3.

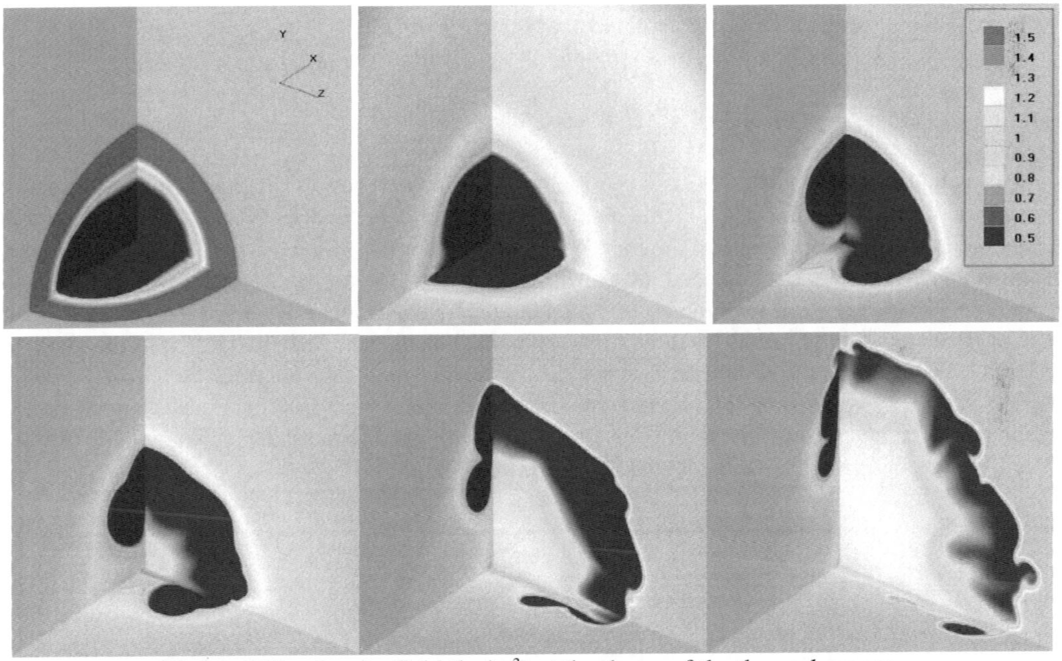

Figure 2 The density field (kg/m^3) at the decay of the thermal cavern, energy 0.5J, time moments 10, 80, 140, 200, 500 and 900 μs.

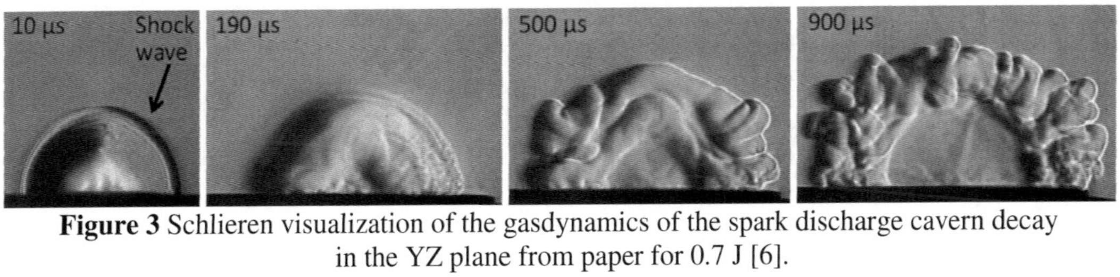

Figure 3 Schlieren visualization of the gasdynamics of the spark discharge cavern decay in the YZ plane from paper for 0.7 J [6].

IOP Conf. Series: Journal of Physics: Conf. Series **1112** (2018) 012026 doi:10.1088/1742-6596/1112/1/012026

In addition to the qualitative comparison, a quantitative comparison of the results of the simulation with previous experiments is performed. Dependences of the coordinate of the thermal cavern boundary (the boundary of the density gradient) on time is obtained, typical position of boundary at time moment 100 μs were 3.6 mm at 60 mJ, 5.2 mm at 200 mJ, and 7.1 mm at 500 mJ. The initial velocity of expansion of the thermal cavern in the first 10 microseconds was 300-500 m/s, and at a later stage the velocity of the hot gas boundary did not exceed 30 m/s. The data obtained in the simulation correlate good enough with the data obtained in the experimental work [6], as it is shown in figure 4.

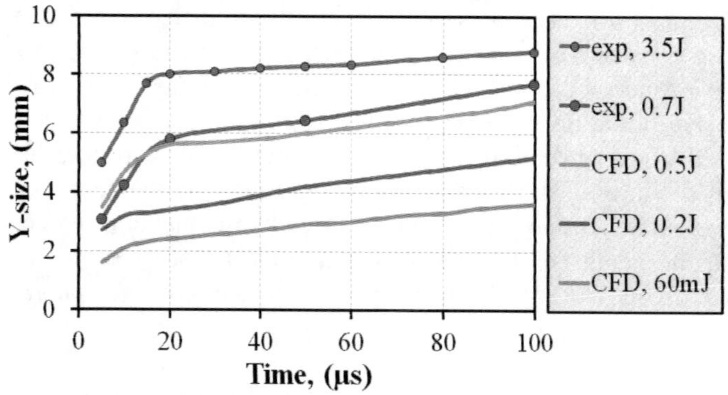

Figure 4 The coordinate of the thermal cavern boundary along the Y axis as a function of time. The results of the experiments were taken from paper [6].

During the analysis of modeling data, it was found that the gas velocity is different from the velocity of motion of the thermal cavern boundary displacement. The gas velocity at 60 mJ energy release in volumetric heat source was 50-100 m/s, and for large values of the energy release it was in the range of 100-200 m/s for a long time about 200 μs. The dependences of the maximum velocity module and velocity components on time are presented in figure 5. It is clearly seen from the curves that the velocity modulus has a second local maximum in the time range of 80-160 μs, depending on the energy. The second local maximum corresponds to the time moment when the jets moving to the center of the spark collide, and a jet flow arises in the YZ plane, directed outward from the center of discharge. As a result, the maximum values of the velocity do not longer correspond to the x-component. In the calculations performed, typically, the maximum value of Vz at a given time is slightly greater than the maximum of Vy. The velocity fields for the planes XY and YZ for two different times - before and after the jet turn - are shown in figure 6.

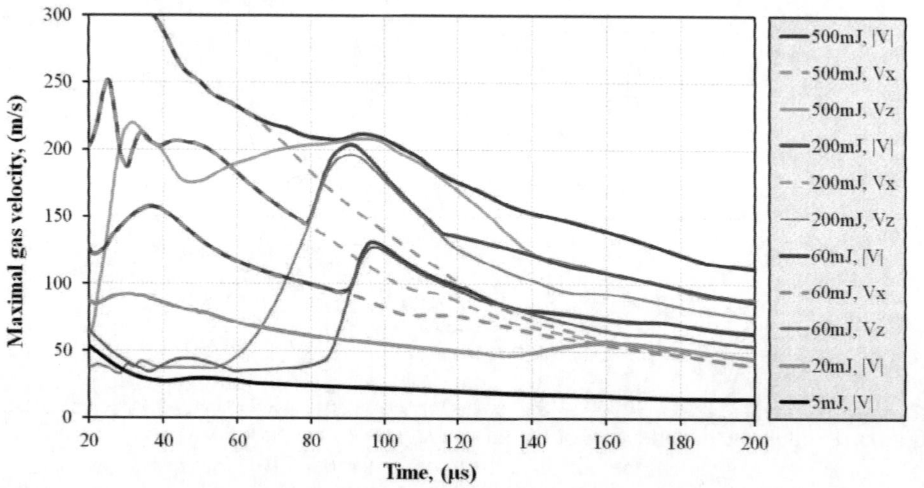

Figure 5 The time dependences of the maximum velocity of the jet at different energy release

IOP Conf. Series: Journal of Physics: Conf. Series **1112** (2018) 012026 doi:10.1088/1742-6596/1112/1/012026

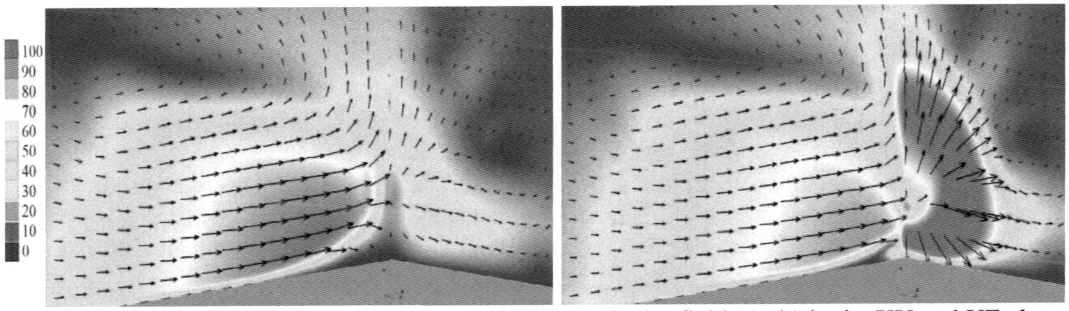

Figure 6 Changing the direction of the jet motion: the velocity fields (m/s) in the XY and YZ planes for the energy release of 60 mJ at the moments of 85 μs and 98 μs

During the analysis the calculation results, it was found that the maximum gas velocity is observed in the low-density region of the thermal cavern. On the other hand, the formation of jet in the hot region also results in the movement of the cold gas, which is sucked into the thermal cavern. As a result, the region of maximum momentum is shifted relative to the region of maximum velocity. To illustrate this effect, fields of the velocity magnitude $|V|$, the density ρ and the specific mass velocity (momentum) of the gas $V\rho$ for the energy release of 200 mJ at the time moment 80 μs are presented in figure 7. For other energy inputs, structures are similar, but are shifted in time. From the presented velocity and density distributions it is clearly seen that they have opposite gradients along X, as a result of that the momentum distribution is uniform enough.

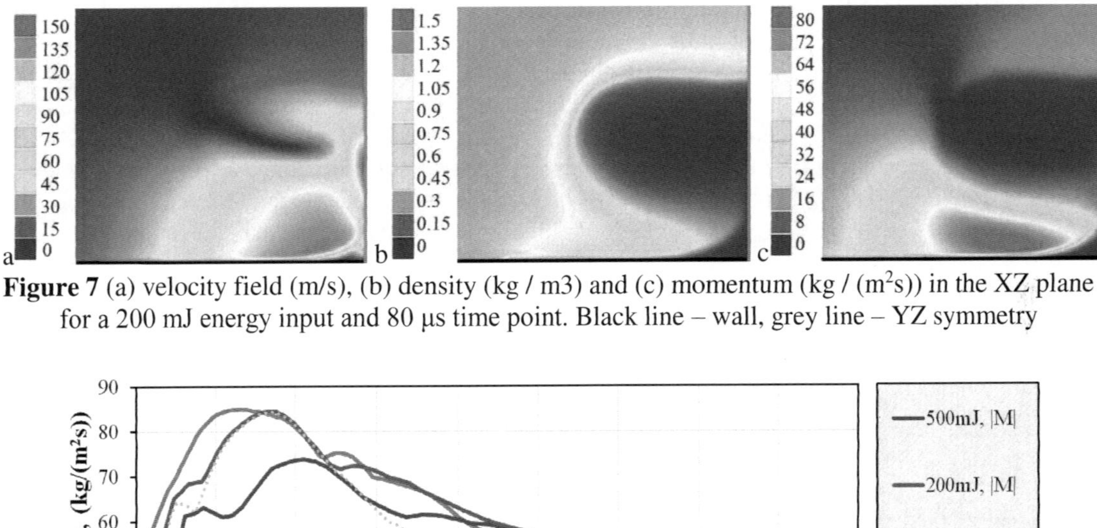

Figure 7 (a) velocity field (m/s), (b) density (kg / m3) and (c) momentum (kg / (m^2s)) in the XZ plane for a 200 mJ energy input and 80 μs time point. Black line – wall, grey line – YZ symmetry

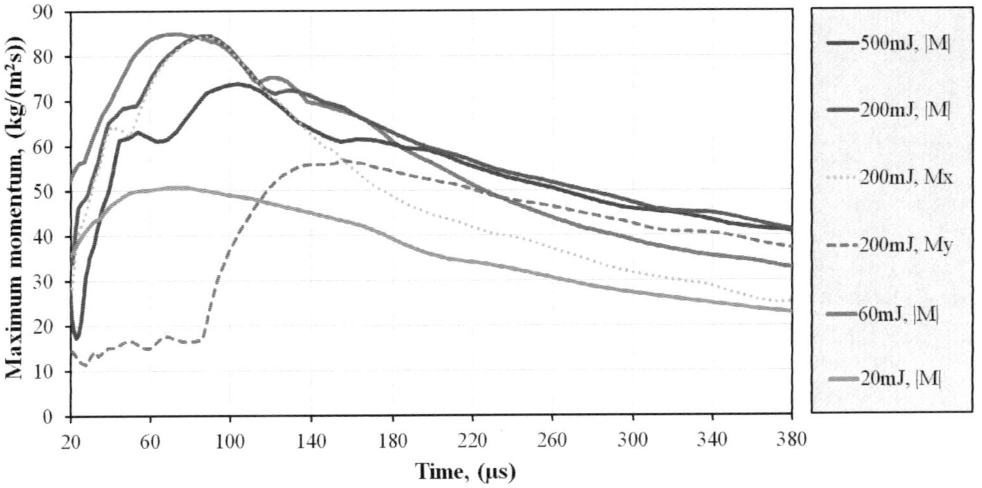

Figure 8 Dependence of jet's maximum specific mass velocity (momentum M) and its components on time

IOP Conf. Series: Journal of Physics: Conf. Series **1112** (2018) 012026 doi:10.1088/1742-6596/1112/1/012026

The study of gasdynamics after spark discharge is caused by an interest to use this type of discharge as an alternative to mechanical and jet vortex generators for solution of problems of flow detachment control and mixing intensification in high speed flows. Therefore, such a characterization as a momentum is of much greater interest than the gas velocity. To estimate the influence of the energy release on the momentum, let us discuss the dependences of the specific mass velocity ρV on time presented in figure 8. The graph shows the maximum values of the magnitude and the components of this quantity in space. It can be seen that the maximum local momentum is achieved not at the maximum energy considered 500 mJ, but at lower values of 60-200 mJ. This can be explained by overheating of the heat cavern: high velocity does not compensate the decrease of the gas density. Thus, with the help of numerical simulation and subsequent analysis of the results it was found that the spark with energy about 60 mJ per 10mm in ambient air should be the most efficient vortex generator for discussed spark geometry and duration of the energy release.

Conclusions

The simulation of the gasdynamic perturbations caused by the decay of a hot channel after the 200 ns pulsed 10 mm long spark discharge (5-500 mJ per 10 mm) arranged near the surface in the ambient air at laboratory conditions was performed using CFD-software package FlowVision 3.10.02. Using numerical simulation, it was shown that pulsed heating of the gas first leads to a rapid expansion of the thermal cavern and subsequent formation of high-speed jets directed from the ends of the plasma channel to its center with the suction of cold gas into the cavern. It was found that, in the time interval 80-180 μs after the discharge, a local maximum of the gas velocity occurs due to the collision of the jets moving to each other and the formation of a radial jet stream directed away from the center of the spark in a plane perpendicular to the axis of the discharge takes place. As a result, the hot gas takes the form of a half of toroid, increasing in time, and the cold gas occupies the central area. The qualitative picture of the hydrodynamics of the hot channel decay as well as the value of velocity of the thermal cavern boundary motion, obtained in the simulation, correlate well with the data obtained in previous experiments. Analyzing the results, it was found that the region of maximum velocity is shifted relative to the region having the maximum gas momentum, since the maximum velocity takes place in a region with a low density. It is shown that, in terms of the generated momentum, the most effective vortex generator is a spark with energy of the order of 60 mJ per 10 mm at discussed conditions.

Acknowledgments

The study of gasdynamic perturbations caused by spark discharge was initiated in JIHT RAS by Prof. Sergey Leonov and authors would like to thank Prof. Leonov for constructive discussions and important comments on the results obtained.

This work is supported by grant of Russian Scientific Foundation № 18-19-00740.

References

[1] Bazelyan E M, Raizer Yu.P 1998 *Spark Discharge* (CRC Press)

[2] Alvarez C E C, Couto G E, Roso V R, Thiriet A B, Valle R M 2018 A review of prechamber ignition systems as lean combustion technology for SI engines *Appl. Therm. Eng.* **128** 107–20

[3] Janda M, Martišovitš V, Hensel K and Machala Z 2016 Study of transient spark discharge focused at NOx generation for biomedical applications *J. Phys.: Conf. Ser.* **768** 012009

[4] Leonov S B, Isaenkov Y I, Firsov A A, Nothnagel S L, Gimelshein S F and Shneider M N 2010 Jet regime of the afterspark channel decay *Phys. Plasmas* **17** **5** 053505

[5] Shurupov M A, Leonov S B, Firsov A A, Yarantsev D A and Isaenkov Yu I 2014 Gasdynamic instabilities during decay of the submicrosecond spark discharge channel *High Temp.* **52** **2** 169-78

[6] Firsov A A, Shurupov M A, Yarantsev D A, Leonov S B and Shneider M N 2013 Mixing Actuation by Unstable Filamentary Discharge *51st AIAA Aerospace Sciences Meeting, Texas, USA, 07-10 January 2013,* AIAA 2013-1188

[7] Firsov A A, Dolgov E V, Rakhimov R G, Shurupov M A and Leonov S B 2018 Mixing enhancement by electrical discharge in supersonic airflow *AIAA Aerospace Sciences Meeting, AIAA SciTech Forum,* AIAA 2018-1195

[8] Leonov S B, Isaenkov Y I, Firsov A A, Yarantsev D A and Shneider M N 2010 High-Power Filamentary Pulse Discharge in Supersonic Flow *48th Aerospace Sciences Meeting, January 2010, Orlando, USA,* AIAA 2010-259

[9] Firsov A A, Isaenkov Y I, Shurupov M A and Leonov S B 2016 Development of plasma actuator based on surface sparks for a buffet control *54th AIAA Aerospace Sciences Meeting, AIAA SciTech Forum,* AIAA 2016-1690

[10] Firsov A, Isaenkov Yu, Leonov S, Moralev I and Soudakov V 2017 Buffet suppression by submicrosecond spark discharge *7th European Conference for Aeronautics and Space Sciences (EUCASS), 3-6 July 2017,* paper EUCASS2017-291

[11] Sidorenko A A, Budovsky A D, Vishnyakov O I, Polivanov P A and Maslov A A 2017 Transonic buffet control by plasma actuator with spark discharge *7th European Conference For Aeronautics And Space Sciences (EUCASS), 3-6 july 2017,* paper EUCASS2017-345

[12] Firsov A A, Moralev I A, Isaenkov Yu I, Leonov S B and Soudakov V G 2018 Suppression of transonic buffet phenomenon by spark plasma actuator *AIAA Aerospace Sciences Meeting, AIAA SciTech Forum,* AIAA 2018-2050

[13] Stepanyan S, Hayashi J, Lovascio S, Stancu G D and Laux C O 2017 Hydrodynamic effects induced by nanosecond sparks in air and air/fuel mixtures *55th AIAA Aerospace Sciences Meeting, AIAA SciTech Forum,* AIAA 2017-1581

[14] Singh B, Rajendran L K, Bane S P, Vlachos P 2018 Characterization of Fluid Motion Induced by Nanosecond Spark Plasmas: Using Particle Image Velocimetry and Background Oriented Schlieren *AIAA Aerospace Sciences Meeting, AIAA SciTech Forum,* AIAA 2018-0680

[15] D'Angola A, Colonna, G, Gorse C et al. 2008 Thermodynamic and transport properties in equilibrium air plasmas in a wide pressure and temperature rang *Eur. Phys. J. D* **46** 129

IOP Conf. Series: Journal of Physics: Conf. Series **1112** (2018) 012027 doi:10.1088/1742-6596/1112/1/012027

Nonlinear dynamics of non-neutral Maxwellian plasma in external trapping field

A R Karimov[1, 2*] **and A A Dementev**[2]

[1]Joint Institute for High Temperatures, Russian Academy of Sciences, Izhorskaya 13 Bd 2, Moscow 127412, Russia
[2]National Research Nuclear University MEPhI, Kashirskoye shosse 31, Moscow, 115409, Russia

[*]Email: arkarimov@mephi.ru

Abstract. Nonlinear dynamics of collisionless non-neutral plasma in an external electrical trapping field is considered. For such system, the time-dependent solution of nonlinear Vlasov-Poisson equations are obtained. Proceeding from this solution, the influence of initial conditions on the dynamics of charged layer is discussed. Conditions leading to the capture of two-dimensional charged collisionless layer in the external parabolic potential are presented.

1. Introduction

There has been a great deal of interest in examining dynamic states of a non-neutral collisionless plasma in an electrostatic or magnetic trap because they are important for understanding the fundamental physical properties of both non-neutral plasma and hydrodynamic systems as a whole [1-6]. In this case, the formation of various dynamic states (so-called collisionless relaxation) is determined by the collective interaction of waves and particles. In particular, in [2,5,7] it has been shown that non-neutral plasma does not reach the state of thermodynamic equilibrium during collisionless relaxation. Unlike conventional thermodynamic equilibrium, the states, in which plasma evolves, explicitly depend on the initial distribution of velocities, particle coordinates, and the external focusing field.

How may it occur? Which is the role of initial conditions and external governing factors? We can indicate a few points. As is known, the dynamics of longitudinal electrostatic waves in collisionless plasma tends to a nontrivial behavior even near the initial equilibrium state (see, for example, [6-10] and references therein). As was shown in [7-11], depending on the initial conditions, one can observe both the O'Neil and the Landau scripta in collisionless plasma and in both cases the initial amplitude of the wave plays the role of a bifurcation parameter which determines the system evolution. Also, in plasma media near the Maxwellian initial equilibrium state, there may exist nonlinear, stable localized distribution functions which realize the energy maximum similar to the steady two-dimensional fluid vortices (see, for example, [5,8,9]). These equilibrium electrostatic structures, so called phase-space holes, may be treated as Bernstein-Greene-Kruskal modes in which a population of charged particles is trapped in a self-created electrostatic potential.

Thus, nonlinearity of Vlasov-Maxwell system and the initial conditions may play the role of the driving parameter which determines the possibility of emergence of quasi-stationary states. This system possesses such a property because it precludes direct particle–particle collisions that tend to

IOP Conf. Series: Journal of Physics: Conf. Series **1112** (2018) 012027 doi:10.1088/1742-6596/1112/1/012027

randomize the particle velocities, and the interaction via the self-consistent electrostatic field cannot change the full system entropy. It means that there are many ways of evolution of the system from the initial state far from equilibrium to a state, where the distribution function practically ceases to change with time. However, one can compare the macroscopic quantities (velocity–space moments of the distribution function) with that obtained from the corresponding fluid models which always are workable near local equilibrium states described by the distribution function of the Maxwellian type [8,9,11,12].

So, it would be of interest to study the dynamics of non-neutral plasma slab via the formation of local equilibrium state formed in the self-generated and external holding fields (see, for example, [8,9,11,12] and references therein). In the present paper, following our previous works [9–12] we are going to study this case in two-dimensional geometry for the cloud of the single charged particles trapping by symmetric quadratic electrostatic field. In order to investigate the evolution of such system, we shall construct exact nonstationary solution of the two-dimensional Vlasov-Poisson equations, which has a local equilibrium form. Using this solution, we will discuss the physical conditions for the realization of local equilibrium states in charged media of the type under consideration.

2. Formulation of the kinetic problem

We consider the dynamics of two-dimensional slab of non-neutral, non-relativistic, collisionless medium constituted by particles with the same mass m and charge q, which is confined in the quadratic potential

$$\Phi_{ext}(x,y) = k_x x^2 + k_y y^2\,, \qquad (1)$$

where k_x and k_y are the constant components of the external focusing parameter in the Oxy plane. Such dependence is typical for Paul, Kingdon and Penning traps. The dimensionless Vlasov-Poisson equations for the one-particle distribution function $f = f(t,x,y,v_x,v_y)$, which describe the electron plasma in the electrostatic approximation, are

$$\frac{\partial f}{\partial t} + v_x\frac{\partial f}{\partial x} + v_y\frac{\partial f}{\partial y} + \left(E_x - \frac{\partial \Phi_{ext}}{\partial x}\right)\frac{\partial f}{\partial v_x} + \left(E_y - \frac{\partial \Phi_{ext}}{\partial y}\right)\frac{\partial f}{\partial v_y} = 0, \qquad (2)$$

$$\nabla \cdot \mathbf{E} = 4\pi n, \qquad (3)$$

where n is the density of particles forming a charged medium, $\mathbf{E} = -\nabla\Phi$ and $\Phi(t,x,y)$ are the self-consistent electric field and potential, respectively. The density, time and space have been normalized by the initial electron density n_0, the inverse plasma frequency $\omega_b = (4\pi n_0 e^2/m_e)^{1/2}$ and the inverse Debye length $\lambda_b = (T_e/4\pi n_0 e^2)^{1/2}$, where m_e the mass, e the charge, T_e the temperature of electrons. The velocity and self-consistent potential are normalized by $\omega_b\lambda_b$ and T_e/e, respectively.

Assume that the system is in the state of local equilibrium. Then we look for the partial solution of time-dependent system of Vlasov-Poisson equations (2)-(3) in the local-equilibrium form

$$f(t,x,y,v_x,v_y) = \frac{n}{\pi T} exp\left(-\frac{(v-V)^2}{T}\right), \qquad (4)$$

where $(\mathbf{v}-\mathbf{V})^2 = (v_x - V_x)^2 + (v_y - V_y)^2$ and $n = n(t)$, $T = T(t,x,y)$, $V_x = V_x(t,x,y)$, $V_y = V_y(t,x,y)$ are unknown functions to be determined.

In view of the symmetry of the slab, we set the additional conditions

$$\left.\frac{\partial\Phi}{\partial x}\right|_{x=0} = \left.\frac{\partial\Phi}{\partial y}\right|_{y=0} = 0, \qquad (5)$$

$$k_x = k_y = k. \qquad (6)$$

The relations (3) and (5) are satisfied by the function

$$\Phi(t,x,y) = \frac{n(t)}{2}(x^2 + y^2), \qquad (7)$$

which corresponds to the self-generated field

$$\mathbf{E}(t,x,y) = -\nabla\Phi = -n(t)x\mathbf{e}_x - n(t)y e_y. \qquad (8)$$

3. Evolution equations

Comparing (1) with (8), taking into account (6), we conclude that for $k > 0$ external and self-consistent fields are oppositely directed: this makes it possible to confine the non-neutral plasma. For this condition, the substitution of (4) into (2) leads to

$$v^{(0)}\left[\frac{\dot{n}}{n}-\frac{\dot{T}}{T}+\frac{\dot{T}}{T^2}V^2-\frac{2}{T}\left(V_x\dot{V}_x+V_y\dot{V}_y\right)+\frac{2}{T}(n-k)\left(xV_x+yV_y\right)\right]+$$
$$v_x{}^{(1)}\left[-2\frac{\dot{T}}{T^2}V_x+\frac{2}{T}\dot{V}_x-\frac{2V_x}{T}\frac{\partial V_x}{\partial x}-\frac{2V_y}{T}\frac{\partial V_y}{\partial x}-\frac{2(n-k)x}{T}\right]+$$
$$v_y{}^{(1)}\left[-2\frac{\dot{T}}{T^2}V_y+\frac{2}{T}\dot{V}_y-\frac{2V_x}{T}\frac{\partial V_x}{\partial y}-\frac{2V_y}{T}\frac{\partial V_y}{\partial y}-\frac{2(n-k)y}{T}\right]+$$
$$v_xv_y\left[\frac{2}{T}\frac{\partial V_y}{x}+\frac{2}{T}\frac{\partial V_x}{y}\right]+v_x{}^{(2)}\left[\frac{\dot{T}}{T^2}+\frac{2}{T}\frac{\partial V_x}{x}\right]+$$
$$v_y{}^{(2)}\left[\frac{\dot{T}}{T^2}+\frac{2}{T}\frac{\partial V_y}{y}\right]=0. \tag{9}$$

This relation is valid for any v_x and v_y. Thus, the coefficients of each power of $v_x^s v_y^l$ should be equal to zero for any l and s. Nullifying the coefficients of v_xv_y, $v_x^{(2)}$ and $v_y^{(2)}$ in equation (9), we obtain

$$\frac{\partial V_y}{\partial x}=-\frac{\partial V_x}{\partial y}, \tag{10}$$

$$\frac{\partial V_x}{\partial x}=-\frac{1}{2}\frac{\dot{T}}{T},\quad \frac{\partial V_y}{\partial y}=-\frac{1}{2}\frac{\dot{T}}{T}. \tag{11}$$

In the simplest case, if T does not depend on the coordinate, from the relations (11) it follows that

$$V_x(t,x)=-\frac{1}{2}\frac{\dot{T}}{T}x+g(y,t), \tag{12}$$

$$V_y(t,y)=-\frac{1}{2}\frac{\dot{T}}{T}y+h(x,t). \tag{13}$$

The equations (12), (13) must satisfy the condition (10). Now we show that this relation will be satisfied when

$$h(x,t)=Cx+D_1(t), \tag{14}$$

$$g(y,t)=-Cy+D_2(t), \tag{15}$$

where C is arbitrary constant, $D_1(t)$, $D_2(t)$ are some functions. Indeed, since

$$\frac{\partial V_x}{\partial y}=\frac{\partial g}{\partial y},\quad \frac{\partial V_y}{\partial x}=\frac{\partial h}{\partial x}\quad\text{and}\quad \frac{\partial g}{\partial y}=-\frac{\partial h}{\partial x}, \tag{16}$$

then relation (10) is satisfied.

Substituting the relations (12) and (13) into equation (9), we can equate to zero the coefficients before x^sy^l for any l and s. Thus, we ensure that this equation is satisfied for arbitrary x and y. As a result, we obtain the relations

$$\frac{d}{dt}\left(\frac{\dot{T}}{T}\right)-\frac{1}{2}\left(\frac{\dot{T}}{T}\right)^2+2n-2k=0, \tag{17}$$

$$\frac{\dot{n}}{n}-\frac{\dot{T}}{T}+\frac{\dot{T}}{T^2}\left(D_1^2+D_2^2\right)+\frac{2}{T}D_1\dot{D}_1+\frac{2}{T}D_2\dot{D}_2=0, \tag{18}$$

$$\dot{D}_1=\frac{1}{2}\frac{\dot{T}}{T}D_1, \tag{19}$$

$$\dot{D}_2=\frac{1}{2}\frac{\dot{T}}{T}D_2. \tag{20}$$

Equations (19), (20) have the following solutions

$$D_1=D_aT^{\frac{1}{2}}, \tag{21}$$

$$D_2=D_bT^{\frac{1}{2}}, \tag{22}$$

where D_a, D_b are constants. For simplicity we put $D_a=D_b=D_0$. In this case we have

$$D_1=D_2=D_0T^{\frac{1}{2}}. \tag{23}$$

By substituting (23) into (12) and (13), for initial moment of time, we obtain

IOP Conf. Series: Journal of Physics: Conf. Series **1112** (2018) 012027 doi:10.1088/1742-6596/1112/1/012027

$$D_0 = V_x(t=0)T(t=0)^{-\frac{1}{2}} = V_y(t=0)T(t=0)^{-\frac{1}{2}}. \quad (24)$$

Thus, D_0 determines the initial distribution of hydrodynamic velocities.

Taking into account (23), from the relation (18) we get

$$\frac{\dot{n}}{n} + \frac{\dot{T}}{T}\left(4{D_0}^2 - 1\right) = 0. \quad (25)$$

The solution of this equation is

$$n = n_0 T^{1-4{D_0}^2}. \quad (26)$$

For simplicity, we put the constant of integration equal to n_0. This solution allows us to exclude the density $n(t)$ from the equation (17), which as a result becomes

$$\frac{d^2u}{dt^2} - \frac{1}{2}\left[\frac{du}{dt}\right]^2 + 2n_0 e^{u\alpha} - 2k = 0, \quad (27)$$

here we use the notation

$$u = \ln T,\ \alpha = 1 - 4D_0^2. \quad (28)$$

By using (12), (13), (26) and (28), we express the parameters in the distribution (4) in explicit form through the function u and parameter α.

$$n = n_0 e^{u\alpha}, T = e^u, V_x = -\frac{1}{2}\frac{du}{dt}x + \sqrt{\frac{1-\alpha}{4}}e^{\frac{u}{2}}, \quad V_y = -\frac{1}{2}\frac{du}{dt}y + \sqrt{\frac{1-\alpha}{4}}e^{\frac{u}{2}}. \quad (29)$$

Thus, proceeding from these relations and (27) we can determine the evolution of the system worked out.

4. Dynamics of a charged slab

The physical properties of a non-neutral plasma are determined by the dynamic properties of the solutions of the equation (27), depending on the initial conditions. In the absence of heating or cooling sources in the system we have to put

$$\left.\frac{\partial u}{\partial t}\right|_{t=0} = 0,$$

which corresponds to immobile non-neutral medium in initial moment of time, and to consider changing of relevant parameters in the local equilibrium distribution (4) for different initial temperatures $u(t=0) = u_0$, the values of the external focusing parameter k and the parameter α.

As is seen from relation (29), the parameter α, associated with the functions n and V_x, V_y, defines the initial distribution of the hydrodynamic velocities. In the case of $\alpha = 1$, the hydrodynamic velocities at the initial moment of time are equal to zero. Therefore, it is convenient to use this value in further calculations.

Figure 1 shows the evolution of the function $u(t)$ associated with density, temperature and velocity by the relations (29) for different values of the focusing parameter $0 < k < 4$ and the initial conditions $u'(t=0) = u(t=0) = 0$, which corresponds to a immobile non-neutral plasma with nonzero initial temperature $T(t=0) = 1$.

In the case of small values $k \leq 1$ (figure 1 (a)), when $u < 0$, the charged layer is expanded accompanied by nonlinear oscillations of the function $u(t)$. As is seen from the above dependences, increase in the parameter k leads to increase in the oscillation frequency. However, according to (27) and (29), these oscillations are most strongly manifested in the velocity field, while the other parameters show insignificantly sensitivity to oscillations of u. The interval $1 \leq k \leq 4$ corresponds to confinement of the plasma. For $k = 1$ the function u, and hence the temperature T and the density n, do not change with time, the velocities V_x and V_y become equal to zero. A further increase in k leads to the emergence of strong nonlinear oscillations in all parameters. For $k > 4$ the distribution $u(t)$ exhibits a singular behavior, which is due to the dynamics of the velocity field.

IOP Conf. Series: Journal of Physics: Conf. Series **1112** (2018) 012027 doi:10.1088/1742-6596/1112/1/012027

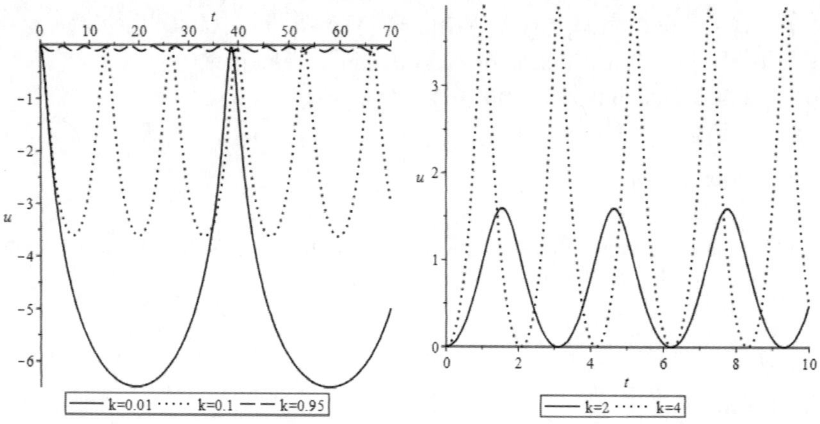

(a) (b)

Figure 1 Evolution of the temperature dependence $u(t)$ for $\alpha = 1$, $u'(t = 0) = u(t = 0) = 0$ under different values of the parameter k.

Thus, from the point of view of plasma confinement, the value $k = 1$ is of the greatest interest, since here the parameters in the distribution (4) remain constant.

For this case figure 2 shows the influence of the initial temperature for $u'(t = 0) = 0$ on the dynamics of the resulting oscillations in the interval

$$-2 \le u_0 \le 4$$

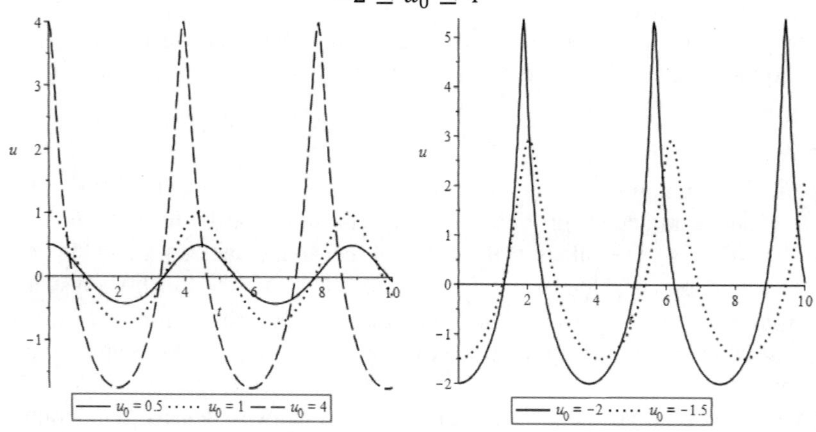

(a) (b)

Figure 2 Evolution of the temperature dependence $u(t)$ for $\alpha = 1$, $u'(t = 0) = u(t = 0) = 0$ under different values of the parameter u_0.

The values of the initial temperature greater than $u_0 = 4$ lead to the emergence of a singularity. This is due to rollover of the velocity profile that is shown in figure 3.

These graphs show that for $u_0 \ge 1$ we can observe plasma confinement, accompanied by the emergence of persistent nonlinear oscillations.

Thus, the dependencies presented in figures 1-3, point out to the existence of a range of k and u_0, in which plasma is confined. For example, we specify a neighborhood near the point $k = 1, u_0 = 0$. Apparently, such states can be identified with stable, long-lived states that can be obtained experimentally if the distribution of the form (4) has already been established in the system.

Now we consider the dynamics of the nonlinear oscillations in the temperature dependence $u(t)$ under the variation of the parameter α.

IOP Conf. Series: Journal of Physics: Conf. Series **1112** (2018) 012027 doi:10.1088/1742-6596/1112/1/012027

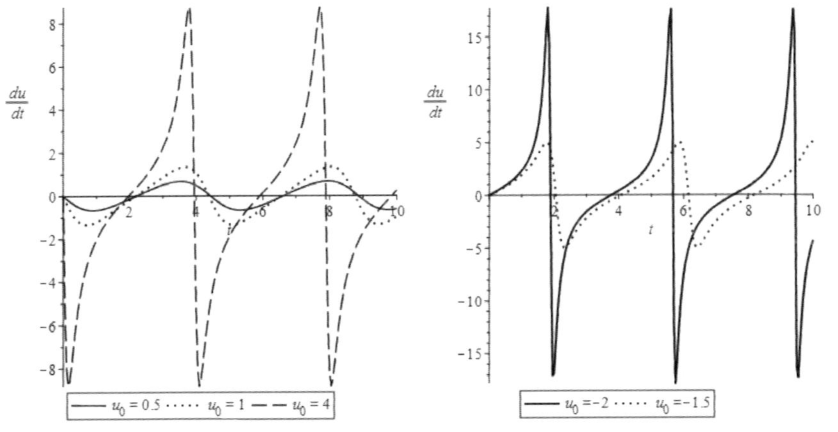

(a) (b)

Figure 3 Dynamics of the first derivative of the function $u(t)$ for $\alpha = 1$, $k = 1$, $u^{'}(t = 0) = 0$under different values of the initial temperatureu_0.

Figure 4 (a) shows the plots of the function $u(t)$ for various values of the parameter $0.4 \le \alpha \le 1$ and fixed initial conditions $k = 1.5$, $u^{'}(t = 0) = u(t = 0) = 0$. As is seen from these dependencies, the amplitude of the oscillations of the function $u(t)$ increases and shows a tendency to a singular behavior when the parameter $\alpha \to 0$. This leads to the fact that the range of values of the parameter k, where the charge layer can be confined, becomes smaller. For example, if $\alpha = 0.4$, this range decreases to $1 \le k \le 1.5$. In the case of $\alpha > 1$, the values of the components of hydrodynamic velocities have an imaginary part.

By setting the value of initial temperature $u_0 = 1$ for $0.4 \le \alpha \le 1$, $k = 1.5$, we obtained the set of graphs presented in figure 4 (b). The comparison of this dependences with dependences in figure 4 (a), makes it possible to investigate the influence of the initial temperature on the dynamics of the oscillations of the function $u(t)$ for various α: we see that the character of the amplitude variation for different u_0 and α can differ. Since for the graphs shown in figure 4 (b), the decrease in the parameter α leads to a decrease in the amplitude of the oscillations of the function $u(t)$, then it becomes possible to select the conditions under which the function $u(t)$ does not change with time. In our case, this is achieved for $u_0 = 1$, $\alpha = 0.4$, $k = 1.5$. Thus, by varying the parameters α and k with initial temperature u_0, for which a decrease in parameter α leads to a decrease in the amplitude of the oscillations of the function $u(t)$, we can find a conditions for the constancy of temperature when the charged layer is confined.

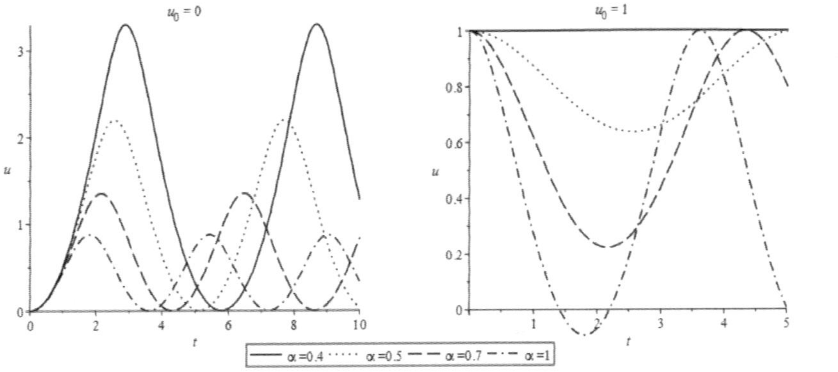

(a) (b)

Figure 4 Influence of the parameter $\alpha = 1$ n the dynamics of the oscillations of the function $u(t)$ for $k = 1.5$, $u^{'}(t = 0) = 0$ under different values of the initial temperature u_0.

IOP Conf. Series: Journal of Physics: Conf. Series **1112** (2018) 012027 doi:10.1088/1742-6596/1112/1/012027

5. Conclusion

In this paper we have considered the dynamics of non-neutral single-component plasma in the external electrostatic field with a quadratic potential for a two-dimensional geometry. As a particular solution of the Vlasov-Poisson equations, we used the distribution of the local-equilibrium form (4), whose macroscopic parameters are described by the dependencies (28) and (29) through the function u(t), which is the logarithm of temperature, and the parameter α. The function u(t) satisfies the equation (27), which describes a series of non-stationary bounded solutions, depending on the initial temperatures, the external focusing parameter k and the parameter α. Analysis of this equation points out to existence of a range of initial conditions, in which nonstationary confinement of a non-neutral plasma is possible.

The two-dimensional model considered here describes a particular case of the evolution of a charged medium, since real systems have a higher spatial dimension, even in the absence of plasma confinement and dissipation processes. On the other hand, our exact solutions, that describe possible strongly nonlinear final states, can give a basic idea of the formation of time-dependent structures in a real system.

Acknowledgments

The work of A.R.K was supported by the Ministry of Science and Education of the Russian Federation under No. 14.575.21.0169 (RFMEFI57517X0169).

References

[1] Davidson R C 1974 *Theory of Nonneutral Plasmas* (New York: Benjàmin)
[2] Dubin D H E and O Neil T M 1999 Trapped nonneutral plasmas, liquids, and crystals *Rev. Mod. Phys.* **71** 87
[3] Werth G, Gheorghe V N and Major F J 2009 *Charged Particle Traps II: Applications* (New York: Springer)
[4] Ghosh K 1995 *Ion Traps* (Oxford: Oxford University Press)
[5] Jin D Z and Dubin D H E 1998 Regional maximum entropy theory of vortex crystal formation *Phys. Rev. Lett.* **80** 4434
[6] Levin Y, Pakter R and Teles T N 2008 Collisionless relaxation in non-neutral plasmas *Phys. Rev. Lett.* **100** 040604
[7] Robinson P A 1997 Nonlinear wave collapse and strong turbulence *Rev. Mod. Phys.* **69** 507
[8] Luque A and Schamel H 2005 Electrostatic trapping as a key to the dynamics of plasmas, fluids and other collective systems *Phys. Rep.* **415** 261
[9] Eliasson B and Shukla P K 2006 Formation and dynamics of coherent structures involving phase-space vortices in plasmas *Phys. Rep.* **422** 225
[10] Lancellotti C and Dorning J J 2003 Time-asymptotic wave propagation in collisionless plasmas *Physical Review* E **68** 026406
[11] Krishan S 1990 Non-linear dynamics of non-neutral plasmas *Plasma Phys. Controlled Fusion* **32** 1209
[12] Amiranashvili Sh, Yu M Y and Stenflo L 2002 Non-neutral plasma column in asymmetric trapping field *Physical Review* E **65** 046402
[13] Karimov A R 2001 Nonlinear solutions of a Maxwellian type for the Vlasov–Poisson equations *Phys. Plasmas* **8** 1533
[14] Karimov A R. 2013 Coupled electron and ion nonlinear oscillations in a collisionless plasma *Phys. Plasmas* **20** 052305
[15] Karimov A R, Yu M Y and Stenflo L. 2016 Properties and evolution of anisotropic structures in collisionless plasmas *J. Plasma Phys.***82** 905820502

AUTHOR INDEX

Institute of Physics
Dirac House, Temple Back
Bristol BS1 6BE UK

ISSN: 1742-6588
ISBN 978-1-5108-7716-0